Robin F.A. Moritz
Edward E. Southwick

Bees as Superorganisms

An Evolutionary Reality

With 76 Figures and 19 Tables

Springer-Verlag
Berlin Heidelberg New York
London Paris Tokyo
Hong Kong Barcelona
Budapest

Prof. Dr. Robin F.A. Moritz
Institute of Biology, Technical University of Berlin
Franklinstrasse 28/29, 1000 Berlin 10, FRG

Prof. Dr. Edward E. Southwick
Department of Biology, State University of New York
Brockport, NY 14420, USA

ISBN-13:978-3-642-84668-7 e-ISBN-13:978-3-642-84666-3
DOI: 10.1007/978-3-642-84666-3

Softcover reprint of the hardcover 1st edition 1992

Typesetting: Best-set Typesetter Ltd., Hong Kong

31/3145 – 5 4 3 2 1 0 – Printed on acid-free paper

Preface

The honeybee (*Apis mellifera* L.) is one of the better studied organisms on this planet. There are plenty of books on the biology of the honeybee for all, the scientist, the beekeeper, and the layman. In view of this flood of publications one is tempted to ask: why does it require another one? The answer is simple: a new one is not required and we do not intend to present a new book on "the honeybee". This would really just add some more inches to the already overloaded bookshelf without substantial new information. Instead, we intend to present a book on the honeybee colony. This of course immediately releases the next question: so what is the difference? Although the difference may look insignificant at first glance, we try to guide the reader with a fundamentally different approach through the biology of honeybees and eusocial insect societies in general. The biology of individual colony members is only addressed when it is necessary to explain colonial mechanisms, and the colony as a whole, as a biological unit, which is the main focus of this treatise. Both of us felt that all current textbooks on bee biology put too much emphasis on the individual worker, queen or drone in the colony. Often it is completely neglected that the colony is a very significant (if not the most significant) biological structure in bee biology. It is only because eusocial insects live in colonies that their ecological and economic impact are of so much more consequence than that of any solitary insect.

The role of the colony has been recognized for a long time giving rise to the term "superorganism" at the beginning of this century. Yet the holistic approach of these early studies to understand complex colonies of social insects failed. It had to fail because nothing was known about communication, social control and the biological processes that rule the colony. The rise of reductionism and the study of individual intracolonial behaviour supplied the data base which forms our current knowledge of the biology of social insects. Though the reductionist approach was, and still is, an important philosophical pathway, it may sometimes constrain our view of biological mechanisms in complex systems. It is correct that without a sound understanding of the simple mechanisms we are bound to fail in understanding the more complex system. Yet this does not mean that because we have chosen to study simplicity in the first place, complexity is of no interest anymore or cannot be studied. The opposite is true. Based on the very detailed knowledge of individual biology we are now in a position to construct a working hypothesis for the more complex situations. It took molecular biology several decades to grasp the link between the genetic code and the physiological correlate on the organismic level. In this book we want to take a similar step from organismic to superorganismic biology.

Insect societies deserve much more attention and study simply based on their ecological and environmental importance and their prodigious biomass (which has been estimated to be as much as one-third of the entire animal biomass in the tropics, Wilson 1985). Colonies of ants, bees, wasps and termites are responsible for considerable turnover of materials, and they provide vast subject matter for analysis.

We have chosen to focus on the colony of honeybees as our central representative of all highly social colonial insects. Being neither an individual nor a population it whets oue appetite for exploration. Our studies of the genetic, physiological, behavioural, and ecological characteristics of bee colonies point to the usefulness of treating it as a biologically recognizable entity. We discuss the convergent strategies that evolution and natural selection has put together to make the superorganism as much a functional unit for gene survival as any individual metazoan.

If this book generates controversy and stimulates students to conduct further work in the field of honeybee biology, it has fulfilled its purpose. It is not intended as a reference book on honeybee biology. It is meant as a new conceptual framework of an "old" biological understanding of the honeybee colony on the basis of modern biological analysis and recent theories. Our book is not the typical monograph of a species, covering all aspects of its biology. Our goal is to present the honeybee as it lives and thrives in its social unit, the colonial superorganism.

This volume would not have been possible without the advice, help and aid we received from several colleagues. In particular we wish to thank M. Ayasse, M. Breed, S.L. Buchmann, R.M. Crewe, M.J. Duchateau, H. Esch, S. Fuchs, J.L. Gould, R. Hepburn, E. Hillesheim, G. Koeniger, N. Koeniger, P. Kryger, M. Makela, D.W. Roubik, J. Schmitz, P. Schmid-Hempel, M. Schwarz, J. van der Blom, and H.H.W. Velthuis for many helpful comments. However, we have not necessarily used all their advice, and any errors that eventually will be found in this volume are fully our responsibility. We are grateful to N. Frisch for preparing the cover figure. Finally, we wish to thank the Alexander von Humboldt Foundation which enabled us to conduct cooperative research over many years. This book evolved from these joint studies in which we became dedicated to the idea that the individual is interesting, but the group forms the central clue for understanding the biology of colonies of social insects.

May 1992

Robin F.A. Moritz
Edward E. Southwick

Contents

Chapter 1 What Is a Superorganism?

Betrachten wir die Termitenkolonie als Individuen höherer Ordnung, so müssen wir sie mit festsitzenden Tieren vergleichen. (K. Escherich 1909, p. 36)

The more complex a society the more simple the individual. (Klaus Jaffe 1990 during a meeting at the IUSSI conference in Bangalore, India)

1.1 From Cells to Metazoan Organisms

Human beings have a very limited capacity to understand complex multidimensional systems and therefore they are in need of models that simplify the degree of complexity. The simplicity of the genetics and the physiology of *Escherischia coli*, for example, allowed for the tremendous progress in molecular biology. Protozoan eucaryotes with chromosomes and a diploid phase already increase the complexity substantially. Genes in the genome can interact and make the analysis of gene expression less simple. Nevertheless, the physiology and genetics of single cells are relatively transparent, and these cells are common experimental organisms in the molecular biology laboratory (e.g. baker's yeast, *Saccheromyces cerevisiae*). The more complex systems must be used whenever questions related to diploid organisms have to be addressed. But what if we want to understand biological mechanisms in more complex organisms that are comprised of more than a single cell? Besides the interaction at the gene level within the cell, we would expect interactions between cells. Even if all these cells contain the identical genetic information, only a part of these genes will be expressed. One cell could well control gene expression in another which would obscure the analysis of individual gene action. The analyses of such interactions at the cell level are particularly well documented in the early development of *Drosophila melanogaster* embryos, where gene regulation of segmentation was studied in detail (e.g. Nüsslein-Vollhard and Wieschaus 1980; McGinnis et al. 1984a,b; Gehring 1987). Nevertheless, we are far from understanding the physiological basis of specialization of cells during development in multicellular organisms.

Even the simplest metazoan organisms, like sponges, have a cellular division of labour, and in the simply structured asconsponge (*Leucosolenia coriacea*), choanocytes line the inner cavity, amoebocytes digest food, chollencytes form the mesochyme structure with neighbouring cells. There is a layer of epithelial cells, and the typical scleroblasts produce the chalk skeleton. The gametes are located in the ascon cavity and develop from archeocytes. The female gametes show amoeboid movement within the ascon and phagocytize on other cells. Semen enters through the pore cells into the ascon and sperms are led to the female gametes by the choanocytes. Apparently some still mysterious interaction causes the somatic cells to remain sterile and usually only the omnipotent archeocytes are able to enter into a sexual reproductive phase (Kaestner 1969; Buss 1987). Asexual reproduction however is common in Porifera. They can bud and form asci or go through fission to form new individuals. They can also form gemmules which are special spore-like structures in which only the totipotent archeocytes retreat during periods of exigency. These gemmules remain in the decaying skeleton. As soon as conditions are favourable, the cells leave the gemmules and recolonize the old skeleton or they are spread by currents of water to new locations.

Clearly, the sponges are living organisms comprised of tens of thousands of individual and specialized cells. Clearly, a colony of honeybees is comprised of tens of thousands of individual organisms. But then the individual sponge shares many more analogies with the bee colony as a whole than with any individual honeybee in the colony. The life history cycle with hermaphroditic sexual reproduction, asexual reproduction via fission, and the survival of periods of dearth in specially adapted structures (gemmula or winterhard cluster), characterizes both the sponge and honeybee colony. These similarities led to the common acceptance of the bee colony as an "organism," a definition that found its way from scientific publications of the middle of the last century even into common current dictionaries. Equating individual insects found within a eusocial group with cells in a metazoan was the early simplistic view of a "superorganism." The functioning social group seemed to resemble an intact organism. Yet the analogy is clearly strained, since the somatic cells of a metazoan body are all genetically identical but the members of an insect society are not (Wilson 1968, 1971; West-

Eberhard 1975; Crozier 1979). The term "superorganism" (or supra-organism) was only rarely used during the 1960s and 1970s and evolutionary biologists were convinced that the death of the concept was a great achievement in increasing our understanding of natural selection (Williams 1966; Ghiselin 1974; Dawkins 1976, 1982; Maynard Smith 1987a,b). West-Eberhard (1975) went so far as to call it "*undesirable*" since it would be understood "*. . . to imply selection at a level higher than the individual . . .*". We agree with Crozier (1979, p. 277) who agrees "*with West Eberhard that readers might draw this implication, but disagree with her that they would thus be led into a biological error.*"

Recently, the concept seems to have had a revival. As Holldobler and Wilson (1990, p. 359) put it nicely: "*. . . old ideas in science never really die out, however. They only sink to mother earth, like the mythical giant Antaeus, to rise again with new vigour*". Indeed, it is the very goal of this book to show that the superorganism concept, although not necessarily a giant, has its merits in many disciplines of contemporary biology. Wilson and Sober (1989) raised the issue of a revival of the concept in the theoretical framework of evolutionary biology. They put forward one very strong argument, that "*superorganisms are more than just a theoretical possibility; they actually exist in nature*" (Wilson and Sober 1989, p. 338). The understanding of the superorganism concept should help to better comprehend natural selection, the behavioural ecology, and the biology of specialized eusocial organisms such as honeybees. We will give ample evidence to conclude that colony-level selection takes place and that the superorganism is a unit of selection as well as of function. We do not want to generate new mystical metaphors around a sort of philosophical holistic view of social life. All approaches which simply accepted the complexity of the social system as something "granted by nature" without questioning the background and the underlying biological mechanisms caused the original demise of the superorganism concept by evolutionary biologists.

In Morris (1976, p. 926) we find that an organism is "*any system regarded as analogous to a living body: the social organism*". Though this may be a common definition, it overlooks a very important issue. The bee colony is composed of individual organisms and therefore the structure as a whole cannot be an organism itself in spite of its overwhelming analogies. In order to acknowledge this fundamental difference, Wheeler (1928) used the term "superorganism," a definition that should describe the strong analogies of a colony to an individual organism. The definition gained wide publicity in the field of sociobiology, although, in analogy to the step from protozoon to metazoon (also population to metapopulation), the term *meta*-organism might have been more appropriate. On the other hand, the definition is historically grown and of such common usage that we stick to the old semantics rather than adding a new definition which would only generate more trouble in the colloquial jungle. As long as everybody knows what is meant by a "superorganism" there is no real need to revise the nomenclature.

1.2 What Makes a Social Group a Superorganism?

Obviously, not every social group fulfils the requirements of a superorganism. Wilson and Sober (1989, p. 339) define superorganisms "*as a collection of single creatures that together possess the functional organization implicit in the formal definition of organism*". Following this track, the most straightforward way to define superorganisms may be to look at the definition of metazoan organisms and simply replace "cell" with "organism." If we look in classical zoological textbooks, we find the following definition for organism: "*Metazoan animals with cells arranged in at least two non-uniform layers and differentiated into somatic and reproductive cells with different functions*" (Kaestner 1969, p. 96). By just replacing "cell" with "organism" we can transform the same definition to something like the following definition for a superorganism: *Superorganismic units with organisms arranged in at least two non-uniform types and differentiated into sterile and reproductive organisms with different functions*. Though this definition helps in specifying what a superorganism is, there are still substantial features missing for a complete picture. A typical characteristic of an organism is that it reproduces and, by doing so, produces another organism of the same type. For the superorganism this means that it should produce superorganismic offspring. This does not imply that each offspring must immediately be a complete superorganism, analogous to vivipary in individual organisms, but it must mature into a reproductive superorganism in the end. The deposition of embryonic stages on a nutritive substrate is the rule in reproductive biology of most organisms.

The term, superorganism, should not be confused with or be used in place of the term, sociality. If social groups are small, natural selection is likely to act mainly upon the individual phenotypes rather than on the group as a whole. If group survival strongly depends on the individual fitness of its members, and if the society cannot buffer failure or loss of some of its members, there is no reason to invoke a definition like superorganism. Superorganisms need a sufficient membership so that the number of organisms involved in a task rather than the individual quality of how a task is performed becomes important.

Another restriction for superorganisms is given if colonies are non-perennial and only fertilized queens survive through a diapause period of dearth. In this case selection will act strongly upon these individual queens. Of course, there will also be fitness differences among the colonies during the active season. Some will produce more sexual off-spring than others yet the highest selection intensity is clearly on the diapausing females. Usually, the queens are produced in excessive numbers and very few are finally successful in establishing a nest in the next season. Once the nests are established they may also die out, but the survival rate is much higher than that of the individual females. The selection intensity on the individual level far exceeds that at the colonial level. At the end of the season natural selection at the colonial level is equally strong for all colonies. They all die and there is no selection for colonial overwintering. Selection among colonies can be particularly strong if they are long-lived and perennial. Therefore, it seems to be a reasonable restriction to only address perennial social groups that fulfil the above requirements as superorganisms.

Within the constraints of the definition, several biological characteristics result that are typical, though not essential, features of superorganisms. The following four traits are particularly often found in superorganisms:

1. *Superorganisms are usually sessile* (with the exception of some highly mobile ant species; for example army ants). The colony is confined to a physiographic nesting site which is commonly occupied by most of the individuals most of the time and where food is stored and brood is reared. Only during extended periods of dearth are these homesites abandoned, and again we find strong analogies to the sponge system discussed above. The fact that the superorganism is sessile does not mean that its individual members are sessile. Just the opposite is true. Foraging social insects can quickly adapt to new foraging conditions and home ranges may vary extensively. Nevertheless, the majority of the workers remains at the nesting site, and the reproductives will only leave for mating.
2. *Superorganisms maintain intraorganismic homeostasis*. In order to maintain a homoeostasis in the colony, food supplies are necessary

to cope with periods of unfavourable climatic conditions. Various ways of food conservation during storage are employed by superorganisms. The most elegant modes of preservation are the fungus gardens of ants and the honey stores of bees. Tightly linked with preservation of food is the evolution of strict nest hygienics. Environmental conditions that favour development of insect larvae are usually also favourable for bacterial development of disease organisms.

3. *Superorganisms are either well armed or highly cryptic.* Because of the stored food supplies, the superorganisms are attractive prey for many predators. To counteract this problem, members of superorganisms are either equipped with effective defensive arms and/or the nests are highly cryptic and located in inaccessible places.
4. *Superorganisms have a large number of colony members* that function as a cooperative unit, and there is no solitary existence except during mating or nest founding. Typically, many thousands of insects are necessary to form a functioning superorganism. This is simply because colony homoeostasis cannot be maintained when only a few workers are available. The larger the colonies, the more typical are the characteristics of a superorganism.

There may be many more characters that are specific for superorganisms than we have mentioned here. In the end, however, only one feature really counts. It makes absolutely no sense invoking such a definition if natural selection does not act upon the superorganism itself. As long as natural selection is only working on individuals we have no need for such an additional perspective. In many of the insect societies, however, selection on the colony level seems to override selection at the individual level. In these cases, in which selection operates chiefly on social units, it seems highly valuable not to abandon the perspective of superorganisms suggested by biologists early in the twentieth century. Reducing our understanding of selection processes on single genes may leave out many exciting aspects of evolutionary biology.

1.3 Diversity of Superorganisms

The study of social insect colonies (ants, termites, wasps and bees) during this century has revealed remarkable social homoeostasis in regulation of nest environment, numbers and kinds of individuals, and behaviour. The ecological conditions that favoured the evolution of superorganisms seem to be most common in the tropical regions of the world. Relatively few superorganisms found their way from the tropical environment to the temperate zones. This may be a result of the extreme physiological costs involved in maintaining a constant high intracolonial temperature in temperate climates. For example, all superorganismic bees but the cave-dwelling honeybees (*Apis mellifera*, *Apis cerana*) are confined to the tropics (Ruttner 1988; Roubik 1989). Thus, the tropics are the primary and probably the original habitat of superorganisms. Apparently, the cave-dwelling *Apis* species only could adapt to the rough winter conditions because of their ability to store large amounts of honey and the sufficient thermal insulation in winter provided by the nest shelter.

1.3.1 Termites

The termites (Termitidae, Isoptera) make up the only non-Hymenopteran insect group that forms superorganisms. They provide a nice example that haplo-diploidy is not a requirement for the evolution of eusociality. Luykx and Syren (1979) and Lacy (1980) suggested that within a colony relationship might be similarly close as in haplo-diploid organisms because of sex-linked multichromosome translocations in lower termites (all families, excluding Termitidea; Syren and Luyks 1977, 1981). Yet Leinaas (1983) and Andersson (1984) argued that the relatedness under such a translocation system may be low and not favourable to the evolution of eusociality. Furthermore, Crozier and Luykx (1985) point out very convincingly that the evolution of termite eusociality is unlikely to have been based on a male-haploid analogy. First, there is considerable variation within and between species in the proportion of the genome included in such translocations. Second, there is only a single translocation in higher termites (Vincke and Tilquin 1978) which then suggests that large-scale sex linkage was not significant for social evolution in their case. Third, Crozier and Luyks (1985) indicated that the evolution of multiple sex chromosomes has taken place much more recently than that of the termites. Finally, male and female workers do not invest more into siblings of the same sex than into the opposite sex, a phenomenon that is clearly predicted by evolutionary theory. The basis for the evolution of eusociality in termites seems to lie elsewhere rather than in an analogy to the haplo-diploid system of the hymenoptera.

Termites are hemimetabolic and the caste system is closely related to developmental pathways (recently reviewed by Noirot 1989, 1990). The individual termite undergoes a series of moults and at each developmental stage a commitment may occur to join a specific caste. Flexibility in the caste system is particularly strong in primitive species (*Kalotermes flavicollis*, *Porrhiontermes simplex*, *Zooteropsis angusticollis*), the determination for a reproductive form (development of wing pads) occurs very late and the differentiation is fully reversible until the final nymphal instar (Miller 1969; Noirot 1985; Noirot and Pasteels 1987). Another major difference from the Hymenopteran societies is that the male termites actively participate in the worker and soldier force. The termite colony has a king which intermittently inseminates the colonial queen. Interestingly, the sexual reproductives do not mate on their nuptial flight. In a particular behavioural sequence of dealation, monogamous pair building, and tandem running, a new nesting site is searched for by the royal couple (Escherich 1909; Lüscher 1951). Only after the nest is established does mating take place. The alates are at enormous risk during this brief period of mating and colony founding. Myles (1988, p. 387) states that "*the age specific mortality rate for dispersing alates is virtually unparalleled for adult animals*". A huge variety of predators ranging from ants to mammals and even man (Deligne et al. 1981) loves the defenceless and tasty alates for breakfast, lunch, dinner or snack in between. For example in *Odontotermes assmuti* 99.5% of all alates are prone to predation. Of the remaining 0.5% incipient colonies, only 12% will survive for more than 3 years. Thus the overall probability of survival to produce alate offspring from dispersing offspring is 0.0006. Thus, there is one surviving alate in 1666 dispersing ones (Basalingappa 1970). The alate production of a colony is necessarily high. Myles (1988) compares alate production and swarming with "*ejaculated gametes when one considers the colony as a sessile superorganism*" (see Chap. 6). The colony will have to grow for many years before the first alates are produced. Skaife (1954) observed a colony of *Amitermes hastatus* which required 4 years after founding to reach "sexual maturity".

Though colony founding by groups of queens has not been recorded, reproduction via colony fission was first described by Grassi and Sandias (1893) for *Termes lucifugus*. Grassé and Noirot (1951) studied colony fission in *Reticulitermes* species. They observed a column consisting of a physiogastric queen, a king, workers and nymphs emerge from the ground. The group moved some distance and then dug itself in. Other groups of non-reproductive termites were observed later, yet they returned into the hole they came from. Harris (1971) mentions similar potential colony foundations by fission for *Syntermes* and *Trinervitermes* in West and East Africa. The phenomonon certainly involves a dramatic behavioural change, because the termites usually are negative phototactic and avoid light. The social fragmentation of the colony depends on the ability to substitute the lost reproductive. Therefore the presence of nymphs in an advanced stage of development seems to be a requirement for colony fission in termites.

Another mode of reproduction is the foundation of a colony by isolation of an already existing one. A precondition is that there are substitute reproductives and that the nest is of a sufficiently diffuse nature to allow parts of it to be isolated. Harris (1971) calls this form of colony founding "budding", yet we will later see that this analogy only partly holds true (see Chap. 6). "Budding" of colonies is frequently found in two East African Rhinotermitidae, *Schedorhinotermes lamanianus* and *Coptotermes amanii* (Harris 1971; Watson and Abbey 1985; Lenz and Barrett 1982; Nutting 1969). These two colonial strategies of reproduction indicate a particularly high level of superorganismic organization. Although colony fission is less common in the termites than in social ants or honeybees, they provide nice systems to study the evolution of colonial reproduction via facultative colony fission.

The impressive above ground nest structures are only found (with the exception of *Coptotermes*) in the more highly developed Termitidae. They house enormous numbers of individuals numbering in the millions in large colonies, often in great vertical structures above ground (as high as 7 m), huge mounds on the ground surface, complex subterranean galleries, or hanging from trees. These colony-built structures dominate areas of the Australian steppe and subsaharan Africa. Tree-dwelling species of termites in South America build solid home structures under large branches (Frisch 1974). Many species, such as those in the genus *Apicotermes*, have extensive subterranean caverns with little evidence above the ground surface indicating the presence of a home site (Schmidt 1964). In this genus, existing in the subsaharan rainforests and savannas, the underground nests are egg-shaped and constructed of excrement from the termites themselves. The nests are complex and species-specific but share some structural detail (Frisch 1974; Schmidt 1964). Most nests have cellular internal structure in storied arrangement with the stories separated by curved, parallel lamellae. These lamellae fuse centrally to form a helicoidal column that serves as a spiral staircase providing communication between stories. Another envelope surrounds the nest with perforations in the inside wall, too small to be used as passageways by the termites. In the 18th century Smeathman (1781) suggested that these openings probably serve as a system to facilitate ventilation, as evidenced by the physics of gas exchange function and pore spacing and capacity for maintenance of ventilation even during extended periods of heavy rainfall.

The ventilatory and thermal properties of nest construction in termites is well documented and is shown best in the compass termites (*Amitermes meridionalis*) of Australia, the species of *Cubitermes* in the rainy tropics, and the mound termites (*Macrotermes bellicosus*) of the Ivory Coast of Africa. The compass termite colonies construct narrow vertical structures with the broad sides facing east and west. This reduces the heat load during the hot midday and warms the nest in early morning and in the evening by absorption of solar radiation on the surface. In the cooler season, the residents congregate at their preferred temperature on the east side in the morning and on the west side in the evening (Frisch 1974). The rainforest-dwelling *Cubitermes* colonies build mushroom-shaped termitaries with overhanging roofs which serve to protect the main structure from the torrential rains. That this shape serves as an umbrella and not a sunshade is shown by other species that exist in arid regions and do not build the mushroom roof portion on their structures.

1.3.2 Superorganismic Ants

The typical ant life cycle shows that most species meet the basic definition of a superorganism. Most ant colonies are initialized by individual foundress queens which rear their first brood on their own with no help from relatives (haplometrosis) or together with other conspecific queens (pleometrosis). The mated queen (= superorganismic zygote) forms the basis of the new superorganism. In contrast to the termite societies, there are many ant species that form reproductive swarms in which mated queens are accompanied by a large number of workers. Comparing this with reproductive behaviour of individual organisms, we have a mix between vivipary and budding. It is clearly different from budding since sexual processes are involved and the new superorganism differes genetically from its parent (like in vivipary).

Some ant superorganisms are less sessile than most other species easily classified as superorganisms. Colony agility is particularly well expressed in some army ants (*Aenictus* spp.) which emigrate more than once a day during the period of peak larvae production. Other ants, however, show the typical behaviour of superorganisms and invest considerable resources in nest construction. *Camponotus herculeanus* nests have been observed to remain at the same location for more than 10 years (Hölldobler and Wilson 1990). The high mobility of the army ants seems to be due to the limits in prey availability. Lack of sufficient food resources is a typical reason for all superorganisms to change their location. Therefore it is not surprising that the actual behavioural patterns in colony migration of the more sessile types is very similar to the agile species (Wilson 1980; Fowler 1981).

Nest structures vary even more widely than in the termites, from the modest passages and cavities of *Myrmecia* in Australia to the woven nests of Oecophylla in tropical Asia. To generalize is impossible and the interested reader is referred to the excellent treatise of Hölldobler and Wilson (1990). Here, we will only give a few examples of the numerous possible dwelling constructions that could be considered. Although ants generally do not hoard food for winter, in the tropics and subtropics where drought rather than cold periodically threatens the colonial existence, some accumulation of food does occur (Frisch 1974). This food, such as seeds, is kept in storage chambers. The carpenter ant, *Camponotus herculeanus*, tends to build in soft wood that is partly decayed. When it builds a nest in a tree, the geometry is determined by the structure of the tree's annual rings. The softer inner portions of each ring are easier to gnaw out leaving the harder parts in place. The workers remove the wood particles. The stronger walls serve to protect the nest cavity. Ants lack the sophisticated systems of temperature regulation found in bees. However, mound building species in temperate latitudes (e.g. *Lasius niger*) take advantage of the mound shape in absorbing solar radiation. Workers maintain the slope of the mound for best absorption when the sun is low on the horizon. Instead of regulating the desired temperature and humidity in a brood area like the bees do, ants carry their brood to areas with the best conditions existing at the particular time.

In temperate climates, most ants have a dramatically reduced activity in winter. The colony temperature drops considerably, and the individual members fall into a chill coma. Ants have only limited capacities for endogenous thermoregulation. Yet, the lack of thermal homoeostasis is not necessarily a contradiction to the superorganism definition. Many organisms, even homoeothermic birds and mammals, survive unfavourable periods or even overwinter with reduced metabolic rates. Some ants, like many of the Formica species, have food stores in specially designed workers (repletes). This gives the colony the option to invest in sexual reproductives before the actual foraging season has started. The queens are mated early in the season and they are carried off by the workers when the colony subdivides later in the year. Clearly, *Formica* species such as *F. polyctena* meet all requirements we expect in superorganisms. An example of a non-viviparous superorganismic ant species may be *Camponotus herculeanus* mentioned above. In this species the colony diapauses. The overwintered workers do not actively increase the colony temperature as *F. polyctena* before May. Mated queens start haplometrotic nest founding by sealing themselves into old beetle burrows. In this "embryonic" superorganism stage the queen lives on her body reserves until the first workers are reared and the colony develops into a mature superorganism.

1.3.3 Bees

Two lineages of bees, the stingless bees (with the genera *Melipona* and *Trigona*) and the honeybees (*Apis* spp.) have evolved superorganismic life cycles in which the only solitary phase is restricted to mating (Michener 1974, Sakagami 1982). Over 150 years ago the honeybee colony was referred to as a "perfect society" (Bevan 1827) and obviously it fits our definition of a superorganism very well. The bees are the only superorganisms with exclusively winged members. This places them in sharp contrast to all other species that form superorganisms. The anatomical constraints that are needed for a flying organism may be a chief reason why the caste system is less diverse than in other social insects and why age polyethism is the basis for division of labour (Lindauer 1952; Rösch 1927). Huge mandibles as defensive weapons of ants or termite soldiers may be possible for ground forces but may hinder in an air defence.

In superorganismic bees, there are only two morphologically different castes in the female sex; queens and workers, and division of labour are achieved via age polyethism (Seeley 1982a; Lindauer 1952). The diet of the bees may be an additional reason for the restricted caste polymorphism in bee superorganisms. Most species are strict vegetarian and only feed on floral carbohydrates and proteins. A necker tank, usually in form of a honey crop, and an efficient device for pollen transport (corbiculae on the hind legs or specialized abdominal hairs) are prerequisites if you want to be a social bee. Some species facultatively forage on dead animals (Chance 1983; Crewe 1985; Baumgartner and Roubik 1989). Only a very few tropical stingless bees, such as *Trigona hypogea* have been shown to be obligatory feeders on decaying animals (Roubik 1989).

The bees, then, with a main focus on the honeybee, will serve as the principal example in our further chapters on the biology of superorganisms rather than the other superorganismic systems. They are the best-studied social insects and substantial data exist on most aspects of their biology including behaviour, physiology, genetics, ecology and reproduction.

1.3.4 The Naked Mole Rat

In the early 1970's Alexander (1974) designed a theoretical eusocial mammal which had obligatory characteristics of living underground with a caste system. Jarvis (1978, 1981) found that the naked mole rat (*Heterocephalus glaber*) exactly met the requirements postulated by Alexander. The question is whether these subterranean mammals also meet the requirements to be considered as another example of a superorganism.

This species of colonial mammal has received considerable recent study because of its physical characteristics and its unique approach to social decision-making among vertebrates which is the only known case to match the eusocial insect superorganism model. Naked mole rats are sightless and hairless mammals that live in large underground colonies in northeastern Africa. Colonies with up to 295 individuals have been unearthed (Brett 1990). The average colony consists of 74 individuals which maintain and defend the colony and assist in rearing the young of the single fertile female and her several male mates (Reeve et al. 1990). They show typical eusocial behaviour including division of labor and reproductive suppression. Although there is only a single reproducing female (Kaayanja and Jarvis 1971), a clearcut multicaste system as claimed by Jarvis (1981) could not be found in follow-up studies (Lacey and Sherman 1990; Payne 1982; Isil 1983). Age polyethism, however, could not be excluded because body size co-varied with both age and behavioural tasks (Lacey and Sherman 1990). Yet Jarvis et al. (1990) indicate that body size is a poor predictor of age which leaves space for further studies in this area.

A general problem in the analysis of social behaviour of naked mole rats may be that they can only be observed in a very unnatural environment of a maze of acryllic tubes. Although the naked mole rat meets many requirements of our definition of a superorganism, there are some critical points that need to be addressed and may conflict with the concept.

1. The adolescent colony size is very small compared to superorganismic insects societies. This is critical, because the response of a naked mole rat colony very strongly relies on individual performance of certain colony members. In large insect colonies mainly the number of individuals involved in a certain task regulates colonial characteristics. The actual quality of the contribution of the individual becomes less significant. It makes little difference in a raid of army ants if there are some workers that do not contribute in an adaptive fashion. However, if a tunnelling mole rat stops to dig, there will be no tunnel.
2. The mode of dispersal of naked mole rats is still unknown. This is a critical aspect because it could easily be that a new colony founded by a single breeding pair needs a very long time to produce a full-size colony. If selection during this developmental period becomes more important than inter-colony selection, then it seems unnecessary to invoke a definition like superorganism for this extraordinary social mammal. Lacey and Sherman (1990, p. 328) therefore correctly pointed out that the eusocial naked mole rat colonies contain elements that can be interpreted as colony level selection (i.e. a "superorganism"), yet there are distinct differences to the superorganismic insect societies and it needs further sudies to reveal the life history in even more depth than achieved so far.

1.4 What Not to Expect from Superorganisms

Although superorganisms may behave like intelligent organisms it certainly seems far-fetched to assume that any member performs particular tasks because it knows the colony has a certain need for it. The search for Maeterlinck's (1901) mystical "spirit of the hive" certainly will be futile. In a group of 2 million individuals, which is common in termite nests (Fenton 1952), it is impossible that every member knows what activity meets the demands of the colony best. A mature colony of three to four million workers of the leafcutter ant (genus *Atta*), occupies three thousand or more underground chambers, and piles the excavated soil in mounds six meters across and more than a meter high. Clearly the queen does not function as a leader in this huge society and there is no central command allocating the tiny workers to the fungus gardens or the three hundred times heavier soldiers equipped with large sharp mandibles for colony defence. Even human society has serious problems in reaching every individual of such a large group number in spite of what we think is a reasonably advanced communication system with all sorts of contemporary media. It is naive to assume that all the termites could possibly achieve detailed information on the status of their colony at any time. It seems much more likely that, based on a very restricted behavioural repertoire, the workers will react to the conditions they meet in their direct environment. The sum of all these local reactions will eventually result in what looks like colony intelligence and a tremendous coordination achievement (Deneubourg et al. 1989). Processing is done by very many "stupid" parallel processors, each one solving its own very little task (see Chap. 4).

Chapter 2 Evolution

It may be said that natural selection is daily and hourly scrutinizing, throughout the world, every variation, even the slightest; rejecting what is bad, preserving and adding up all that is good. (Darwin 1859)

2.1 From Solitary Individuals to Superorganisms: The Theories

2.1.1 Darwin's Dilemma

The basis for the operation of superorganisms is the presence of eusociality, which according to Wilson's (1971) definition implies overlapping generations, division of labour, and the presence of specific castes. The presence of infertile worker castes poses a special problem in evolutionary biology. Darwin (1859) got caught by this when he developed his evolutionary theory. Apparently infertile workers broke the rule of the survival of the fittest and yet social insects were extremely successful in nature and widely aboundant. But how could it be that something non-reproductive could so successfully survive the struggle for existence? "*I will not here enter on these several cases, but will confine myself to one special difficulty which at first appeared to me insuperable, and actually fatal to the whole theory. I allude to the neuters or sterile females in insect communities*" (Darwin 1859).

Darwin focussed his interest on the honeybee, and developed an argument that rescued his theory. In social insects it is not the individual worker which is the most important unit for natural selection, but rather the colony as a whole. The biological unit of a honeybee colony was a well accepted phenomenon in Darwin's time. The "Bien" was a common expression in the German beekeeping literature around 1900 (Jacoby 1949), which specifically denotes the honeybee colony as one ("super")organism. Darwin followed this track and resolved his problem by invoking a very elegant model. He stated that if the workers increase the fitness of the colony, the related queen will produce more offspring, which will be to the benefit of the workers themselves. Darwin perceived the importance of the family (thus a related group) as well as the individual, and therefore made use of both kin and group selection to save his theory of evolution through natural selection in social insects. "*This difficulty* (the one of sterile workers), *though appearing insuperable, is lessened, or, as I believe, disappears, when it is remembered that selection may be applied to the family, as well as to the individual, and may thus gain the desired end*" (Darwin 1859).

Darwin gives a well-taken hypothetical example of cattle breeding, that shows how selection on infertile offspring can take effect via the reproductives. "*. . . I do not doubt that a breed of cattle, always yielding oxen with extraordinary long horns, could be slowly formed by carefully watching individual bulls and cows, when matched, produced oxen with the longest horns; and yet no one ox could ever have propagated its kind*" (Darwin 1859). Here Darwin presents us the kin selectionist approach, in that the genes selected in the offspring are inherited by the parents.

It took evolutionary biologists more than 100 years to grasp the theoretical consequences of Darwin's model. It was not Wheeler's (1911) model of colonies of social insects as superorganisms (which focussed mainly on the colony as the unit of selection, and neglected the individual) that enlightened our understanding of those evolutionary mechanisms which may have formed sociality in insects, but rather the dispute between the group selectionists and individual selectionists [starting in the early 1960s, e.g. Wynne-Edwards (1963) vs Maynard Smith (1964)]. Indeed it was not possible for Darwin to approach these problems with rigour since the fundamentals of genetics and the discovery of male haploidy of Hymenoptera were still to come.

2.1.2 Individual Selection Models

2.1.2.1 Kin Selection

Hamilton (1964a,b) tried to overcome the use of group selection in the design of genetical theories for the evolution of sociality. The male haploid system offers a nice and highly plausible explanation as to why sociality occurs particularly often in Hymenoptera. The only other eusocial group of insects is found in the termites that belong to the phylognetically old insect group of Isoptera. Hamilton's theory is based on the so-called inclusive fitness, J, of an individual which, using the notation of Crozier (1982), can be defined as:

$$J_1 = S_1 + \Sigma G_{Y1} \delta S_{1Y}, \tag{2.1}$$

where J_1 = inclusive fitness of individual 1,
S_1 = direct personal fitness of individual 1,
G_{Y1} = relatedness of individual Y to 1,
δS_{1Y} = effect of individual 1 on direct personal fitness of individual Y,
$\Sigma G_{Y1} \delta S_{1Y}$ = indirect individual fitness of 1.

The inclusive fitness is the sum of classical individual fitness and the fitness gained through the reproductive success of other related individuals. The latter fitness component, which is also called indirect fitness, depends on the relationship between the two individuals. The more closely related the two individuals are, the more gains one has from the the direct fitness of the other. If the indirect fitness is large, it might be selectively advantageous for one organism to reduce its own direct fitness and invest in the fitness of a related individual. Such an investment will incur costs (c) for the donor and provide benefits (b) for the recipient individual. Whenever b/c is larger than the inverse of relationship, 1/r, the donor should invest in the fitness of the recipient. Hamilton's rule, therefore, is as follows

$$b/c = k > 1/r, \tag{2.2}$$

where b = benefit to recipient,
c = cost for donor,
r = relationship,
k = benefit-cost ratio.

The above measure of the relationship actually represents the pedigree coefficient of relationship, G, (see Sect. 7.3.2.1) since r is usually used as Wright's symmetrical coefficient of relationship. A directionality in the measure of relatedness is very important for the operation of the model.

The basis for Hamilton's theory was the single mating of the queen heading a colony. Under multiple mating, the average worker relationship decreases rapidly and weakens the argument why the male haploid system should particularly facilitate the evolution of sociality. Boorman and Levitt (1980) focussed on the problems by using $k > 1/r$ in population genetical models. They found that recessive altruist traits, in most cases, have an advantage over dominant ones. Furthermore, they found a wide range of possible polymorphic systems where the potential altruistic allele does not go towards fixation in spite of $k > 1/r$. They also found various fitness conditions under which polyandrous matings favour altruism more strongly than single matings. Boorman and Levitt's models, however, are one-locus systems and do not consider pleiotropic gene effects. The results of models considering pleiotropic effects of large numbers of genes are substantially different. In the quantitative genetic models of Yokoyama and Felsenstein (1978) and Cheverud (1985), results were consistent with Hamilton's $k > 1/r$. Generally under weak selection and slow Darwinian evolution $k > 1/r$ works well (Hamilton 1987; Michod 1982; Boyd and Richardson 1980). Nevertheless, it is obvious that Hamilton's heuristic rule bears many hidden genetic problems not all of which are yet resolved.

2.1.2.2 Assured Fitness Returns

An extension of Hamilton's (1964a,b) ideas was presented by Queller (1989) in the so called head-start hypothesis. He argued that the workers have a reproductive head start, because there is already related brood in the nest. A little additional investment of a single worker can cause a brood member to become a reproductive individual. The solitary female, however, must survive through the entire egg to adult cycle in order to produce fertile offspring. This clearly is a disadvantage for the solitary strategy whenever the fitness of a solitary female (expressed as the product of the probability of survival s, the number of offspring b, and the relatedness r) is less than that of a worker. Thus there is a selective advantage for sociality if the following is true

$$r^*s^*b^* > rsb, \qquad (2.3)$$

where s = probability of a solitary female to reach the age of first production of fertile offspring,

b = number of offspring produced, given a solitary female survives,

r = genetic relatedness between mother and offspring,

s* = probability of a worker to reach the age of first production of fertile offspring,

b* = number of offspring produced,

r* = average individual relationship between the workers and the brood they rear.

Using this model, Queller (1989) obtained huge fitness advantages for sociality, and proposed that social insects should be able to produce between 5.3 and 17.4 times as many offspring as their solitary counterparts. Gadagkar (1990) found some flaws with this theory that do not invalidate the approach but reduce the power of sociality. His main point is the assumptiom of $s^* = 1$. Queller assumed that the probability for a worker to survive until the first attended brood becomes fertile is 1. Gadagkar (1990) correctly points out that this assumption is unlikely to hold true, because the "*contribution to any worker's fitness should clearly be in proportion to contribution towards the rearing of each offspring; otherwise full fitness for rearing a larva gets assigned to several workers on account of the same larva*" (Gadagkar 1990, p. 18). Gadagkar (1990) modifies the head-start concept further by redifining b^* precisely as the number of individuals that result from the efforts of a worker provided she survives the entire brood cycle. Figure 2.1 shows how b^* and b relate to each

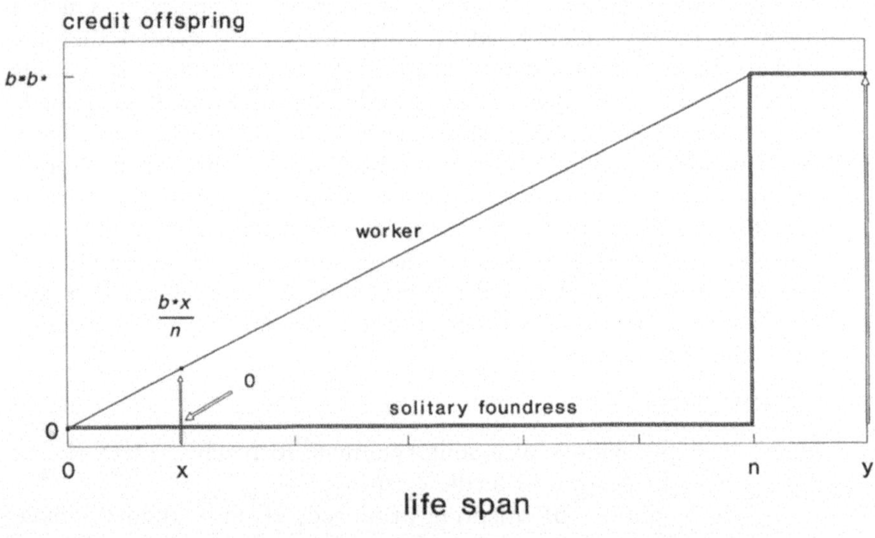

Fig. 2.1. The relation between b (returns for workers) and b^* (returns for solitary foundress). The solitary foundress has to survive the developmental time of the brood (n) to get any returns. If she dies beforehand, for example at time x, she will have no offspring. Workers dying at time x, however, can get partial credit for rearing offspring (b^*x/n), because the colony as a whole will survive until time n. In this model a longer life span (y) does not increase the number of offspring because only a single synchronously produced batch of brood is considered

other. The major conceptual addition is, however, the redefinition of s* which is a function of a survivorship curve and can be expressed as

$$s^* = \sum_{i=1}^{n-1} p_i(i/n) + \sum_{i=n}^{\infty} p_i \tag{2.4}$$

where p_i = proportion of workers with a life span of i days,
n = developmental time from egg to adult in days.

Obviously, this more realistic definition reduces the dramatic advantage postulated under the head-start hypothesis. Using survival data from an empirical example of a primitively social wasp (*Ropalidia marginata*), Gadagkar found that workers would break even with solitary foundresses if they are either related to the brood by at least a scant $r^* = 0.14$, or if they rear more than 28% of the brood to the adult stage, compared to that raised by solitary females.

The assured fitness concept builds on the uncertainty of solitary females producing fertile offspring. The workers gain from partial investment to the brood and even if they die before the brood has reached the adult stage their investment is not lost because others continue the brood rearing. The assured fitness returns model becomes particularly attractive over Hamilton's (1964a,b) haplodiploidy model if the survival of workers is low. Only if the developmental time of the brood is less than 2 days does the concept of assured fitness returns not promote sociality. To our knowledge there are no social insects that have a developmental time of less than 2 days.

2.1.2.3 Parental Manipulation

The theory of parental manipulation has been mainly propagated by Alexander (1974). But again Darwin had thoughts on parental manipulation in social insects: "*I believe that natural selection, by acting on the fertile ants or parents, could form a species which should regularly produce neuters*..." (Darwin 1859). The basis for the theory is that a dominant parent, for example a queen, manipulates its offspring to maximize its own fitness. Though this is a plausible mechanism for the evolution of sociality, it has one unattractive drawback. It does not provide any explanation why sociality occurs particularly frequently in Hymenoptera. The rationale is similar to that derived for kin selection. An offspring individual A may increase its own inclusive fitness by using fitness units from a sibling B. According to kin selection theory this pays off if

$$\delta W_A > -\delta W_B G_{BA}, \tag{2.5}$$

where δW_A = fitness gain of A,
δW_B = fitness loss of B,
G_{BA} = relationship of B to A.

Under the parental manipulation theory, a parent will only favour such action if it is to the benefit of its own inclusive fitness ($\delta W_A > \delta W_B$ irrespective of the relationship between A and B).

Parental manipulation does not necessarily mean that the parent physically dominates its offspring. Especially in the case of the honeybee colony, the queen seems to have control via a genetic programming over her offspring (see Chap. 4). In bumble-bees or social wasps, however, parent offspring conflict often results in physical fighting (see Sect. 2.2.3.3.).

2.1.2.4 Mutualism and Reciprocal Altruism

The mutualism model was introduced into sociobiology of insects by Lin and Michener (1972). It is based on a system in which both partners gain from an apparent altruistic behaviour. Thus

$$\begin{aligned} J_1 &= W_1 + \delta W_2 G_{21} - \delta W_1, \\ J_2 &= W_2 + \delta W_1 G_{12} + \delta W_2 \end{aligned} \tag{2.6}$$

where J_1 = inclusive fitness of individual 1,
J_2 = inclusive fitness of individual 2,
W_1 = direct fitness of individual 1,
W_2 = direct fitness of individual 2,
$\delta W_2 G_{21}$ = indirect fitness of 1 though offspring of 2,
$\delta W_2 G_{12}$ = indirect fitness of 2 through offspring of 1.

If the indirect fitness gain is positive for both individuals, the mutualistic trait will be favoured by natural selection. Since both consitions

$$\delta W_1 G_{12} - \delta W_2 > 0 \text{ and } \delta W_2 G_{21} - \delta W_1 > 0$$

need to be met, the mutualism theory needs more preconditions than the two other models. Nevertheless, Boorman and Levitt (1980) showed in various one-locus models that after reaching a critical freqency, β_{crit}, an allele for altruism can be easily established (Fig. 2.2). Furthermore, once sociality is established it is much less likely to reverse to solitarity than to rise in the first place (Table 2.1).

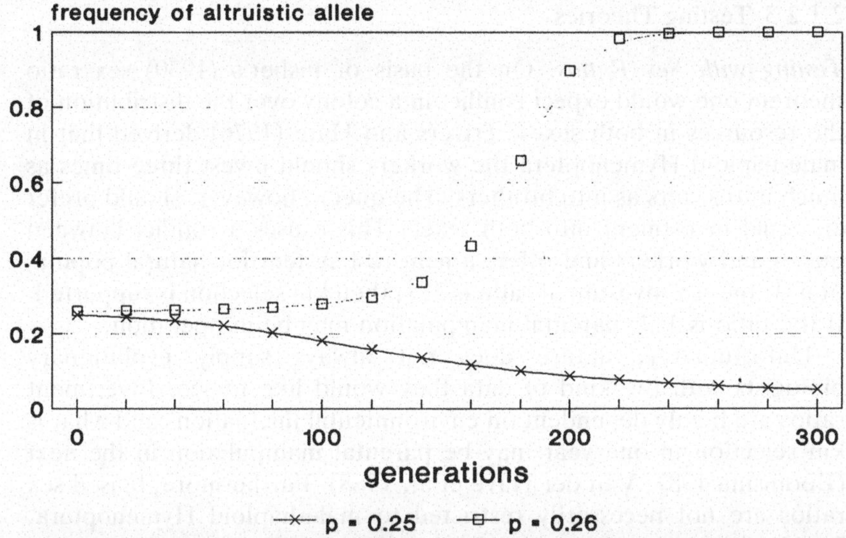

Fig. 2.2. Fixation of an allele for altruism. As soon as the frequency, *p*, for the altruistic allele is larger than 0.26, the allele will go toward fixation and is unlikely to be counteracted by drift or mutation. (After Boorman and Levitt 1980)

Table 2.1. The likelihood of fixation and reversion from sociality for a dominant allele for altruism (Boorman and Levitt 1980). L = size of social group, N = population size

L	N	Probabilities for a change from solitary → social	social → solitary
10	100	4.22×10^{-2}	4.0×10^{-4}
10	500	5.42×10^{-2}	8.1×10^{-7}
10	1000	5.09×10^{-2}	2.1×10^{-10}
100	500	9.83×10^{-2}	2.9×10^{-7}
100	1000	12.60×10^{-2}	2.1×10^{-11}

One basis for the fixation of such altruistic alleles is through random drift effects which are particularly strong in small populations. Mutualistic mechanisms may be especially important in primitively social bees. West-Eberhard (1978) found such mutualistic evidence in Metapolybine wasps.

2.1.2.5 Testing Theories

Testing with Sex Ratios. On the basis of Fisher's (1930) sex ratio theorem one would expect conflict in a colony over the distribution of the resources in both sexes. Trivers and Hare (1976) derived that in male-haploid Hymenoptera the workers should invest three times as much into sisters as into brothers. The queen, however, should prefer an equal investment into both sexes. This causes a conflict between queen and workers and offers a neat test system for natural population. If the sex investment ratio is 3:1, then kin selection is supported. If the ratio is 1:1, parental manipulation may be in operation.

Unfortunately, nature does not always supply evolutionary biologists with the kind of data they would like to see. Investment ratios are highly dependent on environmental fluctuations and what is kin selection in one year may be parental manipulation in the next (Boomsma 1987; Van der Have et al. 1988). Furthermore, biased sex ratios are not necessarily restricted to male-haploid Hymenoptera. Local mate competition may lead to female-biassed sex investment ratios (Alexander and Sherman 1977). In social spiders, female-biased sex ratios have been found (Aviles 1986, 1987) which are less likely to be affected by kin selection. Another complication may arise if both, parental manipulation and kin selection, operate simultaneously. As in most biological problems, there is no good reason why one should assume an either-or model. There are wide areas of fitness parameter overlap for both theories as shown by Starr (1979). Under polyandry of the queen and/or polygyny, the difference between kin selection and parental manipulation becomes even less clear since the investment ratio tends towards unity under kin selection with an increasing number of matings (Charnov 1978; Starr 1984; Moritz 1985b; Gadagkar 1985b). Kin recognition might counteract this problem such that members of particular subfamilies reveal intra-colonial nepotism.

In ants, the dry-weight ratio of the male and female sexuals provides a useful estimator of investment ratios in field studies (Nonacs 1986). In honeybees, however, the estimation of the sex ratio is less clear. Thousands of drones are produced, but only few queens. A strongly male-biased investment ratio? Not necessarily, as Macevicz (1979) or Page and Metcalf (1984) argue. They consider the workers which accompany the leaving queen during swarming or those workers that stay in the colony with a sister queen as investment in the female sex. Based on this argument they obtain an investment ratio close to 1:1. Certainly, the numeric sex ratio of queens and drones is further from being 3:1 than 1:1, weakening the arguments of kin selection and reducing the likelihood for realized evolutionary worker strategies (see also Chap. 6). Generally, testing both hypotheses by sex ratio studies seems rather cumbersome since sound data can be obtained only with repeated measurements of populations over years (Moritz et al. 1987c).

Testing with Kin Recognition. Another possibility to test both theories is indirect, using kin recognition. Individuals which are able to discriminate kin from non-kin, have a selective advantage under kin selection. Indeed, it seems unlikely that kin recognition systems would have evolved simply by chance. On the other hand, it may not always be easy to dicriminate true kin recognition from highly sophisticated nestmate recognition (Moritz 1988a, Chap. 6). And again, both parental manipulation and kin selection could be simultaneously in operation, so that the presence of kin recognition would support kin selection but not reject the parental manipulation model.

2.1.3 Group Selection

2.1.3.1 Classical Group Selection

Besides the efforts to explain the evolution of sociality on the individual basis, several attempts have been made to construct models on the basis of group selection. The major drawback of invoking group selection is that one of its preassumptions is what actually should be selected for, that is, groups. The goal of an evolutionary theory should be to explain the step from individuality to a social group. If the model requires groups before sociality has been evolved there is certainly a prerequisite that severely weakens the arguments of group selection theory in social evolution.

Nevertheless, once groups (social or non-social) are formed there is certainly no plausible reason why natural selection should not operate at the level of groups. The idea of groups as units of selection in the evolution of sociality was implicitly invoked in evolutionary models of the early twentieth century. Most of them, however, were heuristic descriptions of how the change from solitary life to sociality could have taken place. Legewie (1925a,b) misinterpretated observations on the reproductive biology of *Lasioglossum malachurum* in that he assumed the second brood female offspring would be produced by parthenogenetic laying workers and not by the queen. Although workers can produce males, Stöckhert (1923) observed earlier that it is only the foundress queen that produces female offspring in the colony, and he explicitly pointed out that Legewie's (1925a,b) to-be-published work was in error. History proved Stöckhert's criticism to be correct and his concept found repeated confirmation in later studies (Michener 1974). Nevertheless, based on his error, Legewie developed a very appealing theory in which ecological constraints on reproducing solitary insects cause offspring with a reduced fertility. His idea was that this offspring can only produce fully fertile reproductives if they cooperate. Cooperation may be a very simple and an almost automatic mechanism, mainly achieved via errors in brood care. Legewie (1925a) claimed repeated evidence for such mistakes in the sweat bee *L. malachurum* (Fig. 2.3). Because the first brood offspring workers of overwintering females nest all in the same cavity, they deposit their pollen loads and food supplies in any cell (not only the self-constructed one). Legewie's theory, though clearly based on misinterpretation of his observations and completely meaningless for *L. malachurum*, offers a nice alternative model to kin selection because it predicts that natural selection should operate against "own brood cell" recognition within the nest. The contents,

Fig. 2.3. Overwintered female of *Lasioglossum malachurum*: about to make a mistake? (Photo M. Ayasse)

rather than previous visits to the cell, are important for further provisioning or egg laying. The system will be evolutionarily stable if nests with highly mistaking females produce more successful reproductives than nests in which each female supplies only its own brood cell. Thus the frequency of mistakers in a nest determines its fitness, which clearly is a group selectionist approach as we will see below. Though his model does not fit the case of *L. malachurum* with a clear-cut reproductive hierarchy (although it may be applicable to drone production by laying workers), it fits nicely the case of *Exoneura bicolor* where more than one female founds a nest (Schwarz 1987, 1988a) with progressive feeding of the brood. There is no evidence of any intranidal kin recognition in this species (Schwarz 1988b).

Other models were less precise (Roubaud 1916; Wheeler 1918; Rüschkamp 1921) than Legewie's and focussed on trophallaxis as the central cue for the evolution of sociality. Here too, the social group was supposed to be a central unit in evolution. The major drawback of these and other theories before the 1960s was that they were descriptive and tried to find a tightly linked chain of biological examples from solitarity to sociality. No explanations were offered how sociality could have evolved on the basis of natural selection and genetic variation. The main question, Darwin's dilemma, of why and not how an individual should subordinate its direct fitness to another, remained unanswered.

Wynne-Edwards (1962) developed an evolutionary theory in which the group is the unit of natural selection and individuals play no significant role. Under this model, there is no dilemma, because the direct fitness of any individual in the group is insignificant and only the group's fitness is important. Yet, the theory needs a large array of preconditions to be met before sociality can evolve in a population of solitary individuals. The critical part is that a possible selfish trait will spread successfully in a population of altruists and cause their extinction. Only in small populations with large genetic drift and a minimal migration could social populations remain social. Furthermore, populations with social groups must be much more fit than the solitary populations. It was these preassumptions that caused severe criticism of group selection models (for a review, see Starr 1979).

2.1.3.2 Competitive Group Selection

A modified group selection theory was independently developed by Matessi and Jayakar (1973, 1976) and DS Wilson (1975, 1977), called *competitive group selection* because various subgroups in the population (trait groups) compete for reproductive success. The trait groups are the units of selection which produce individuals for a dispersal phase. In this dispersal phase, the individuals randomly regroup into new trait groups. The fitness of a trait group depends on the frequency of altruists in the group. The more altruists, the higher the fitness. Though the frequency within the group is critical for the group fitness, this must not be confused with classical frequency-dependent selection. The overall frequency of an altruistic allele in the population is unimportant for the fitness of the groups (Fig 2.4). Although this model has been said not to apply well as a general

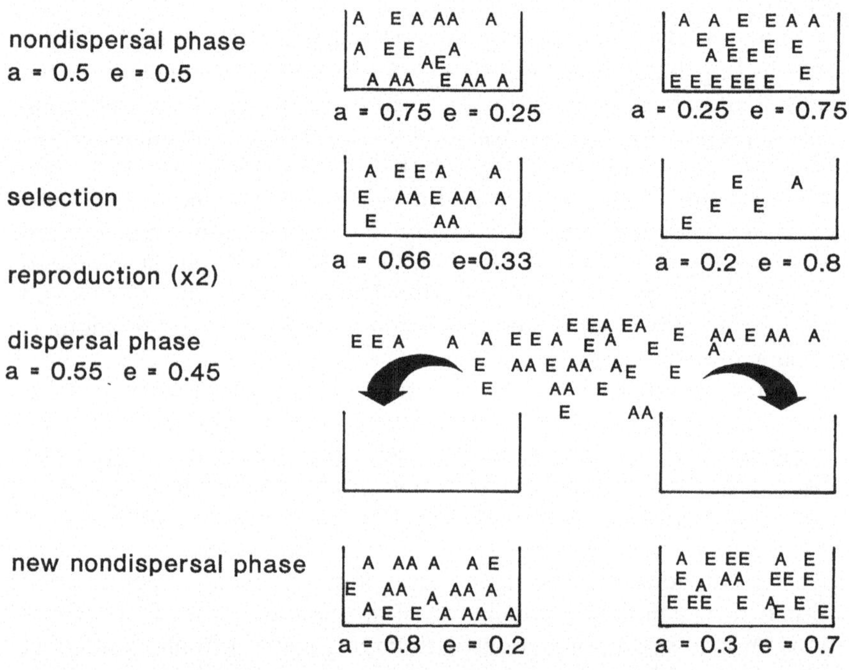

Fig. 2.4. Numerical example for the competitive group selection model (after Wilson 1975). Initially the frequency of altruists (*A*) and egoists (*E*) are equal (0.5; in this example 20 individuals of each type). However, there is stochastic variance among subunits in the population (the so-called trait groups in the non-dispersal phase). Groups with a low frequency of altruists have a reduced rate of individual intragroup survival (selection). This causes an increase in the overall altruist frequency in the population. Both surviving altruist and egoist individuals reproduce equally (x2) and regrouping occurs at random after a dispersal phase. Selection will again act upon the trait groups in the new non-dispersal phase

theory for the evolution of sociality in insects (Matessi and Jayakar 1976; Starr 1979), it fits nicely for some features of recent natural selection in honeybee populations. The difference is that the "trait groups" (= colonies) of honeybees are related and therefore we have an overlap of group and kin selection, probably operating synergistically. What is good for the group may also be good for the cooperating individual's inclusive fitness. However, there remains a testable difference. Under group selection, the recognition of one's own investment or that of kin is selectively neutral or even detrimental, which is not so under kin selection.

The following scenario may illustrate this. Suppose a primitively social bee returns with a load of pollen to the nest. There are two options: either deposit it as fast as possible to increase the gross pollen intake, or search for its own cell to invest only in her own future offspring. The latter may be advantageous for the direct individual fitness but not for the indirect or group fitness. The nest as a whole may have a reduced pollen intake because it takes the bee longer to handle the pollen in the nest. Schwarz's (1988b) observations on *E. bicolor* fit in this scenario since he could not observe any preference of brood by the bees. On the other hand, the lack of recognition does not rule out kin selection as an important mechanism of natural selection. If there is no genetic variation for a recognition system, it cannot be selected for even if kin selection is the exclusive evolutionary force. Yet, a complete lack of kin recognition mechanisms would be surprising, in the light of the fact that so many potential recognition labels have been experimentally found in the social insects. The problem with kin recognition seems to be whether the observed variability in kin recognition has any significance in natural selection (see Chap. 8).

2.1.3.3 Colony-Level Selection

Recently the colony of social insects as a unit of selection regains popularity in evolutionary theory (Owen 1986, 1989b; Moritz 1989; Seeley 1989). Looking at honeybee colonies, with their superorganismic organization, this certainly is highly plausible. Every bee breeder performs colony level selection when he decides from which colonies to rear queens to produce the next generation. The honey production, definitely a colony trait, usually is his major selection criterion. Natural selection is likely to operate in a very similar way, but will not restrict itself to group characters. It may operate simultaneously at all three possible levels and thus also via direct and indirect individual selection. With increasing complexity of the social organization of a colony, the group fitness is likely to gain in relative selective importance compared to the individual fitness. Group selection is undeniably of crucial importance for superorganisms, such as the honeybee colony.

2.1.3.4 Testing Theories of Group Selection

The documentation of the presence of group selection is not an easy task. What is selectively advantageous to the individual may also be advantageous to the group, thus producing the same phenotype. Even traits, detrimental to the individual but beneficial to the group, do not tell us much about selective mechanisms. Kin selection, as we have seen, easily explains the reduction of the direct fitness of an individual. Only if a trait that increases individual fitness but decreases group fitness is found, can we find evidence of actual group selection. The only example for honeybees is given in the Cape honeybee, *Apis mellifera capensis*, where individually fit workers cause poor colonies (Hillesheim et al. 1989; Moritz 1989). However, before we deal with this problem in detail, it is necessary to understand some genetics of superorganisms. We therefore will return to this problem in detail in Chapter 8, after having dealt with the genetics (Chap. 7).

2.2 From Solitary Individuals to Superorganisms: The Evidence

2.2.1 Sociality in Bees

The majority of bees live in a solitary or primitively social organization, far from representing a superorganism. Unlike in ants, where only eusocial species are known, in bees a variety of naturally occurring social organizations can be studied. Based on this biological background it is intriguing to develop a phylogenetic system similar to the level of social organization of bees. Sociality, though extremely efficient once developed, seems to be hard to evolve. From all bees, only the Meliponini, with the genera *Dactylurina*, *Lestrimelitta*, *Melipona*, *Meliponula*, and *Trigona*, and the Apini, with the only genus *Apis*, have developed colony structures that fit the definition of superorganism. In principle, following the above theoretical framework, there are three possible phenotypical "routes to eusociality" (Michener 1958; Wilson 1971; Lin and Michener 1972). One is via aggregation of nests, tolerance of conspecifics, and joint nest use termed the "semisocial route" (Schwarz 1988b). Another is via matrifilial colonies where the offspring aids her mother, the "familial route". Finally, there could be a direct one from solitary to primitively social life. In bees, however, there seems to be little evidence for a "direct route" from solitary females to social groups. The other two concepts, however, can be supported with behavioural evidence in various species of primitively social bees.

2.2.2 *Solitary and Subsocial Bees*

Those interested in the general biology and systematics of these bees are referred to Michener (1974) for a global discussion or Westrich (1988a,b) for a more detailed view of the bees of central Europe. Instead of listing all known species, we will focus on only a few representatives to look at possible social interactions between individuals that might have lead to superorganisms, and suggest possible evolutionary pathways (semisocial or familial) which they are following.

2.2.2.1 *Megachile rotundata*: The Individualist

Among the leafcutter bees, *Megachile rotundata* is the best investigated species mainly because of its commercial use as pollinator in Alfalfa (Bohart 1972; Parker and Torchio 1980). The common name indicates a general feature of these bees. They cut preferably soft leaves, which they use as nest material. The nest consists of a linear tunnel in which one brood cell is located behind the other. Megachilids are solitary bees although they frequently are found in aggregations. Aggregations seem to persist because emerging bees tend to return to their native nest sites (Michener 1974). Though *M. rotundata* is gregarious, there are no reports on sociality, i.e. two females provisioning one common nest. Each female constructs its own nest, interactions among females are extremely rare, and parent offspring interactions are generally lacking since this is a univoltine species. Each cell is mass provisioned with pollen before the egg is laid. One female provisions 8 to 12 brood cells on the average (Westrich 1988b). In every provisioned cell an egg is placed on a food ball of pollen and nectar (Fig. 2.5). After oviposition, each cell is sealed and the larva develops to a prepupal diapause stage with a cocoon in which it overwinters. The female deposits fertilized female

Fig. 2.5. Female *Megachile rotundata* placing a leaf piece in the nest hole. (Photo by W.P. Nye)

eggs in the first constructed inner cells and male eggs in the outer cells of each tunnel. The sex ratio depends strongly on the length and size of the nesting hole, but generally tends to be male-biased by 1:2 (females to males, Westrich 1988b).

Although there is no trace of sociality in this species, we can see that groups can form irrespective of social interactions. Simply the necessary restriction and the patchy distribution of nesting sites can cause individuals to aggregate. Indeed, it seems mandatory to have some mechanism to assemble individuals into groups before social interactions can occur. Such interactions can even evolve without any primary adaptive value, if, for example, a few females start making errors in provisioning cells. Looking at the current technique of commercial leafcutter beekeeping, such errors actually are likely to be frequent. Up to 3000 nest tunnels on a 50 × 60 cm square are provided per unit (Parker and Torchio 1980). Twelve of these units are placed in single shelters, thus involving as many as 36 000 bees in search for their cells. Let two bees enter the same cell, and let one decide to wait instead foraging and we have reached the first step towards social interactions. The error of the one bee would form the basis for social interactions. Communality and errors of individual bees therefore may be one pathway to sociality via the semisocial route without the establishment of a dominance hierarchy. If nesting sites are clustered and rare, competition may add an additional and selectively adaptive dimension to interactions between the bees in search of a nest. Fights among nests may occur, resulting in reproductive heirarchies. Apparently for *M. rotundata* both mechanisms are not significant because the bees maintain their solitary life history strategy in spite of many opportunities to become social.

2.2.2.2 Carpenter Bees: the Hierarchists

The biology of large carpenter bees has recently been reviewed in detail by Gerling et al. (1989) and Michener (1990). Many Xylocopine bees are univoltine, nest in dead wood, and construct a nest consisting of 3 to 15 cells. Though females of species in the temperate zones are mainly solitary nesting, social interactions among females occur depending on the environmental conditions. Whenever nesting sites are rare, there is strong competition for existing nest sites and nest defense becomes increasingly important. This, in combination with food shortage and robbing of nests, promotes a prolonged life of the mother. She may have to guard her offspring until emergence, with only brief foraging flights. As a result, the first step toward a matrifilial society is accomplished. If two females fight for one nesting site, the loser might decide not to leave the nest and instead functions as a guard. *Xylocopa pubescens* (Fig. 2.6) commonly found in the Middle East reveals such signs of drastic interactions among females competing for nesting sites. During nest establishment, physical fights determine a hierarchy between two competing females. The loser will become subordinate and will not produce offspring. It will only be involved in guarding (Velthuis 1987; Velthuis and Hogendoorn 1989; Van der Blom and Velthuis 1988). However, the chance of reproduc-

Fig. 2.6. X-ray photograph of *Xylocopa pubescens* nest. The foundress mother, one offspring daughter, various larval and pupal stages and pollen provisions can be recognized. Individuals with large heads represent males, those with smaller heads are female individuals. (X-ray photo J. van der Blom)

tion for the subordinate is not zero. For example, the dominant bee may die and the subordinate bee may produce her own offspring later on. Depending on the environmental conditions, the reproductive capacity of the social nests may be lesser or higher than that of solitary nests, which maintains a balance between social and solitary strategies. Sociality might be selected for via kin selection if the two females are closely enough related. However, since relationship data are lacking, it is unclear whether other selective forces may also lead to the social nests. For example, if poor ecological conditions cause severe constraints in nesting site availability, the social nests could be more common, whereas under favourable conditions there would be no "need" for the bees to share nesting sites. Nevertheless, here we clearly have a case where a species is on the doorstep to sociality. They have options to change from individual to social nesting depending on environmental situations. The individuals have become "plastic" concerning their nesting behaviour. If this phenotypic plasticity is determined by genotype – environment interactions, then natural selection can cause the spread of genes for dominance and subordination in the population.

A further step towards sociality has been done in another Xylocopine bee, *X. sulcatipes* (Gerling et al. 1989). Here also, nests can be founded facultatively by more than one female with the establishment of a dominance hierarchy. The subordinate females forage and the dominant female stays guarding in the nest. In some cases we find a clearcut matrifilial dominance hierarchy with the dominant mother as guard and the daughter as forager. If the mother dies or is expelled by the workers to found a new nest, a hierarchy develops among the sister workers with one female taking over the role of the egg layer. Although carpenter bees provide us with a nice example for primitive sociality, it does not yet explain how sociality evolved, since all of the above-mentioned theories can plausibly fit the system. Nevertheless, it clearly provides us with an example for the "familiar" route mechanism towards sociality.

2.2.3 Primitively Social Bees

2.2.3.1 *Exoneura bicolor* Smith: The Mutualist

The most unusual feature of allodapine nests is the lack of cells and the progressive feeding of the brood which is reared in a common burrow. They typically nest in dry stems but also readily occupy empty burrows of beetles in wood. Because the brood is not enclosed in sealed cells, the nest-founding female is the only possible physical protection against predation. Furthermore, the foundress must at least survive until the brood reaches the post-feeding stage. Both factors may facilitate the evolution of social structures in the nest. Michener (1985) called these, at least temporarily matrifilial nests, *subsocial* groups.

Sometimes, more than one laying female can be found (Michener 1971; Sakagami and Maeta 1985). Maeta et al. (1985) found almost 10% of all nests in an artificial population of *Braunsapis sauteriella* to be multiply founded. The phenomenon of multifemale nests is particularly common and well studied in an Australian allodapine bee, *Exoneura bicolor* (Fig. 2.7). Schwarz (1987) found 2.7 females (n = 119) per nest on the average. In contrast to closely related species (e.g. *Exoneurella lawsoni* Michener 1974), no dominance hierarchy could be found and all foundresses were mated and had equally enlarged ovaries (Schwarz 1986, 1988b). Co-founding seems to be a general feature of the genus *Exoneura* since Schwarz (1988a,c) found multiple foundresses in two other species (*E. richardsoni* and *E. bicincta*). The lack of a hierarchy is significant. All foundresses lay eggs in a common clump at the nest bottom and Schwarz (1988b) argues that kin discrimination of the foundresses is extremely unlikely. Since the larvae are progressively fed and not mass provisioned,

Fig. 2.7. Two adult females of *Exoneura bicolor* attending brood in the nest. (Photo M.P. Schwarz)

it should be advantageous for the foundresses to feed only their own brood, and even cannibalizing the brood of other foundresses (cheating) might be a selective advantage. The potential for such nestmate parasitism seems to be considerable in view of the parasitic *Inquilina* cuckoo bees (Wilson 1971). These bees enter the nest, and deposit eggs, but do not engage in brood rearing. Michener (1977) found that all parasitic species are very closely related to their hosts. Cannibalism would not pay off if inclusive fitness is larger under mutualism. Thus, closely related nest foundresses or large mutualistic benefits counteract nest parasitism. Indeed, Schwarz (1987) found that nest foundresses are more closely related than random samples of the population. Furthermore, nest parasitism has not been observed in *E. bicolor*, which fits nicely the model of mutualistic behaviour.

Could there be a possible advantage in not recognizing one's own brood? Schwarz (1988b) offers a plausible explanation on the basis of predator and parasite pressure. Since the brood is progressively fed, the nest indeed is quite vulnerable to raiding ants. A permanent guard at the nest entrance may prove to be most helpful to avoid predation. Thus, instead of wasting time by discriminating between own and foreign brood (which still is related anyway) it may pay off to feed the brood as fast as possible and then guard the nest. But certainly there are also other plausible advantages in communal nests if task performance is coordinated among nestmates. Such a system could be explained with colony level selection and the competitive group selection theory. If, furthermore, nest foundresses are related, there may be no need to discriminate among brood within the colony because the indirect part of the inclusive fitness is high enough, in any case, to facilitate $k < 1/r$, and inclusive fitness theory may supply another plausible theory to explain the evolution of communal nesting in *E. bicolor*. One theory, however, can definitely be excluded. Parental manipulation has clearly no significance because both parents died a season ago and have no obvious impact on the behavior of their mutualistic offspring.

However, *E. bicolor* is not an entirely mutualistic species. Mutualism is only expressed in the nest-founding phase. Offspring females that survive until winter may re-use the parental nest. In these overwintered nests the bees are semisocial with a very rigid reproductive hierarchy (Schwarz and O'Keefe 1991). The order of eclosion determines which female becomes the dominant one in the group. Ovarian differentiation may even occur between females that eclosed the same day. This hierarchy may be maintained over periods of more than 5 months. The reproductive female inhibits ovarian development of her nestmates via pheromonal cues (O'Keefe and Schwarz 1990; Schwarz et al. 1987). No ovarian inhibition has ever been found in newly founded multifemale nests (O'Keefe and Schwarz 1991).

It becomes obvious that the evolution of sociality in this organism is not a simple "either-or" model. Blows and Schwarz (1991) showed that inbreeding did not contribute to between-colony genetic variance as predicted under kin selection theory (Hamilton 1972, 1975). Relatedness in a population studied in Victoria (Australia) was not high enough to account for the presence of worker behaviour. Joint action of individual and group selection seems most likely to bridge the gap between plain solitary and social behaviour. As soon as the group is formed, selection operates on it. The formation of the group in a newly founded nest underlies plain individual selection since every female can choose to join or establish a nest of itself. The system of mutualistic cooperation, however, seems to be the exception in bee societies. Usually, as in other Allodapine bees, there are always clear-cut hierarchies with one dominant female producing most of the offspring (Michener 1974). A typical example may be the

Australian *Braunsapis* species in which a single foundress dominates over her offspring (Michener 1962).

2.2.3.2 Halictine Bees: Getting Social

Among the sweat bees, the biology of *Lasioglossum zephyrum* is the best studied (Batra 1964, 1966; Michener 1974). Its life history is typical for halictine bees. In the spring, single females found a nest (haplometrosis) by digging a small burrow. Occasionally other females may intrude after the initial nest has been excavated and a polygynous colony will be established (Michener 1974). The nest initially is 20–25 cm deep with two to seven brood cells (Fig. 2.8). Typically, there are nest aggregations, since suitable habitats are rare (vertical heavy soil banks, Batra 1966). The structure of such aggregations and the general colony composition may be different in diverse habitats (Kukuk 1989). After founding the nest, the queen supplies the cells with food and deposits an egg in each cell (mass

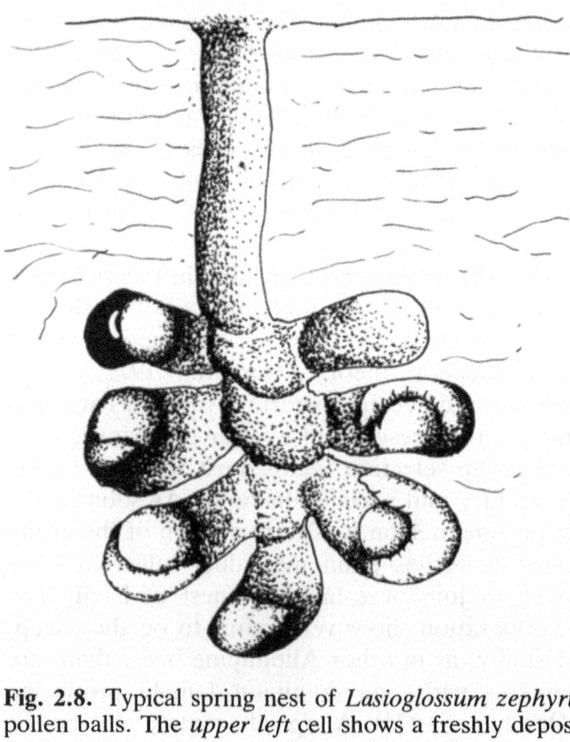

Fig. 2.8. Typical spring nest of *Lasioglossum zephyrum*. The larvae feed on pollen balls. The *upper left* cell shows a freshly deposited egg

provisioning). The brood cells are not sealed which allows the queen to remove faeces from the cells. The nest entrance is closed until the first brood, which consists exclusively of workers, hatches. The workers construct further cells, forage and defend the nest. The overwintered queen often dies in mid season and is replaced by her offspring, leading to a semisocial colony (eusocial if the queen survives). The offspring of the new queen is more likely to mate than workers of the old queen (Greenberg and Buckle 1981). Though sometimes more than one reproductive female can be found (Michener 1974), particularly in large colonies, only a single dominant female is the functional queen (Batra 1964; Breed and Gamboa 1977). Often it is difficult to discriminate between gynes and workers (Michener 1974). In autumn the number of males and gynes is increased and the colony dies as the gynes for overwintering are mated.

What are the selective forces that drive *L. zephyrum* towards or from sociality? In a rigorous experiment, Greenberg (1979) found support for the kin selection theory. He used the guarding behaviour of Halictine workers as a bioassay. The guard of *L. zephyrum* controls incoming bees whether they are nestmates or not. Greenberg (1979) reared several lines in the laboratory and could produce workers with different degrees of relationship. In over 1500 experiments he exposed the guard bees to intruding workers of various relationships (ranging from $r = 0$ to $r = 0.88$) from different colonies and found that the closer the relationship between intruder and guard, the more likely the guard would allow the intruder to pass. The conclusion was that the guards used a genetically based discrimination system and that environmental cues were less important, because the bees were from different nests. Buckle and Greenberg (1981) showed that learning certain labels from the nestmates in the adult stage was crucial for the discrimination ability. In experimentally mixed colonies (three workers each of two different lines) the guard accepted bees from different nests, but related to one of the two lines present in the tested nest. Bees related to either line were accepted at the same rate irrespective of the genotype of the guard. A further experiment showed that the guard learns its discrimination template from the other nestmates. In experimental colonies made up of five workers of one line and with one odd bee of another, the odd bee would not accept its own kin but rather bees related to the other nestmates. Apparently, learning is involved in kin recognition in *L. zephyrum* when the guard forms its template to discriminate the labels carried by the entering bees. These labels, however, are genetically determined and under natural conditions the system operates with minimal mistakes (Buckle and Greenberg 1981). The learning component and the "errors" made under experimental conditions do not imply that the discrimination ability is meaningless for natural selection. The skill of kin recognition will always affect natural selection irrespective of how it is achieved.

Although the potential for kin recognition might be a symptom of kin selection (Breed 1989), other explanations are also plausible for explaining the discrimination ability. Particularly in Greenberg's experiments, where nestmate recognition was tested, discrimination could equally well be a result of colony-level selection instead of kin selection as argued by Moritz and Hillesheim (1990). Furthermore, Smith (1987) found that colonies of *L. zephyrum* with closely related workers were more productive than those with unrelated ones, indicating possible selection at the colony level. Indeed, the genetic structure at the population, the aggregation and the intracolonial level might be of utmost importance if we want to study how social evolution operates in the sweat bee. At the population level, Crozier et al. (1987) found a weak but significant differentiation among various aggregations in Kansas, whereas Kukuk et al. (1987) found no such phenomenon in populations along five out of six streams in New York state. In spite of this partial contradiction, both of these studies and a pilot study of Greenberg (1982) seem to indicate that most individuals remain near their natal nests and only very few of them migrate. This is confirmed by looking at the nest aggregation structure. Crozier et al. (1987) found related "neighborhoods" within nest aggregations, a phenomenon predicted by Kukuk and Decelles (1986) on the basis of behavioural observations. Kukuk (1989) shows that early season colony initiation occurs in clusters and not at random. These features increase the relationship between potentially interacting individuals and therefore make kin selection potentially important. However, how high is the relationship in the nest? Polyandry or intraspecific nest parasitism as found in subtropical *Halictus ligatus* (Packer 1986a,b) might again reduce the intranidal relationship and therefore make kin selection more unlikely.

Polyandry and polygyny seem to play only a minor role in *L. zephyrum*. Electrophoretic studies indicate an average intranest relationship of $G = 0.7$ (Crozier et al. 1987) or $r = 0.72$ (as defined by Queller and Goodnight 1989) (Kukuk 1989) for the second brood produced by the non-overwintered female. Initially there is more variation in the nest. The first brood has a smaller (but not significant) estimate of the average intranidal relationship ($r = 0.43$ to

0.57, Kukuk 1989). The increased genetic variability may be due to nest usurpation or doubly mated queens (Kukuk et al. 1987). Anyway, the relationship among nest members is certainly high enough to allow kin selection. However, the sole presence of relationship does not exclude other selection mechanisms. Lin and Michener (1972) suggest that the semisocial evolutionary route is more likely than the familial route to sociality in halictine bees, yet they admit that "*some forms almost certainly have taken the familiar route*."

Again, we get no conclusive clue to exclude any of the above theories on the evolution of sociality. Individual selection models are plausible because of a high intranest relationship and a matrifilial hierarchy structure, and group selection models are plausible because the genetic make up of the colony determines its productivity. Likewise, it seems difficult to imaging which mechanistic pathway evolution took to shape sociality in halictides. The level of social complexity reached in *L. zephyrum* is already high and it is difficult to find specific traits that discriminate the possible routes to sociality.

2.2.3.3 *Bombus*: Controlling Environment

Bumblebees provide an example which adds a further dimension to social nesting in that they have partial control over the nest climate (Heinrich 1974). In spring, nest initiation starts with single overwintered queens building a small nest (Fig. 2.9). The queen will produce some eight to ten eggs, which she deposits on a pollen ball. She will also construct a honeypot to store nectar. During the development of this first brood the queen is mainly involved in keeping the brood warm. She actively regulates temperature and is able to maintain brood nest temperature above environmental temperature in the cold. Bumblebees are typical bees of the cooler and cold climate and only very few species are found in tropical habitats (for example *Bombus atratus*). Bumblebees maintain a brood temperature of 25 °C at an ambient temperature close to freezing even without any insulative material (Heinrich 1974). The energy spent during incubating can be as high as during flight with a full load of nectar (Heinrich 1979b). Overheating, however, may pose a problem to bumblebee nests since there are no reports on active cooling mechanisms. Since bumblebees are adapted to the cold, there may be little adaptive need for sophisticated cooling behaviour.

After the first brood emerges, the colony grows more rapidly and the queen restricts herself to egg laying. With further colony growth

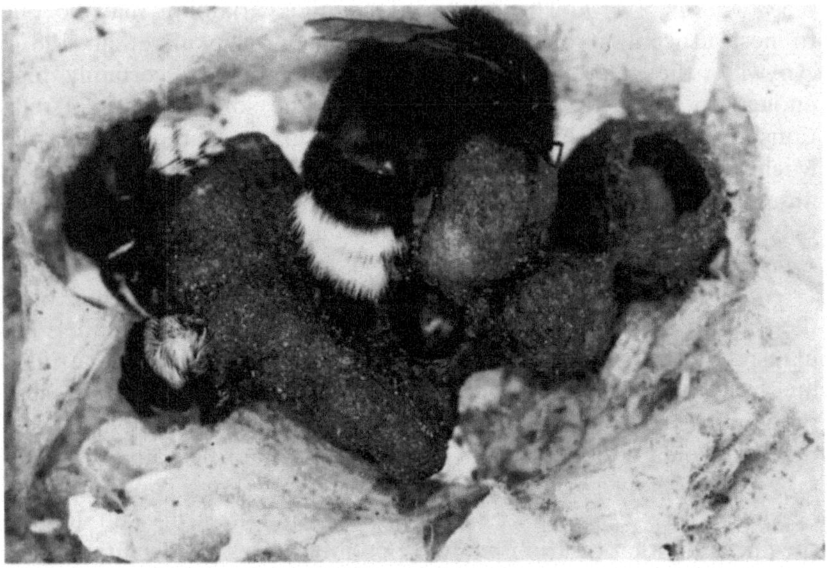

Fig. 2.9. *Bombus terrestris* queen warming a brood cell in an early stage nest. (Photo M.J. Duchateau)

she starts the production of drones (Michener 1974; Pomeroy and Plowright 1982; van der Blom 1986; van Doorn and Heringa 1986). After reaching a so-called "competition point" (Duchateau and Velthuis 1988, 1989; Duchateau 1989) the dominant position of the queen is challenged. Plowright and Plowright (1990) hypothesize that the queens can monitor their egg-laying and begin with drone production as soon as the first female eggs can be reared to queens. When the colony produces female reproductives, the position of the queen in the nest is clearly critical. Only a few eggs and larvae will reach adulthood due to excessive reciprocal oophagy (Sladen 1912; Richards 1977). Laying workers are physically competing with the queen among the production of males and more than 50% of all drones can be worker laid (Owen et al. 1980; Owen and Plowright 1982). In this stage, since only queens and drones are produced, the nest degrades rapidly. After mating the queens prepare themselves for hibernation.

The communication system of bumblebees shows a major difference to the sweat bees and forms a further step toward a superorganism. In contrast to *Lasioglossum*, workers of attacked bumble bee colonies alarm nestmates and defend their nest against attacking individuals (Heinrich 1979b). Although communication apparently has reached a more complex level, the colony as a whole is far from reaching a true superorganismic level. Information processing is mainly restricted to individuals. Typical for bumblebee behavioural ecology is its foraging behaviour. The workers make foraging decision based exclusively on their own field experience. Workers rapidly learn to visit nectar- and pollen-rewarding flowers (Heinrich 1976, 1979a). Although, the bees are individually optimizing, the colony as a whole may not be. There is no communication among the workers for rich nectar and pollen sources. There is no trophallaxis among adult nestmates which potentially could form a basis for communication of food quality. The nest economy of the bumblebee colony looks like an accumulation of characters of individual workers, and colonial characteristics merely represent the arithmetic mean of the individual characters of all nest members.

Reproductive hierarchies are similar to those of *Lasioglossum*. Dominant females "rule" mainly via physical strength (Röseler and Röseler 1977; Van Honk and Hogeweg 1981; Van Doorn and Heringa 1986). Van Honk et al. (1981) showed that a single *B. terrestris* queen can control only up to 80 workers. Body size is a major determinant in polyethism. Surprisingly, the small workers preferably remain in the nest, and the large workers join the foraging force, in spite of their high individual reproductive chances in the colony. It is mainly chance and the location of the brood cell that determines body size of the imagines. Plowright and Jay (1968, 1977) and Röseler (1970) argue that those larvae that develop more closely to the periphery of the communal batches receive less food then those more favourably placed, causing a wide variance in body size. Also temperature could play an important role. Nevertheless, division of labour is extremely flexible and workers dedicated to nest construction may shift to foraging, for example if all foragers are experimentally removed.

In addition to the physical dominance, queen pheromones have been shown to be active in various species. Mandibular gland secretions suppress worker egg laying in *B. terrestris* (Van Honk et al. 1980) and hinder worker ovary activity (Röseler et al. 1981). Pomeroy (1981) found further evidence for queen pheromone activity in six other bumblebee species. Van Doorn (1988) suggests that the retinue behaviour observed in *B. ignitus* (Katayama 1971) or *B. terrestris* (Röseler and Röseler 1977; Van der Blom 1986; Van Doorn 1988) may be elicited by queen pheromones. However, the pheromonal control and regulation system is far less sophisticated than in superorganismic insect colonies where the queen can "rule" without any physical power but rather "drugs" her worker force into submission (see Chap. 4).

It is not only the less developed communication system that makes a difference between a bumblebee colony and a superorganism. Another major factor is still lacking. Bumblebee colonies "die" at the end of each season. After the sexuals are produced, the virgin queens mate and overwinter in the soil. The colony degrades until the last workers leave the nest and die. There is no continuity of the colonial structure, and the superorganism term is not validated in this case. Colonies of the tropical *B. atratus*, however, seem to come close to functioning superorganisms since they are perennial. But here too, we still have an extended solitary phase in which selection exclusively operates on the individual during the nest-founding phase of queens. Bumblebees have reached a level of social organization that makes it an arduous organism in which to study the critical evolutionary steps from solitary to social life. Colony efficiency is an important unit in bumblebee fitness, contributing an additional factor to natural selection. Nevertheless, the bumblebee colony may serve as a typical intermediate example on the way from social individual organism to superorganism. The study of bumblebee colonies should be particularly rewarding if one is interested in intracolonial conflict between individual and colonial "interests."

2.2.4 Superorganismic Bees

All highly eusocial bees, according to the definition of Michener (1974), can be addressed as superorganisms. They have long-lived colonies, and an intracolonial homoeostasis which is independent of environmental variation. Food is stored in designated areas of the nest and the workers use glandular secretions to feed the brood. Division of labour is extremely well developed and typically pheromone regulated. Nest-founding is done within a social context. Actually, the queens are always in the company of workers except during mating. Another important difference to the primitively social bees is the ability of superorganisms to abscond (although rare in stingless bees). Under adverse environmental conditions the colony can leave the old nesting site in search for a new one (see Chap. 5). Colonies of primitively social bees have no option to change nesting sites. Those foundresses that choose a poor site are doomed to failure. The superorganismic bees are all in the subfamily of Apinae which is divided into the stingless bees of the tropics (Meliponini) and the old world honeybees (Apini).

2.2.4.1 Stingless Bees

All eusocial stingless bees within the two major genera *Melipona* (Fig. 2.10) and *Trigona* of tropics come close to or actually do meet the criteria of the superorganism definition stated in Chapter 1. The perennial colony forms an intact entity with brood and stores in combs. Nest climate is controlled by fanning or active heating. Nest climate control seems to be best developed in *Trigina spinipes* where the brood chamber stays about 35 °C (33.3–36.2 °C) under ambient temperatures ranging from 8.2 to 30.3 °C (Zucchi and Sakagami 1972). The cooling of the nest seems to be a problem. *Trigona denoiti*, a minute ground-nesting species in the African Highveld, prevents overheating during the hot midday by a reduced nest ventilation (Moritz and Crewe 1988b). At night-time the stale inside nest air is exchanged with cool fresh outside air. Overheating during daytime is further prevented via insulation of the nesting site. The nests can be up to 1 m deep in the ground where soil temperature remains very constant (Fletcher and Crewe 1981).

Fig. 2.10. *Melipona quadrifasciata* worker guarding the nest entrance

Social organization of stingless bees can be extremely complex and colonies have been claimed to exceed 180 000 individuals in some *Trigona* species (Lindauer and Kerr 1960). A distinct worker caste can be discriminated from the queen. The solitary phase in most species is reduced just to the mating flight. These mating flights can be extremely short since the males hover in front of the nests and the virgin queens are inseminated as soon as they leave the nest entrance. This is the one and only time the queens are not within a group of workers. New colonies are established via colony fission and even here the queen is guided by escort bees to the new nesting site which has been located and constructed by workers before the new queen moves in. A once established colony as a physical unit may have an extremely long lifetime though they certainly are not immortal as stated by Wilson (1971). Genetically, their lifetime is limited by the longevity of the queen. A new queen, even though a daughter of the old one, changes the genetic make-up of the colony, which, depending on the male's genotype, may have drastic effects on the further fate of the colony.

Colony Structure. A stingless bee colony with a long-lived queen is less genetically variant over time than a honeybee colony which changes its genetic make-up every swarming season. In stingless bees the old physiogastric queen remains in the nest and the virgin queens leave for new nesting sites. Furthermore, the monandrous mating system of many stingless bees reduces genetic variation within a colony. The monandry hypothesis is based on semen counts in freshly mated queens (Kerr et al. 1962). Furthermore, Kerr and Nielsen (1966, 1967) found support for the monandry model from phenotypic segregation, and da Silva et al. (1972) who noticed that the queen was unable to remove the mating sign (male genitals and seminal vesicles) during the mating flight.

Though monandry offers all features that favour the evolution of sociality according to inclusive fitness theory, several stingless bee species are know to be polyandrous. In *Lestrimelitta*, even laying queens can be mated by several males in the nest (Sakagami and Laroca 1963). Also gravid *Trigona* queens are attractive to swarming males outside the nest and readily mate (Michener 1974). It is hard to conclude from plain behavioural observations and experiments whether polyandry or monandry is a derived character in stingless bees. It is trivial that polyandry has originated at one time from monandry, but the case of *Melipona* monandry could equally well be a return from a previously polyandrous mating system. The intracolonial relationship in the polyandrous species is not known, still it is expected to be high enough to make kin selection operational. A strong impact of kin selection on evolutionary processes in the stingless bees does not exclude other selection mechanisms, which may be equally effective.

Chromosome Data. The phylogeny within the stingless bees has been investigated in some detail with cytogenetical studies. Kerr and co-workers (Kerr 1972a,b; Kerr and Silveira 1972; Silveira 1971, 1972) give data on chromosome numbers in stingless bees. Kerr and Silveira (1972) combined the chromosome data with morphological, physiological and ethological data from various studies (Cruz-Landim 1967; Kerr 1969; Kerr and Esch 1965; Michener 1944) and constructed a phylogenetic tree. Generally, the chromosomal set of Trigonid species is considered to be more derived than that of Meliponids. Kerr and co-workers regard a chromosomal set of $n = 8$ as the basic number of chromosomes in stingless bees. Species like *Melipona quadrifasciata* or *M. marginata* with $n = 9$ presumably arose via polysomy. *M. quinquefasciata* with $n = 18$ is thought to be derived via a duplication of the chromosome number. A vast number of *Trigona* species has a chromosomal set of $n = 17$. Kerr and Silveira (1972) suggest that this may be due to Robertsonian fusions of two chromosomes. The same may be true for those *Trigona* species with $n = 14$ or $n = 15$ which could present fused forms of an ancestor with $n = 16$. Some modifications of this tree have been suggested by Pompolo et al. (1989). Two *Plebeia* species were found to have $n = 17$ instead of $n = 18$, as stated in Kerr and Silveira (1972). However, even in the light of these additional data that included a C-banding analysis, the model is highly tentative because of the small number of species (48) karyotyped and the lack of variation in chromosome morphology (Crozier 1977). As long as the individual chromosomes cannot be identified, any thoughts on Robertsonian fusion or duplication remain speculative.

Fossils. The fossil record provides additional information that shows that much more data are needed before detailed phylogenies should be discussed. Recently a Meliponine bee fossil was found in the New Jersey amber ($96–74 \times 10^6$ years before present, Michener and Grimaldi 1988). The specimen is clearly a *Trigona* type stingless bee, very similar to *T. cilipes*, which is abundant in the Amazon Basin and the Guianas today (Fig. 2.11). The age of this fossil *Trigona* is most surprising, since *Trigona* is considered to be a highly social species.

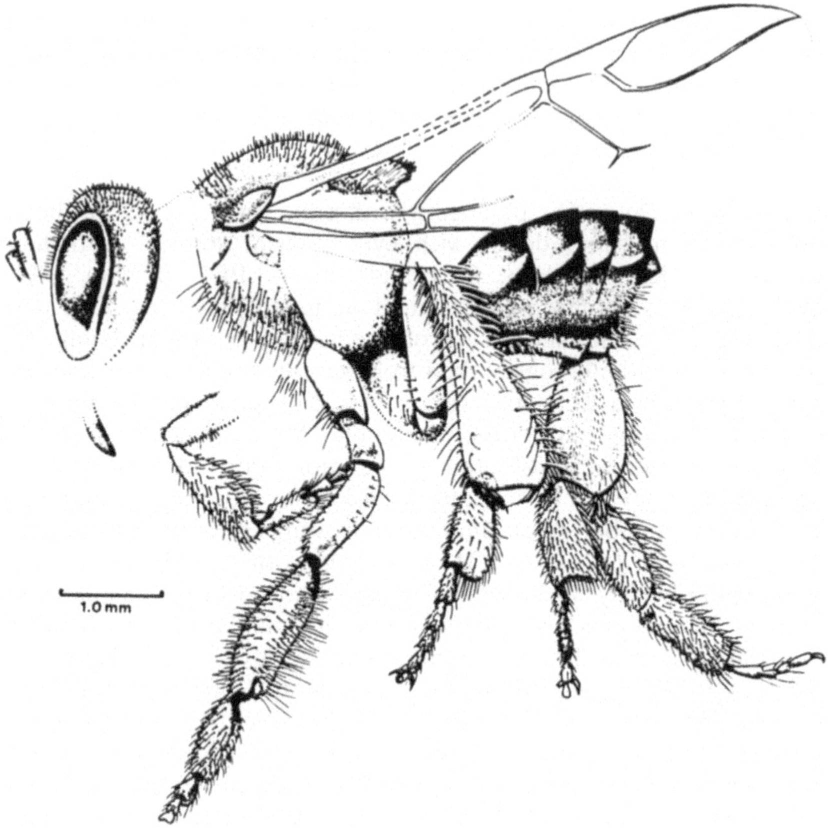

Fig. 2.11. Fossil *Trigona prisca* worker found in Jersey amber (94–74 × 10^6 years B.P. (Modified after Michener and Grimaldi 1988)

Since the fossil, *T. prisca*, was a worker, obviously sociality had been evolved by then. Bees are assumed to be co-evolved with the flowering plants (Michener 1979). Given that the first angiosperms appeared in the early Cretaceous (130 × 10^6 years B.P.; Cerpet and Taylor 1985; Crane et al. 1986), this leaves only 50 × 10^6 years for the Trigonids to evolve from a solitary prey-catching insect to a nectar- and pollen-feeding bee living as a superorganism. Furthermore, there is a surprisingly long stasis in evolution if the bees some 80 million years ago are almost identical to one of the most advanced superorganisms presently known. On the other hand, this may just reflect the tremendous effect of the co-evolution between flowering plants and their pollinators for the speed of evolution on the one hand and the rise and success of sociality on the other. Recently, the date of the above fossil has been challenged (Michener pers. comm.) because of associated fossil insects that are likely to be of more recent origin, but all other information still supports the late Cretaceous age to which the amber was originally attributed. Certainly the fossil shows that we are far from understanding the true story of Meliponine evolution.

Differences Between Stingless Bees and Honeybees. Notwithstanding the wide common basis of the biology of stingless bees and honeybees, there are distinct differences that suggest important divergences in the evolutionary pathway of both taxonomic groups. Winston and Michener (1977) conclude from the divergence in form and function of many biological features that the Apinae are not an offshoot of the Meliponinae but they evolved independently. The similarities are therefore considered as convergent evolution due to similar selection pressures.

The striking differences between the two superfically similar superorganismic bees have been recently reviewed by Engels and Imperatriz-Fonseca (1990). The most obvious difference, of course, is the lack of a sting in the stingless bees. But this has little to do with their social organization because even without a sting they can fiercely defend their colonies (for example *Trigona tataira* Kerr and Costa Cruz 1961). The next most evident difference is in the nest architecture. Vertical construction of combs which is found in very few stingless bees (for example *Dactylurina staudingeri*, Darchen and Pain 1966) is the rule in honeybees, and they construct the same type of cells for food storage and brood. In most stingless bees, brood cells are very different from the large honey and pollen pots. Another fundamental difference between honeybees and stingless bees is the mode of brood care. Honeybees feed their larvae progressively, whereas stingless bees mass-provision their cells. More subtle differences can be found in the caste determination system. Queens of Meliponine bees seem to be determined by a genetic two locus system (Kerr 1950, 1973, 1974), whereas honeybee queens can be reared from any female larva just by feeding it sufficient royal jelly. The reproductive biology is not only different on the behavioural level (for example drones mate the virgin queens just in front of the nest in Meliponine bees) but also the role of laying workers is very different in both groups. The nursing workers of stingless bees are generally fertile and contribute substantially to the number of drones produced by a colony (Sakagami et al. 1965; Beig 1972; Bego et al.

1983). Also, trophic eggs are laid that serve mainly as food for the queen (Akahira et al. 1970; Engels and Engels 1977) a phenomenon never found in honeybees. Long ago, Spinola (1840) observed that swarming is very different between these groups. In stingless bees the workers first establish a filial nest before the virgin queen moves in. The old physogastric queen remains in the nest. In honeybees the old queen leaves the hive while not-yet-emerged virgin queens remain in the colony. In stingless bees, swarming and colony fission is a long-lasting process with extended phases of polydomy (Engels and Imperatriz-Fonseca 1990), whereas in honeybees swarming has explosive dynamics (see Chaps. 5 and 6). While honeybees use a dance language for communication, stingless bees have evolved a scent marker system. Kerr (pers. comm.) believes that the scent trail of the Meliponine bees is even more efficient than the dance language of the honeybees in dense rainforest habitats.

The necessarily condensed and fragmentary collection of biological differences between stingless bees and honeybees very clearly shows that the similarities between the two groups are rather superficial indeed. Comparative studies on stingless bees and honeybees (Schwarz 1948; Winston and Michener 1977; Engels and Imperatriz-Fonseca 1990) generally come to the conclusion that the route to eusociality was very different in these groups of bees.

2.2.4.2 Evolution Within the Genus *Apis*

The genus *Apis* is comprised of several species that nest in the open (*A. dorsata*, *A. florea*, *A. andreniformis*) and those nesting in hollows (*A. mellifera*, *A. cerana*, *A. koshevnikovi*). Although open nesting has been suggested for *A. laboriosa*, yet the final confirmation that it is not just a subspecies of *A. dorsata* is still lacking (Ruttner 1988). Several attempts have been made to clarify the phylogeny within the genus *Apis*. Nevertheless, there is still no phylogenetic system that is free of contradiction. Generally, the study of evolution of the honeybees will contribute only a little to the understanding of the evolution from solitary individuals to superorganisms. We will restrict ourselves to a brief overview of the diversity within the genus *Apis* since Smith (1991) covers this topic in extensive detail. The reader with an interest in evolutionary biology is likely to be disappointed in the paragraph to come because we loose the tight grip of connecting theory and evidence. This is because honeybees confront us with an extremely complex and highly developed society in all species. Of course, we can study natural selection in the honeybee and the impact of sociality on selective processes in populations (we will deal with this in Chap. 8), yet on an evolutionary time scale we are restricted to establishing phylogenetic trees with the routine technology we know from the study of individual organisms. The basic structural framework of a honeybee colony is nearly identical in all species of the genus. Variability is clearly much less than in the other genera discussed so far. One reason for this limited degree of variability seems to be the very young phylogenetic age of the genus *Apis*.

Fossils. Although the fossil record of insects is vast in the Baltic amber, there are no reports on any *Apis*-like insects. Therefore, the genus is believed to be much more recent than the stingless bees. *Electrapis* is thought to be one of the possible links between stingless bees, bumblebees and honeybees (Winston and Michener 1977). However, in light of the potentially great age of the stingless bees (see above) it seems to be unlikely that the species can form a true link. The oldest *Apis* fossil, *Synapis henshawi* (Cockerell 1907) dates back only to 25–22 $\times$ 10^6 years B.P. Zeuner et al. (1976) classified the specimen as a subgenus of *Apis*. More recent specimens from the upper Miocene (Sarmatian 12 $\times$ 10^6 B.P.) were found in southwest Germany and were classified as *Apis armbrusteri* (Zeuner et al. 1976). The fossils reveal a clear trend in their wing venation pattern. Ruttner (1988) observes the elongation of the longitudinal venation and a migration of the third cubital toward the wing base. In a biometrical analysis, Ruttner et al. (1986) found that the fossil wing venation patterns are most similar to recent *A. dorsata* wing structure. Using a principal component analysis on 16 wing characters, the data of *A. armbrusteri* and *A. dorsata* are grouped in the same cluster (Fig. 2.12). Ruttner suggests that *Apis dorsata* is just the "mega" edition of *A. armbrusteri*. The similarity between both groups was already stressed by Armbruster (1938) and Maa (1953). The fossil evidence indicates that *A. dorsata* is the ancestral outgroup in the honeybees. Does this fit with data coming from other disciplines?

Behaviour. There were several attempts to clarify the phylogenetic relationships within the genus *Apis* on a behavioural ecological basis. Buttel-Reepen (1906) considered open nesting primitive in contrast to cavity nesting. However, Koeniger (1976) stressed that there is no convincing evidence against assuming cavity nesting to be more primitive than open nesting. It could be an adaptation of an initially cavity-nesting honeybee to the tropical environment. Indeed all trop-

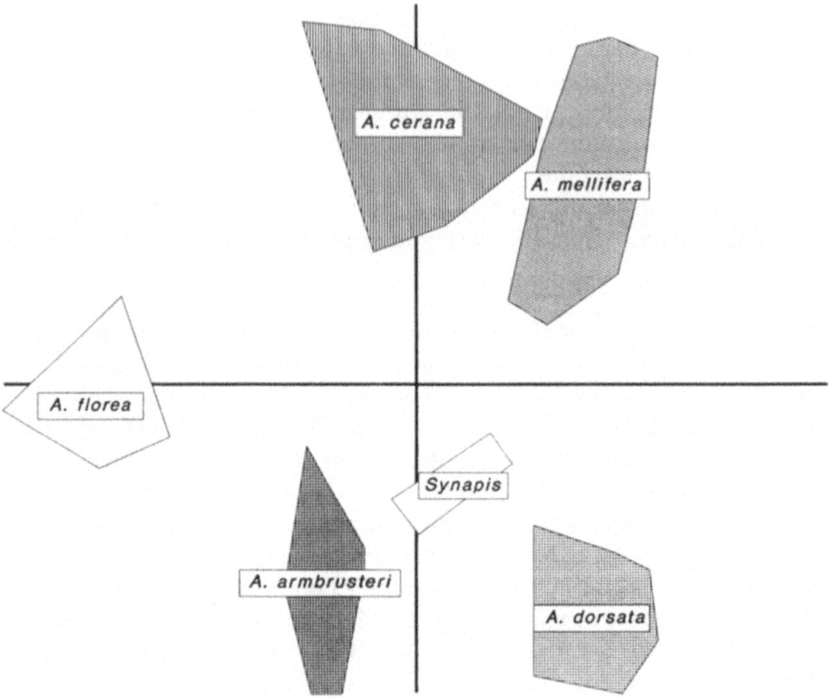

Fig. 2.12. Plot of the first two factors of a principal component analysis on 16 wing characters of *A. armbrusteri*, *A. dorsata*, *A. mellifera*, *A. cerana* and *A. florea*. (Based on data of Ruttner 1988)

ical stingless bees are cavity-nesting species as are the bumblebees. The nesting in burrows and hollows is actually a typical feature of even non-social and primitively social species. There seems to be no really convincing argument why open nesting should be a primitive character in bee biology.

Another approach was taken in a comparative study of the communication system of the various *Apis* species by Lindauer (1956). He found a clear cut sequence *A. florea* – *A. dorsata* – *A. cerana*/*A. mellifera* in terms of memorizing and transposing the position of the sun from the horizontal to the vertical plane. At first glance, the evidence seems highly convincing. *A. florea* dances horizontally on top of the nest directly indicating the position of the food source. *A. dorsata* can only dance if the sun or at least parts of the sky are directly visible. The cavity-nesting species memorize the sun's position and communicate in the dark. Subsequent work, however, makes this picture less clear. Free and Williams (1979) and Gould et al. (1985) found that *A. florea* is quite able to dance on vertical combs, thus transposing the angle between sun and food source to the vertical plane. Furthermore, the communication dances of *A. florea* actually imply a more complex information processing, because non-celestial marks with the memorized sun position are both used for orientation (Dyer 1985a). Thus again, we can identify *A. dorsata* as the most ancestral type. The two cavity-nesting species are almost identical in their communication system and *A. florea* may have the most advanced system.

The phylogenetic proximity of two allopatric cavity-nesting species is particularly obvious when we look at the mating behaviour of *A. cerana* and *A. mellifera*. Both are very recent species, since drones of one species readily mate with queens of the other species whenever they are kept in the same area (Ruttner et al. 1973). Even cross-fertilization is fully functional and there is normal embryonic development until the blastula stage. Then, however, the cells disintegrate and further development of the hybrid tissue is aborted (Ruttner and Maul 1983). The third cavity-nesting species, *Apis koshevnikovi*, does not interbreed with sympatric *Apis cerana* (Koeniger et al. 1988; Tingek et al. 1988; Rinderer et al. 1989; Ruttner et al. 1989).

The five sympatric *Apis* species of Asia have a variety of mechanisms to prevent interbreeding. The morphology of the mating apparatus is very different, which should prevent any semen transfer in potential mismatings (Ruttner 1988; Wongsiri et al. 1990). Furthermore, the times of mating activity have been shown to be distinctly separated between *A. dorsata*, *A. cerana*, *A. florea*, and *A. koshevnikovi* (Koeniger and Wijayagunasekera 1976; Koeniger et al. 1988). A virgin queen on a mating flight will only meet its own conspecific drones, since the others are waiting in the nest for their time to come. There are no reports yet on flight times of drones and queens of *A. andreniformis* (Wongsiri et al. 1990) but one would expect similar mechanisms as for the other species.

Karyotyping Does Not Help. Deodikar et al. (1959) found that *A. mellifera* and *A. cerana* have the same set of n = 16 chromosomes, a figure which was repeatedly verified (Sharma et al. 1961; Fahrenhorst 1977a,b; Hoshiba et al. 1981). Chromosomal studies (Deodikar and Thakar 1966; Thakar and Deodikar 1966), which claimed that the open-nesting species had a chromosome set of n = 8 in contrast to n = 16 in the cave-nesting species, were readily shown to be false by Fahrenhorst (1976a,b). Previously, Kumbkarni (1964) had reported on a chromosomal set of n = 16 in *Apis dorsata*. Since the karyograms of all four species contain only a single large mediocentric chromosome, the number of n = 16 cannot result from polyploidy as argued by Deodikar and Thakar (1966) or Kerr and Silveira (1972). Four species within the genus have a uniform chromosome number of n = 16. *A. koshevnikovi* and *A. andreniformis* have not yet been studied. So far the cytogenetical approach does not give us any information on the phylogeny within the genus because of a complete lack of variability in chromosome number.

Biochemical Evidence. In spite of the rapid development of molecular techniques and their use in evolutionary biology, very few studies related to honeybee evolution have been published. Jordan and Brosemer (1974) studied the GC (guanine-cytosine) content of the two cavity-nesting species and *A. florea*. Though *A. mellifera* and *A. cerana* were very similar, all three species had a wide GC-distribution and there was no clear separation. Sperlin et al. (1975) found that 60% of the *A. mellifera* DNA can form heterohybrid DNA with *A. cerana* DNA. Since only these two species were tested, no phylogenetic relationships within the genus can be derived from these data. Moreover, the DNA divergence was surprisingly high regarding the close similarity of the two species, and a better documentation of this phenomenon seems desirable. Kreil (1973, 1975) used an amino acid sequence analysis on the 26 amino acids of the melittin of *A. mellifera*, *A. cerana*, *A. florea* and *A. dorsata*. Melittin is the major protein in the poison of the bee. He found no sequence differences between *A. cerana* and *A. mellifera*. The amino acid sequence of *A. dorsata* melittin was changed at three sites. Finally, there were five amino acid exchanges each between *A. florea* and *A. dorsata*, and *A. florea* and *A. cerana*/*A. mellifera*.

In an isozyme study on four *Apis* species (*A. dorsata*, *A. cerana*, *A. mellifera*, *A. florea*) Sheppard and Berlocher (1989) added information but were also unable to contribute clear phylogenetic evidence. In contrast to the above data set, their analysis favours *A. florea* as outgroup. Yet, in a cladistic analysis they find unrooted trees supporting all possible combinations of the four species. Obviously, further work, also including the two recently re-recognized species *A. koshevnikovi* and *A. andreniformis*, is needed. The picture is still very incomplete, because neither a sound cladistical analysis nor a profound molecular data set is available to give reliable support to the above hypothesis.

Further evidence is lacking, and studies like those of Smith (1990) producing molecular data on the DNA level promise to be enlightening in our understanding of *Apis* evolution. A first attempt was done by Garnery et al. (1991) by using sequence data of 270 bp in the CO-II region of honeybee mtDNA. They concluded from distance estimates that the speciation of *A. cerana* and *A. mellifera* took place about 6 million years, and the trifurcation *florea – dorsata – cerana/mellifera* about 9 million years ago (Fig. 2.13). However, the standard errors of their estimates are still substantial and may affect both the phylogenetical time-scale and the topology of the tree.

Biometrical Analyses. Most work on honeybee phylogeny is based on external morphological characters. This is convenient because museum samples can be analyzed and there is no need for the systematicist to expose himself to live honeybee colonies which may cause inconvenient situations in the case of a colony defense. Ruttner (1988), however, very convincingly shows that it is not sufficient to base systematic or even phylogenetic analyses of honeybees on a few museum samples. Accurate field sampling of a large number of colonies is essential to obtain a reliable data base. Biometrical data on variation of characters among workers of a colony, among colonies, among populations and finally among species give a clear demographic picture. Five of the six known species can be clearly discriminated, based on the morphometry of 16 wing venation angles. *A. dorsata* is most similar to *A. armbrusteri*, and the cave-nesting species are well separated from both *A. dorsata* and *A. florea*. Although Ruttner (1988) does not provide a tree analysis in his monograph, it seems clear that the cavity-nesting species are more similar than those nesting in the open. The recently described open-

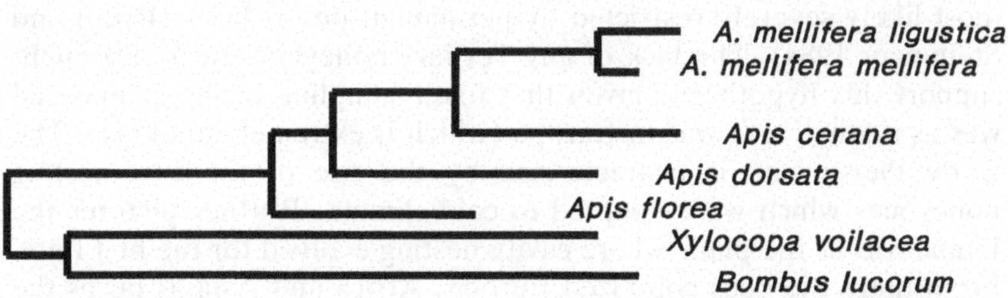

Fig. 2.13. Phylogenetic parsimony tree of four *Apis* species based on sequence data of the CO-II-region of mtDNA. The tree is rooted by two outgroups *Bombus lucorum* and *Xylocopa violacea*. (From Garnery et al. 1991)

nesting *A. andreniformis* (Wongsiri et al. 1990) is, however, very similar to *A. florea* which may leave *A. dorsata* as the "ancient" outgroup.

Zoogeography and the History of Honeybee Colonization. In spite of the above problems, Ruttner (1988) provides a radiation of the subfamily Apinae and a history of the zoogeography of the subfamily, which will be briefly discussed here. According to Ruttner (1988), the major forces in honeybee evolution are the temperature constrains under the various climatic conditions. The story begins in the early Miocene (24–23 $\times$ 10^6 years B.P.) when the first honeybees appear in (at that time) tropical central Europe. In the warm climate, *Synapis*, a presumably open-nesting species, survived well. However, some 11 million years later the climate became cooler, but it still was temperate to subtropical with an annual mean of 16 °C and 1500 mm of annual rainfall (Gregor 1982). These conditions would have allowed the presence of an *A. dorsata*-like honeybee. *A. armbrusteri*, the abundant bee of that period is at least morphologically very similar to *A. dorsata*. In the Pliocene, open air nesting became impossible and such species could only survive in tropical south Asia. Europe was free of honeybees and most likely cavity-nesting species were not evolved, because otherwise they might have survived the cold climate like other thermophilic animals and plants. A major problem arises with the ancestral position of honeybees in Africa. Why did those free-breeding *A. dorsata* type honeybees not colonize the African continent? Ruttner (1988) argues that during the Eocene and Oligocene, the Mediterranean Sea was much wider than today and separated Africa from Europe and Asia. However, in the Late Miocene the straits of Gribaltar were closed, forming a connection between Europe and Africa. Nevertheless, the exchange of fauna was most likely severely restricted by permanent desert belts (Roegl and Steininger 1984). The lack of any Tertiary honeybees in Africa might support this hypothesis, given that fossil sampling of insect material was as intense as it was in Europe (which is extremely unlikely). The Early Pleistocene is characterized by the rise of the cave-nesting honeybees which were adapted to cold climate. Ruttner pictures the Himalayas as the place where cavity nesting evolved for the first time. From there the bees colonized Europe, Africa and Asia as far as the climatic conditions allowed. Although it may be speculation whether the first cave-nesting bee lived in the Himalayas, it is well supported that *Apis mellifera* colonized Europe and Africa (via Arabia) and *Apis cerana* remained in Asia.

Chapter 3 Physiology

. . . it is a common rule throughout nature that the same end should be gained, by the most diversified means. (Darwin 1859)

The thousands of individuals living together in colonies of social insects function as integer entities which show coordinated group responses to a variety of stimuli. The organized responses only rarely resemble the summation of responses seen in individuals outside the social framework. The physiology of the individual ant, termite or honeybee that is a part of the unified whole superorganism is not remarkably different from that of other single insects, whether they are social or solitary (e.g. Rockstein 1978, 1987; Blum 1985; Kerkut and Gilbert 1985). Yet the social group itself often displays a "group physiology" which we discuss, using the honeybee colony, in this chapter, and which is very different from that shown by individuals isolated from the group. Physiological functions of the social unit are adaptations toward solving problems of homoeostasis that involve novel and fundamentally different mechanisms, than those utilized to solve the same problems in normal intact metazoans.

3.1 Development

Development of the honeybee colony is closely tied with the queen and her fecundity. A colony's success is especially dependent on a large population of workers to carry out the support tasks of brood and queen care, house keeping and foraging as well as thermal balance in winter and summer.

3.1.1 From Egg to Adult

Individual honeybees develop through complete metamorphosis from egg to larva to pupa to adult within the cells of the wax comb. The single fertile queen lays about 1000–2000 eggs each day into empty cleaned cells (Nolan 1975). Nest conditions, such as little available empty comb or swarming preparations, reduce the egg-laying rate (Allen 1956; Fell et al. 1977; Seeley 1985).

After 3 days of embryonic development the larvae emerge from the eggs (Nelson 1915). During the first few days, all honeybee larvae (L1) are fed exclusively royal jelly (see below) secreted by the mandibular glands in nurse workers, after which honey and pollen are added to the worker larvae diet. The details of embryogenesis and how differentiation occurs are well documented (Fleig and Sander 1986; Fleig 1990). Hormones, especially the terpenoid juvenile hormones (JH) secreted by the corpora allata (Robinson 1987a,b; Sasagawa et al. 1986), are known to play an important part in larval development and in division of labour of adult worker bees.

All brood are given royal jelly until mid-L4 (see Table 3.1) and those which are fed royal jelly after mid-L4 stage will develop into queens (Dietz and Lambremont 1970; Hartfelder 1990). Queen development is promoted by the stimulatory effect on hormone production by feeding on royal jelly. Most brood are destined to be workers and are subsequently fed only honey until pupation. In the final larval stage (L5), larvae feed for a time followed by a spinning phase. During the feeding phase of L5, the JH titer is increased.

As brood develops and grows, it is compelled to shed its exoskeleton and synthesize a new one five times (Bertholf 1925). This moulting is precipitated by ecdysteroids produced mainly in the prothoracic glands. The steroids bind to intracellular receptors inducing transcription of specific genes which initiate the moulting

Table 3.1. Development of honeybee brood

Brood stage	Queen	Worker	Drone
Egg	1st day	1st day	1st day
Larva			
L1-instar	3rd day	3rd day	3rd day
L2-instar			
L3-instar			
L4-instar			
L5-instar			
Cell is capped	8th day	9th day	10th day
Pupa			
Morphogenesis			
Adult emerges	16th day	21st day	24th day

process (Hill and Wyse 1989). JH promotes retention of juvenile characteristics until pupation which commences when the silk gland produces the cocoon. At pupation the tissues are exposed to ecdysteroids with little or no JH, and metamorphosis ensues. The additional hormones, bursicon and eclosion hormone, are necessary for hardening of the new adult cuticle and are required for neural generation of movements used by the new adult to emerge from the sealed pupal cell. In the prepupal stage, the cell is sealed with a wax cap by house bees. The pupae metamorphose but do not moult in the sealed cells and emerge as young adults about 2 weeks later with their size largely dependent on the dimensions of the cell in which they were reared (Abdellatif 1965).

The honeybee provides the only example in bees in which queen development is more rapid than worker development (Hartfelder 1990). Growth of the larvae depends on quality and quantity of the food provided by the workers and on the timing of provisioning (Wirtz and Beetsma 1972; Wirtz 1973; Beetsma 1979). Hartfelder argues that the rapid queen development in honeybees may be a result of the movement of relevant endocrine events to the beginning of the last larval instar instead of at the end of the instar. This would place the events in a different temporal position with respect to the metamorphosis regulating endocrine activity. The more rapidly developing queen has an advantage over other queens as she is the first to emerge and can kill those still sealed in their cells. Natural selection would then favour rapid development of the queen larvae but sufficient time is still required for completion of morphogenesis in the pupal stage.

Drone cells are usually constructed at the periphery of the brood area (Free 1967; Taber and Owens 1970) and may be prone to cold temperatures (Allen 1963; Page 1981). Drones have a slower development, which may be an adaptation or consequence of the temperature fluctuations (Spivak et al. 1991). Clearly, the variability of drone developmental time is much larger than that of queens or workers (Moritz and Jordan 1991).

3.1.2 Caste

The only castes in bees are queen and worker (Michener 1974). In only two families of social bees have distinct castes been described; the Apidae and the Halictidae (Michener 1974; Hartfelder 1990). Juvenile hormones, food, and pheromones act in concert to control caste development and behaviour within the colony. For example, the undeveloped ovaries of workers are associated with low JH titer from inactive corpora allata brought about by exposure to queen contact pheromones. These are licked off the queen and picked up on the body surfaces of her attendants and passed on to other workers by contact, antennation and trophallaxis (see Chap. 4). No data are available on the endocrine mechanisms involved in regulating caste differentiation in the eusocial species of halictines, although the endocrine regulation of caste development has received attention in stingless bees (subfamily Meliponinae) (Velthuis 1976; Velthuis and Velthuis-Kluppel 1975; Sakagami 1982). Figure 3.1 illustrates events relative to caste polymorphism and endocrine and trophic events in honeybees and stingless bees. It is seen from the figure that control mechanisms of caste development are affected by progressive and differential feeding. A caste specific peak in JH content at the start of the feeding period (L5) takes place soon after the nutritional switch in L4 (Rembold 1986). Queen development in honeybees is due to her exclusive feeding on royal jelly. During her very early development from a fertilized egg, the queen larva is reared primarily on a food containing hypopharyngeal and mandibular gland secretions of attendant workers. This royal jelly differs in this way from food fed to workers. The developing queen larva receives ten times more biopterin and pantothenic acid and three times as much sugar in her food as does a worker larva (Hanser and Rembold 1960). She also is fed a larger amount and more frequently. Lindauer (1952) found that queen larvae are visited ten times more often than worker larvae by nurse bees. The high sugar content in royal jelly may act as a phagostimulant (Goewie 1978b; Asencot and Lensky 1988) resulting in the faster growth and larger body size of queen larvae. The diet induces high levels of juvenile hormone (JHIII) in queen larvae starting on about the third day of development throughout the larval feeding period (Wirtz 1973; Hagenguth and Rembold 1978; Lensky et al. 1978; Rembold 1986). The high JH titer at this critical age prompts development into queens rather than workers (Wirtz and Beetsma 1972). Hartfelder (1990) suggested that the nutritional state of the larvae may be related to these hormone modulations in two general ways: First, hormone change or their relative balance could result from specific hemolymph components (which are reflective of caste-specific composition of larval food), or, second, larval size may be the critical factor affecting hormone levels. The first hypothesis invokes the "determinator" concept (von Rhein 1933; Rembold 1976; Wittmann and Engels 1987) requiring a specific substance or balance of substances in the larval food that induces caste control. Although

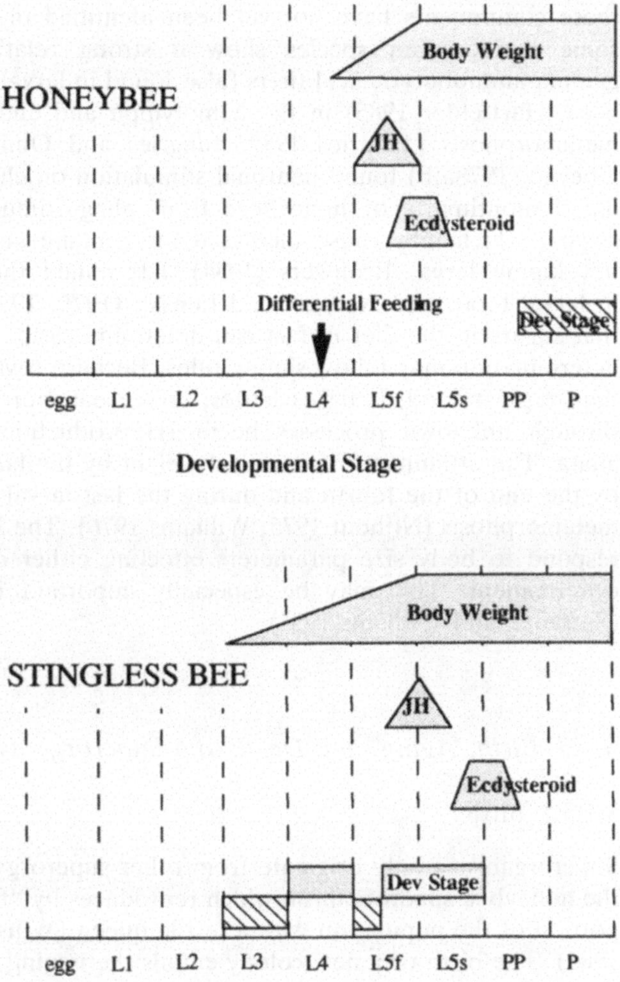

Fig. 3.1. Caste polymorphism, endocrine and trophic events in developing bees. Honeybees (Apis mellifera) compared with stingless bees (Scaptotrigona postica depilis). L1 through L4 designate 1st through 4th larval instars; the 5th instar is further divided into f, the feeding phase, s, the spinning phase, PP, the prepupal and P, the pupal phase. Different modes of larval feeding (mass provisioning in the stingless bees and progressive and differential feeding in honeybees). The mass provisioning brings on an earlier increase in body weight of queens (grey area) and differences in developmental stages (indicated by grey where queen > worker and hatched where worker > queen). Caste-inducing endocrine events are later in stingless bee development. (Redrawn and modified from Hartfelder 1990)

these components have not yet been identified in larval bee food, some lepidopteran species show a strong relationship between gamma-aminobutyric acid titers (also found in larval food of stingless bees, Hartfelder 1985) in the hemolymph and endocrine events of metamorphosis (Jungries 1982; Jungries and Omilianowski 1980). Goewie (1978a,b) found neuronal stimulation on chemoreceptors on larval mouthparts of honeybees from phagostimulatory effects of sugars, which may cause increased food intake and growth of the developing larva. Brouwers (1984) determined the specific sugars in larval food, and Asencot and Lensky (1976, 1984, 1988) showed that sugars in the diet in fact can determine caste. Therefore, caste determination may follow simple rules. Because royal jelly is sweeter than regular food, it tastes better, larvae eat more, grow fast, and through unknown processes begin JH production in the corpora allata. The attainment of a critical weight by the larvae is important by the end of the fourth and during the last larval stage for proper metamorphosis (Nijhout 1975; Williams 1976). The JH synthesis may respond to body size parameters effecting either queen or worker development. This may be especially important in stingless bees (Safranek and Williams 1984).

3.1.3 Birth, Aging, and Death of a Superorganism

3.1.3.1 Birth

Superorganisms only originate from other superorganisms, shown in the honeybee superorganism which reproduces by "fission" of a large portion of the population with a fertile queen (Winston 1980; Seeley 1985). The birth of a new colony entails the issuing off of the parent swarm, leaving part of the colony with its new queen (or queen cell) at the established nest site. Once the parent swarm locates and occupies a new nest site, workers rapidly adjust to organized division of labour and establish a fully functional colony just as they do in the colony left behind at the original nest (Rothenbuhler and Page 1989). A somewhat different process is seen in the stingless bees (Meliponini), in which a new home site that is found by searchers is prepared for occupation, and a swarm with a young gyne goes to the new nest (Michener 1974). After flying on her nuptial flight and mating once, the young gyne returns to the new nest and brood production begins.

3.1.3.2 Aging and Age Variation

The colony is comprised of workers varying from freshly emerged to about 50-days-old during the foraging season (Lindauer 1952, 1954; Bühlmann 1985; Fig. 6.7) with extended age to many months in the non-foraging season. During a strong nectar flow the workers life-span may average less than 16 days. The profile changes during the winter in temperate regions where foraging is not possible, with a high proportion of very old bees. The dependence of worker health and age on their protein diet has been repeatedly demonstrated (Haydak 1935; Maurizio 1950; Groot 1951; Haydak and Dietz 1965; Haydak 1970).

Honeybee colonies are perennial and the same nest site may be occupied for at least 40 years (Buchmann 1989). It is not the same genotypic colony that occupies the nest site for this extended period, but successive generations, with changes in the queens every year or two. These occur as the old colonies leave and newly born ones stay at the original site. Ages of different colonies are dependent upon the date of occupation of a particular site and the date of their queen royalty. Some colonies could be only a few weeks or months old and die out, while other colonies may theoretically be several years old. It is the "genetic integrity" maintained by the fertile queen that is important in determining age of a superorganism. In fact, a newly established colony can be quite old if a prime swarm with an old queen moves into a new location.

It is well known that adult *A. m. mellifera* workers show polymorphism for age, living longer in winter than in summer (Bodenheimer 1937; Lindauer 1952). Adult longevity varies from summer individuals living 14 to 50 days (median <17 days) after emergence, to winter-emerged workers' lives averaging 154 to 170 days or more with a record of 350 days (Bühlmann 1985; Sakagami and Fukuda 1968; Schmidt-Hempel and Wolf 1988). The seasonal differences exist even though the individuals are progeny of the same fertile queen mother. Apparently there is strong gene–environment interaction, that is, environmental factors affect genes that control the life-span of worker bees. A number of biological properties show seasonal differences including JH, lipid, glycogen, hemolymph sugar, protein reserves, but their relationships to aging and life span are uncertain (Flury 1982; Shehata et al. 1981; Robinson 1991). Longevity and age-related activities may be genetically programmed (Rockstein 1972) but are also affected by nutrition, state of brood rearing in the colony, onset of foraging and other colonial characteristics (Neukirch

1982; Omholt 1987). Expansions in colony size are at the sacrifice of individual life-spans (Fukuda and Sekiguchi 1966).

Nearly all organisms have a number of physiological characteristics that show decline with age, such as the decline in physical performance and in maximum oxygen consumption ($\dot{V}O_2$). Metabolic rate is an integrator of many physiological functions and is a good indicator of whole organism "physiological fitness" at any age. Life-span and aging seem to be closely tied to metabolic expenditure (Adelman 1978). There is evidence from several insect species (Ragland and Sohal 1973; Trout and Kaplan 1970) that there may be fixed quantities of some vital substances such as enzymes or substrates that are used up, thereby shortening life-span in metabolically more active animals. Flies, for example, that are restricted from flying live twice as long as those allowed to fly. This has given support to the "rate of living" theory (Sohal 1975, 1985, 1986). This theory assumes a specific energetic "bank account" in each individual organism. The composition of such a bank account is unknown but may consist of specific metabolically important enzymes or substrates that are not replenished when used up. When the account is used up, that individual dies. In fact, experimental data do show that a worker bee's life-span is proportional to its time spent in the nest (Neukirch 1982; Ribbands 1952). Once they forage, they rapidly use up the account in about 2 weeks, regardless of their length of time previously spent in the nest (Winston and Katz 1982). Thus, a longer hive period would simply be added on to the fixed foraging time for total life-span. Although the correlation exists, the cause-effect is unknown. Neukirch (1982) suggests that the hive period may be instrumental in the longer-lived winter bees which may never get out to forage. However, a portion of the very long-lived winter individuals actively maintain the warm central cluster temperatures (at 35 °C) by shivering, which can take as much metabolic energy as flying (Southwick 1985a).

Normal decline in several enzyme activities occurs with increasing age in honeybees and other insects (Gunn and Gatehouse 1979; Harrison 1985). Seasonal experiments on bees show little difference in muscle enzyme concentrations or enzyme activities in winter bees older than 120 days when compared to summer bees no older than 30 days (Smugor 1991).

Honeybees, active in the nest or foraging in the field, exhibit a wide variety of behaviours dependent on their physical, social, and temporal surroundings. One pattern of behaviour highly dependent on the colony age profile and of particular physiological interest is age polyethism. Observations by Rösch (1927, 1930) revealed that there is a set of tasks performed by individual workers. These tasks are dependent on age but cannot be treated outside the context of the social structure (Seeley 1982a). Within a normal, queen-right colony, activities of young bees (0–10 days old) occur mostly in the central nest area (Fig. 4.1). These include cleaning cells, feeding brood, capping brood, and attending the queen. Older workers (15+ days old) operate nearer the periphery of the colony, receiving and storing nectar, packing pollen, and finally leaving the nest on foraging trips. Other activities occur throughout the colony structure with middle-aged bees (5–20 days old) grooming and feeding nestmates, shaping comb and ventilating. However, all of these individual tasks can be and are performed by various aged workers if the circumstances require it. Removing a number of workers all of the same age, for example, causes others to step in and take over their tasks (Lindauer 1952). The removed workers of honeybees when placed in an empty nest with brood and stored food will assume all the necessary roles. Some roles even require substantial changes in the physiology of the individual. For example, wax production by abdominal wax glands cannot be performed by the older workers with inactive wax glands (e.g. foragers >20 days old or freshly emerged bees <5 days old) (Rösch 1930; Free 1961; Lindauer 1961; Kolmes 1985). Yet, in this social context, within a few days some of the bees' wax glands will become active and this task resumed, suggesting that colony requirements influence temporal activities (Haydak 1930; Sakagami 1953; Seeley 1982a). If all bees removed are foragers, some of the remaining workers will stop foraging and become house bees, incubating the brood and performing other necessary house functions. The mechanism of conversion to individual tasks is likely to be a result of local phenomena, just as in a normal colony. When a worker encounters brood needing feeding, it responds by seeking pollen and honey.

3.1.3.3 Death and Immortality

A superorganism is a living entity and just as metazoans die, a colony may also die. Colony death may be brought about by disease, pests or parasites, starvation, predators or disasters. In these cases, all the individuals in the colony expire because they are killed directly (e.g. eaten by a predator). Of course, if all the constituent individuals are dead, the superorganism is dead.

Colonial death must be considered from the genotypic standpoint because, although a colony is phenotypically perennial in any particular nest site, it is not genotypically immortal. In fact, a colony is a constantly changing entity, with its individual components being different today than they were yesterday or will be in another few days. This is brought about by the disappearance and death of individual bees and their replacement by newly emerging bees, but all from the single mated queen mother. This is similar to the continuous process of replacement of cells in metazoans which also are not, of course, immortal. Countless cells die daily in any mammal or bird (e.g. human erythrocytes survive only about 120 days and die at the rate of 2 million *each second*; Keeton and Gould 1986), yet the intact animal lives on as a single genotype and phenotype. Cell replacement in the metazoan is via newly produced cells, with the same genotypes (except for mutations) as those which are being replaced. The daily rate of turnover of red blood corpuscles in a human amounts to nearly one percent of the total cell population (i.e. about 2×10^{11} RBC's replaced each day of the total 2×10^{13} RBC's in the body). Extending consideration of cellular life-span and turnover, we see that blood within a single organism has become composed of totally new and different cells about every 4 months. Thus, a single metazoan is not truly the same organism today as it was a year ago, or last week, or will be tomorrow. Yet, we have no problem considering the animal with its new components as the same whole organism through time because it retains its genotypic integrity. In the honeybee superorganism, the component turnover rate is nearly the same as found for red blood cells in humans when examined during the active foraging season (Avitable 1978). Using egg-laying rates and rates of population growth, one can estimate that about 300 to 400 workers succumb each summer day in a typical colony of 35 000 (Gary 1960). Although we caution against applying too much emphasis on numerical analogies, here, too, we see a daily turnover of about one percent, and again in 4 months the components would be completely replaced but the superorganism would remain intact with only minor fluctuations of the colony genotype (see Chap. 6).

Death of a superorganism occurs when the genetic integrity is lost. Whenever the genetic make-up is substantially changed through the loss of a queen, for example, the resulting superorganism with its new queen is not the same as the original one. The new queen will have mated with other drones and the newly emerged workers are not genetically similar to the old workers. The genotype of the superorganism is completely different and comprises a new organism, although this may not be manifested in the colonial phenotype. Thus, death of a colony results from any form of queen loss. The queen normally is replaced every 3–4 years as her egg-laying vigor becomes reduced; however, it can happen that the queen will die before a replacement queen has been produced to take her place. In such circumstances, the workers will raise emergency queens from young larvae. However, if no queen can be raised, the colony is doomed to die out completely as the workers age. Some workers may develop their ovaries because of the lacking queen pheromones, but all these unfertilized eggs are destined to become drones. There are records of some successful requeening from laying workers, especially in the Cape honeybee, *Apis mellifera capensis* (Onions 1912; Anderson 1963). In these colonies, laying workers are capable of producing viable eggs that parthenogenetically develop into diploid females. Even in this special case the colonies are not immortal because the genotypic compositions change. Worker offspring compose a genetic clone and the father drone contribution to the colony is lost (i.e. variability due to polyandry of the queen is lost). Eventually a new queen will be reared which will be mated by a new set of drones, giving rise to a new colony (see Chap. 6).

3.2 Glands and Secretions

The superorganism has functional "tissues" and "glands" composed of various groups of bees or individual bees which release various secretions resulting in physiological and behavioural changes within the organism. One of the most important and potent "glands" is the queen herself. She actively secretes a blend of chemicals that have hormonal effects on her worker force. Other "glands" include specialized groups of workers, drones, and brood. The true glands of each individual worker bee include the individual endocrine system, and those secreting substances externally having their effect on other nestmates within the superorganism.

3.2.1 *Endocrine System*

The glands and their secreted products inside the individual insect include: salivary (saliva), hypopharyngeal (royal jelly), corpora allata (juvenile hormone), corpora cardiaca (lipid and carbohydrate mobilizing hormones), brain and ganglia neurosecretory cells, and various digestive and excretory glands including the Malpighian tubules (excrete nitrogenous wastes and salt), and reproductive organs (Table 3.2). The corpora cardiaca serve as sites for release of neurosecretions synthesized in the brain (Hill and Wyse 1989). Here we discuss the various endocrine glands only in the context of their influence on the social organism in the appropriate sections of the text. Further details of endocrine anatomy and function in individuals are available in general insect physiology textbooks such as Blum (1985).

3.2.2 Exocrine System

Pheromones are metabolically produced chemicals released to the outside world by one individual of a species population which elicit specific physiological, developmental and/or behavioural responses when detected by another individual of the same species (Karlson and Butenandt 1959; Blum 1985; Hoar 1983; Prestwich and Blomquist 1987). In honeybees, these chemical messengers are of particular importance for the stability, homoeostasis, and integration in everyday functioning of the colony and they are reviewed by Blum (1985) and Free (1987). The glands that secrete products to the outside of the individual bee include: abdominal Nasanoff gland (scent pheromone), sting gland (alarm pheromone), Dufour's gland, mandibular glands (especially in the queen), hypopharyngeal gland, head and thoracic labial glands, tergal glands, koschevnikov's gland, Arnhart and tarsal glands (foot print) and also the abdominal wax glands. More than two dozen chemical substances are of pheromonal importance in honeybee colonies (Lensky et al. 1984, 1987; see Table 4.1). Sealed brood is recognized probably because of their odour. This results in their attraction to bees which incubate them (Koeniger and Veith 1983). Besides those actively secreted substances, even the wax produced by the bees and used to construct honeycomb has distinctive odours which influence bee behaviour (Free 1987). Wax may take on as well a variety of associated odour components from flowers, and other natural substances with which forager bees have contact.

As the major impact of the chemical environment inside the superorganism is at the level of local behaviours, the importance of pheromones is discussed more completely in Chapter 4.

Table 3.2. Some important glands of individual honeybees (see Table 4.1 for pheromone producing glands)

Glands	Function
Salivary glands	
Thoracic	Watery secretion dissolving sugars
	Nectar processing enzymes
Head	Oily secretion (wax manipulation?)
Hypopharyngeal gland	Production of larval food
	Nectar processing
	Glucose oxidase
	Invertase
Corpora cardiaca	Molting hormone
Corpora allata	Juvenile hormone
Mucus glands	Mucus on endophallus in drones
Spermathecal gland	Spermathecal fluid in queens
Wax glands	Production of wax scales

3.3 Feeding and Digestion

3.3.1 Nutrition

Honeybees, by nature, are vegetarians and generally consume only sweet flower nectars and ripe pollen. They derive their total nutrient requirements from these two resources. Pollen found in individual flowers is the only food source of protein (Snodgrass 1956; Schmidt and Buchmann 1985). Only under extreme conditions of starvation, lack of pollen resources, or under the influence of disease or pesticides do workers show cannibalistic behaviour toward developing larvae (Haydak 1935; Myser 1952; Rothenbuhler 1964a,b). Also diploid drone larvae are cannibalized (see Chap. 7).

Flower nectars are commonly composed of disaccharides, and hexoses dissolved in water (Wykes 1952; Percival 1961), with minute quantities of other substances such as amino acids depending on the floral species (Baker and Baker 1973; Southwick 1986b). Nectar concentrations are quite variable ranging from almost no sugar in flowers filled with dew or rain water, to over 80% sugar with average nectar concentration found in the floral standing crop at about 40% (Beutler 1953; Oertel 1944; Southwick et al. 1981). Salivary secretions of invertase break down the food from disaccharides into the simpler monosaccharides found in honey. These monosaccharides are utilized in energy metabolism, as described later under Section 3.7.1.2. Carbohydrate reserves in individual honeybees are present as glycogen (stored in tissue) and trehalose (in the haemolymph). These reserves can readily be transformed into glucose (Chippendale 1978) which provides the main energy source for muscle activity in honeybees (Storey 1985).

Many quantitative data are available on whole colony food consumption, mostly based on weight loss during periods of confinement in managed hives. A managed colony consumes about 60 kg of honey during a year and requires about 19–25 kg honey to survive north temperate winters where the coldest month averages −4 to +7 °C (Farrar 1952, 1960; Dyce and Morse 1960; Johansson and Johansson 1969) and 7–10 kg in more Mediterranean climates where the average temperature remains above 10 °C (Morse 1956). The few data available indicate that wild colonies store much less honey, amounting to 11 kg or less (Seeley and Morse 1976).

3.3.2 Food Exchange and the Meniscus Effect

The colony has been qouted to have a "communal stomach" (Wilson 1971; Gould 1988; however, see Korst and Velthuis 1982) manifested by the rapid and continuous food exchange among colonial members (Free 1959; Gary 1975, reviewed in Moritz 1992). The nature of this food exchange was demonstrated by studies utilizing radioactive tracer labelled food (Nixon and Ribbands 1952). The mechanism of social food sharing is through mouth-to-mouth trophallaxis which occurs when one bee shares its food with others in its immediate vicinity. The transmission of food from one worker to another is initiated by one of them begging or offering food to the other. The begging bee extends her tongue toward the head of an offering bee which is usually older (Free 1959) or behaviourally dominant (Korst and Velthuis 1982). The offering bee regurgitates food from her honey stomach and offers it as a droplet on her spread mandibles. The begging bee then laps up the droplet. The exchange usually lasts only a few seconds, rarely exceeding 20 s (Istomina-Tsvetkova 1953, 1960). Throughout the exchange, both participants show active antennation. Often trophallaxis is only ritualized and no exchange of food occurs (Van der Blom 1991). This implies communicative functions in the behaviour, including transport of queen substance (Naumann et al. 1991), and the establishment of worker hierarchies (Korst and Velthuis 1982).

When one considers the physiological processes followed by multicellular vertebrate organisms under starvation conditions, the following scenario develops. First, carbohydrates circulating in the blood are utilized (Hill and Wyse 1989). When these are exhausted, carbohydrates stored in the liver and muscles will be catabolized, followed by breakdown and metabolism of body fat. Only at this point in starvation will the body begin to attack the organismal protein. Then, gradually, the entire organism will grow weaker as the cells obtain less and less of their required nutrients, resulting in death of the whole body. Groups of cells (tissues) do not become cannibalistic under these conditions, consuming other cells to enhance their own survival. They generally do not somehow hoard the available food supply either, but instead the supply is shared among all cells, and they succumb in a general effect. Yet, if a number of the whole organisms, each made up of these same cells, was placed in a group, for example 1000 or 10 000 vertebrates (or even people), which was then placed under starvation conditions, individuals would be unlikely to share what they had. Instead, they would hoard whatever food was available, and as conditions worsened, even resort to cannibalism. The honeybee colony functions like the individual whole organism, not at all like a group of individuals, when under starvation conditions. Because of their trophallaxic behaviour, individual workers share whatever food there is remaining down to the last drop (of honey). There is no evidence that they form hoarding subgroups, but instead they all starve together.

If it is a set of cool temperature conditions that brought about the starvation situation because of the inability for even an edge of the superorganism to move to the food stored in the comb, then those workers located centrally are most likely to survive until a rise in temperature allows movement. This results from the starved bees occupying the outer portions of the colony structure succumbing first to the chill coma resulting in immobility and ultimately death, while those centrally located are protected by the insulating properties of their comatose compatriots. It has been noted that under a prolonged winter cold spell, if even just a small bit of the edge of the cluster (i.e. a few bees) can reach the food, the entire cluster is saved. This can be likened to the physical properties of fluid surface tension making a meniscus of nectar which crawls up the side of a floral corolla tube allowing a feeding bee that has a proboscis only long enough to reach just the edge of the fluid to suck up nearly all available in the nectar reservoir. Essentially the same effect is observed in the winter feeding cluster. In the meniscus feeding effect, the outermost bees in contact with the food source imbibe the honey into their honey stomachs. They then turn around and through trophallaxis provide the food to nearby neighbours which continue the process until all 20000 to 40000 individuals in the cluster have obtained a supply of food. Trophallaxis is essential for survival during clustering as food is nearly always confined to the outer portions of the comb with the inner comb cells being utilized for brood rearing.

3.3.3 Pollen Stores

Winston (1987) reviews the evidence for pollen collection activity related to needs for brood rearing. The number and proportion of pollen foragers is strongly correlated with the amount of eggs and larvae present in cells in the brood area, and further, the proportion of workers collecting pollen increases with increasing egg-laying rates by the queen (Cale et al. 1968; Filmer 1932). Doubling the brood in a colony triples the pollen collection within 24 h (Free 1967; Winston 1987).

Fresh pollen brought into the nest by foragers is packed into wax comb cells by the carriers themselves. They use their forelegs and mandibles to manipulate the pollen load from their corbiculae into comb cells. The bees also add some acid secretion which prevents germination of the pollen grains. The pollen is generally stored in cells adjacent to the brood where it is easily accessible for feeding the developing larvae (see Chap. 4).

3.3.4 Making Honey

House receiver bees accept the nectar loads from foragers and fill empty cells or "top off" cells already containing nectar in the wax comb. The unripe honey is left in open cells with the surface exposed to the drying hive environment. Workers may remove nectar from cells to ripen it by producing a thin film between their extended probosces and mouth parts exposing large surface areas for water removal in the dry air (Gary 1975). They may increase the air movement across the droplet film by fanning their wings to achieve a 20% water concentration. Ripe honey cells are capped with wax, although they are frequently broken into whenever food is needed.

3.4 Respiration

3.4.1 Respiration in Individuals

Individual bees breathe through tracheal tubes which open through the integument via spiracles. These tubes are on the order of 2 µm in diameter, and branch into tracheoles measuring 0.6 to 0.8 µm in diameter for 200 to 400 µm in length, when they taper to blind ends of 0.2–0.5 µm diameter (Bailey 1954). They also connect to large air sacs in the abdomen, thorax, and head. The details of breathing gas exchange in individual honeybees resemble those of many other insects and can be found in any textbook on insect physiology, such as Wigglesworth (1984).

3.4.2 Respiration in Colonies

The metabolism of individual bees in the colony requires adequate exchange of respiratory gases in the nest cavity of the superorganism. Circulation of the atmosphere is maintained by fanning behaviour of some of the bees. Often the nest cavities have only one entrance (e.g. in hollow trees) so the fanning must function in a way that assures outflow of stale air and inflow of fresh air in a way similar to eupnea in vertebrates. Ventilation activities of bees have been observed in acrylic nest chambers each with only a single 2-cm diameter entrance (Southwick and Moritz 1987b). The bees form legions of fanners in fixed positions about the bottom and sides of the nest cavity, and even outside the entrance on the landing platform. One to three fanners orient themselves in the entrance hole. The side fanners circulate the inside air, but the behaviour of the entrance fanners is the crucial factor in determining the direction of air flow into and out of the nest cavity. When these few bees fan with their abdomens pointed outward at the entrance, stale air flows outward. This is indicated by an increase in temperatures at the nest entrance resulting from the warm nest air, drawn across thermocouples placed in the entrance hole, and a simultaneous decrease in oxygen content and increase in carbon dioxide content in the sampled air stream. Whether air moves into or out of the cavity is determined solely by the few bees located directly in the entrance hole.

Normal eupnea in vertebrates is a negative pressure type in which inspiration is active and expiration passive. Instead of a muscular thoracic pump, such as utilized by intact mammals, the honeybee colony actively changes direction of air flow by the behaviour of entrance fanners. Natural nest cavities may have rather large or even several entrances allowing for one way ventilation. In mammalian organisms, regulation of respiration is determined by centrally co-ordinated effects of carbon dioxide receptors and the need to cool. Similar regulation is effected in ventilation of bee nest cavities through social group coordination, showing another facet of colony-level homoeostatic control. In our study of small colonies confined in artificial cavities with a single opening, normal ventilation during the day showed an average respiratory frequency (RF) of 3 "breaths"/min. Normal tidal volumes (TV) averaged 147 ml, with a maximum TV of 462 ml (a yawn?). This results in an average minute volume (MV) of 0.42 l/min. These data are well within the range reported for intact vertebrates such as cats (RF = 15 breaths/min, TV = 25 ml and MV = 0.38 l/min) and marmots (RF = 8 breaths/min, TV = 22 ml, MV = 0.17 l/min). In the honeybee colony respiration activity was depressed at night (RF of 0.4 breaths/min, TV remained the same, periods of apnea occurred). Similar ventilatory behaviour has been shown in cave-dwelling social stingless bees under field conditions. Ground-nesting *Trigona denoiti* and tree-nesting *T. gribodoi* in South Africa colonize cavities with only small entrances (Moritz and Crewe 1988b).

3.5 Circulation

3.5.1 Haemolymph Circulation

Within individual bees, haemolymph fills the each bee's body cavity (haemocoele) in an "open" circulation system. One circulatory vessel moves haemolymph forward along the midline of the back by peristaltic pulsations of the tubular dorsal heart. The haemolymph travels forward through the convoluted aorta, leaves the open-ended vessel in the head and percolates backwards bathing all the tissues. It carries no respiratory pigment and seems to be unimportant in the transport of respiratory gases. It has important roles in transporting hormones, and functions in the mass transfer of heat by movement of warm haemolymph from one region to another. Under the temperature stress of flying at high air temperature, the circulating haemolymph is utilized in removing excess metabolic heat from the active thorax. During such flight, haemolymph flow to the head is enhanced and heat is lost by evaporative cooling of regurgitated honeycrop contents held on the tongue (Heinrich 1980b). At low air temperatures

(<5 °C), there is also more haemolymph circulated to the head, which keeps it warmer by mass transport of heat from the heat-producing thorax (Crailsheim 1985). Under non-stress temperatures at rest, the haemolymph pressure in the body cavity may be less than the surrounding atmospheric pressure. The rigid cuticle prevents collapse. The haemolymph hydrostatic pressure increases during movement, and in drone mating causes endophallus eversion and ejaculation (Koeniger 1984, 1987; Koeniger and Ruttner 1989).

3.5.2 Circulation in the Colony

There is a special form of circulation within the superorganism. To qualify as a true circulation scheme, the process must fulfil three complex functions as defined by metazoan physiologists (Beck 1971).

The circulatory system must:

1. serve as the body's main transport system in delivering nutrients from storage depots to distant groups of cells, hormones from their secretory points to cells elswhere, and carry wastes away;
2. play a leading role in regulation of water balance and distribution, and temperature regulation; and
3. be responsible for essential defence mechanisms.

Of course, there was no intention or thought about social organisms, yet in the honeybee colony all three of these functions are carried out by individual worker bees serving as carriers. They transport a mix of pheromones and food that is moved rapidly among members of the group through trophallaxis and body contact. Although slow in comparison to vertebrate blood flow, the transport of these substances via trophallaxis is effective in getting to each of the workers just as a vertebrate blood circulatory system permeates tissues to within a few cell diameters of each cell.

The circulation of workers also functions in the transport and dispersion of semiochemicals throughout the colony, such as queen-produced surface pheromones (Ferguson and Free 1980; Naumann et al. 1991). The importance of the workers in the retinue of the queen acting as "messenger bees" (Seeley 1985) moving through the colony is described in Chapter 4. Those messenger bees have surface contact with many other workers after intimate association with the queen. In this fashion, the pheromones on the queen's body surface are circulated quickly and constantly through the colony, providing an atmospheric molecular background important in maintenance of stability and monogyny. This also fits nicely with the finding that the queen substances show a gradual decline in their activity with a half-life of about 20 min (Juska 1978). Such pheromone deactivation is probably mediated through enzymatic mechanisms and ingestion (Pain and Barbier 1981; Naumann et al. 1991).

The single queen status found in most highly social insects results from competition between queens, and worker cooperation in eliminating supernumerary queens. Wilson (1971) suggested that a "queen-worker complex" may have evolved so that such queens would be removed as have the least familiar genetically determined pheromonal bouquet circulating within the superorganism. Working with fire ants (*Solenopsis invicta*), Fletcher and Blum (1983b) demonstrated that queen recognition by workers is dependent on unique quantitative blends of pheromones produced by queens (see queen recognition in Chap. 8). They hypothesize that the quantity of pheromonal blend circulating in the colony is maintained within some optimal range very much like the narrow range in blood titres exhibited in vertebrate blood. Deviations outside this range cause worker responses that tend to restore the level to within optimal limits. Queenlessness causes a great pheromonal deficit which results in workers accepting or producing a replacement queen (Butler 1957). Two-queen colonies collect more honey, but this is probably due to the simple increase in the population rather than to any increase in circulating queen pheromones (Barker and Jay 1974; Free 1987). Presence of additional queens may increase the circulating pheromonal level above some threshold which might cause workers to show aggressive behaviour toward some queens. A key factor in this hypothesis is that maintenance of monogyny is the effect of pheromonal quantity acting at the level of the individual on queens, and at the level of the colony on workers (Fletcher and Blum 1983a,b), and queen recognition is the quantitative result of the circulating pheromonal blend. Different queens produce different amounts of components of the pheramonal blend, and the amounts produced are correlated with their age and fecundity. Queen size and fecundity are lower in polygynous social insect colonies than in monogynous ones.

3.6 Colonial Excretion and Water Balance

The efficient removal of waste, dead bees, mould, pathogens and debris from within the superorganism is achieved through individual cleaning and hygienic behaviour, and via circulating undertaker bees (Rösch 1927; Perepelova 1928; Lindauer 1952; Ribbands 1953; Sakagami 1953; Seeley 1982a; Visscher 1983). The undertaker bees, comprising 1%–2% of the worker population, specialize for a few days in collecting and removing dead adults, flying as far as several hundred meters from the nest, and dropping them (Gary 1975; Visscher 1983). The hygienic actions result in maintenance of a "clean" internal environment (Rothenbuhler 1964a,b; Morse 1972). No defecation occurs within the nest, even during periods of nest confinement in non-foraging seasons of dearth, including the extended cold period during a northern winter. Nitrogenous wastes from protein catabolism build up and are retained in distended recta, until a mild day allows for flight when evacuation occurs outside the hive during brief cleansing flights.

Effective water balance in the superorganism involves precise control of humidity surrounding the bees as well. The problem of water balance is enhanced in the winter when excess water vapour from respiration condenses on the cold surfaces within the nest cavity. While bees in the warmer core are dehydrated, those on the periphery are hydrated (Omholt 1987). Of course water may be circulated among individuals by trophallaxis, but nevertheless it is likely that the whole cluster water accumulation will increase with decreasing temperature, with no way to get rid of the excess. Omholt (1987) describes a model in which brood rearing in the winter is a strategy of reducing water excess problems. This model is based in part on the decreasing humidity and decreasing evaporation from the centre of the cluster radially outward, based on Fick's law of diffusion and assuming stable conditions in a spherical geometry [Eq. (3.1) after Omholt 1987]:

$$\int_0^r 4pr^2 D(r, T_a)\, Z(r, T_a)dr = -4pr^2 G(r, T_a)\, dW(r, T_a)/dr, \qquad (3.1)$$

where T_a = ambient temperature outside the cluster surface,
$D(r, T_a)$ = bee density per unit volume at a distance, r, from the centre,
$Z(r, T_a)$ = rate of evaporation for each individual bee,
$G(r, T_a)$ = coefficient of diffusion of water molecules in air,
$dW(r, T_a)/dr$ = humidity gradient.

The cluster humidity gradient determines the water density at any specific point. Omholt (1987) shows that the radial distribution of humidity would be as plotted in Fig. 3.2. This results from differential temperatures maintained in the cluster (see below) and the

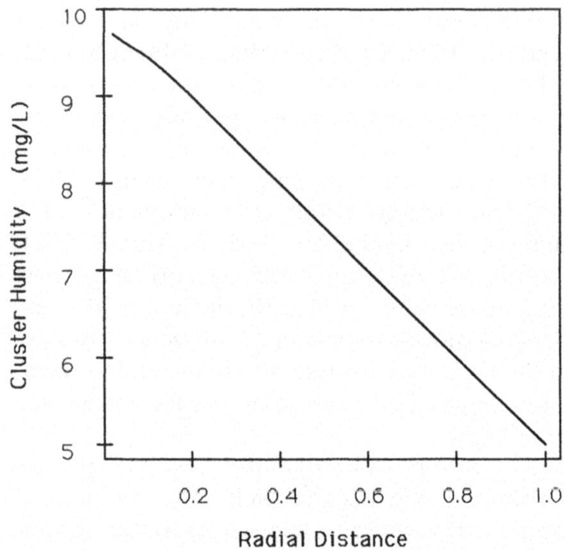

Fig. 3.2. Radial distribution of humidity in a honeybee colony. Moving out from the core of the colony, the water content of the air falls nearly linearly with distance. (Redrawn from Ohmholt 1987)

increased density of bees in the outer portions of the cluster. As ambient temperature decreases, the core humidity will be further reduced by simple diffusion of water molecules outward. The bees in the core will become dehydrated while the bees in the outer periphery will become more and more hydrated. The rate of accumulation of water for the whole cluster is given by Ohmholt as:

$$A(T_a)_{cluster} = \int_0^{R(T_a)} 4pr^2 D(r, T_a)\, A(r, T_a)\, dr. \quad (3.2)$$

This model shows the strong dependence of water accumulation on the size of the cluster and on environmental air temperature, with large clusters and low temperatures causing the greatest water load. Bees may change places within the cluster because of temperature preferences or motivation by hunger, thirst or water excess (Omholt 1987). The bees located in the inner dehydrating core could move to the outer hydrating periphery to take on water and those on the outside could move in to dehydrate. However, as long as there are honey stores in contact with the cluster extremity, there is considerable food exchange by trophallaxis from the food stores at the periphery inwards (Furgala 1975). Such exchange of food by trophallaxis, which could include food dilution by the outer bees, resulting in some water being passed inwards, would decrease the need for within-cluster movements. In fact, under low temperature conditions, there is little movement (Southwick 1987a). Food and water transport within the colony is supported by data on rectal weights (dry and fresh) which show no difference between bees located at the centre or periphery (Hallund 1956). Even with such transport, with the high rate of metabolism, water accumulation for the superorganism as a whole would still increase with falling air temperature.

Brood rearing in winter, then, not only must be timed for colony reproduction (Seeley and Vischer 1985), but has survival value for the superorganism with a water excess problem (Moebus 1980). The production of glandular brood food removes some of the water surplus from the bees. Also, the mere presence of increased brood area results in increased core temperature maintenance which allows for greater evaporation efficiency at that location. At locations in the nest cavity outside the cluster, where surface and air temperatures are quite low, evaporation rate is slow and would not be a significant route of water loss as it would be under summer conditions.

The model is supported by data collected on winter bees. If an individual bee's storing capacity is about 40 mg of water (Altmann and Gontarski 1963), in an air temperature of 0 °C, based solely on surplus water problems, a cluster of 15 000 bees is predicted by Omholt's (1987) model to begin brood rearing 43 days after their last cleansing flight (during which they have their last chance to easily lose water). This is clearly in line with the field evidence given by many observers (Avitable 1978; Jeffree 1956; Moebus 1980 and Wedmore 1947). By assuming a total water content in the brood of 85%, and a larval mass just before cell sealing of 140 mg, this cluster would be expected to rear about 12 350 brood in the time period from mid-November to the end of February (Omholt 1987). Avitable (1978) found average brood production for colonies in the north-eastern United States where the mean temperature was a little below freezing in the winter time period to be 11 300 brood for colonies with young queens. The brood rearing is accompanied by an increased core (brood area) temperature to about 35 °C (Himmer 1927; Simpson 1950, 1961; Kronenberg and Heller 1982; Southwick 1988). At this temperature, evaporation from the bees increases substantially as the saturation deficit of the cluster air increases (see Sect. 3.7.1.1).

3.7 Metabolic Physiology

Oxidative metabolism of the superorganism is an integrator of the combined physiological functions of all the bees in the social group. Even small perturbations in their physiological state are reflected by measurable changes in their overall metabolism (Southwick and Moritz 1985). Rates of oxygen use and carbon dioxide release provide reliable data for calculating metabolic rates (Parhon 1909; Corkins 1930; Free and Simpson 1963; Heusner and Roth 1963; Southwick and Mugaas 1971; Rothe and Nachtigall 1989). These measures have been particularly valuable in clarifying mechanisms of temperature regulation of colonies, groups and individual bees as described below.

3.7.1 Temperature Control

Individual honeybees fall into a cold comatose condition at temperatures below about 11 °C and cannot revive without an external source of heat (Free and Spencer-Booth 1960; Heinrich 1987; Southwick 1989; Goller and Esch 1990). Yet, originating in the Asian tropics (Ruttner 1988), *A. mellifera* has expanded into north temperate regions where it maintains a perennial nest as a direct result of its remarkable social thermoregulatory capacity. The honey bee is the only insect that survives winter cold by keeping warm as a social group (Seeley and Heinrich 1981). The social interaction in a colony of honeybees combined with their physiological responses to low environmental temperatures allow them to capitalize on early flowering spring plants, and give this species unique advantages allowing it to occupy numerous ecological niches that would not otherwise be available.

Heat production for temperature regulation in the bee colony is one of the most studied homoeostatic mechanisms in superorganisms (Gates 1914; Phillips and Demuth 1914; Himmer 1927; Lindauer 1954; Simpson 1961; Owens 1971; Southwick and Mugaas 1971; Heinrich 1981a,b, 1985; Ritter 1982; Southwick 1983a, 1985a,b,c). The picture that emerges from employing hundreds of temperature probes in honeybee colonies (Owens 1971) is a thermal profile with high stable temperatures in the central core area (where the brood is located) and concentric isotherms of lower temperatures at distances from the centre (Fig. 3.3). The range of environmental conditions which is tolerated by full-size colonies is illustrated with two studies.

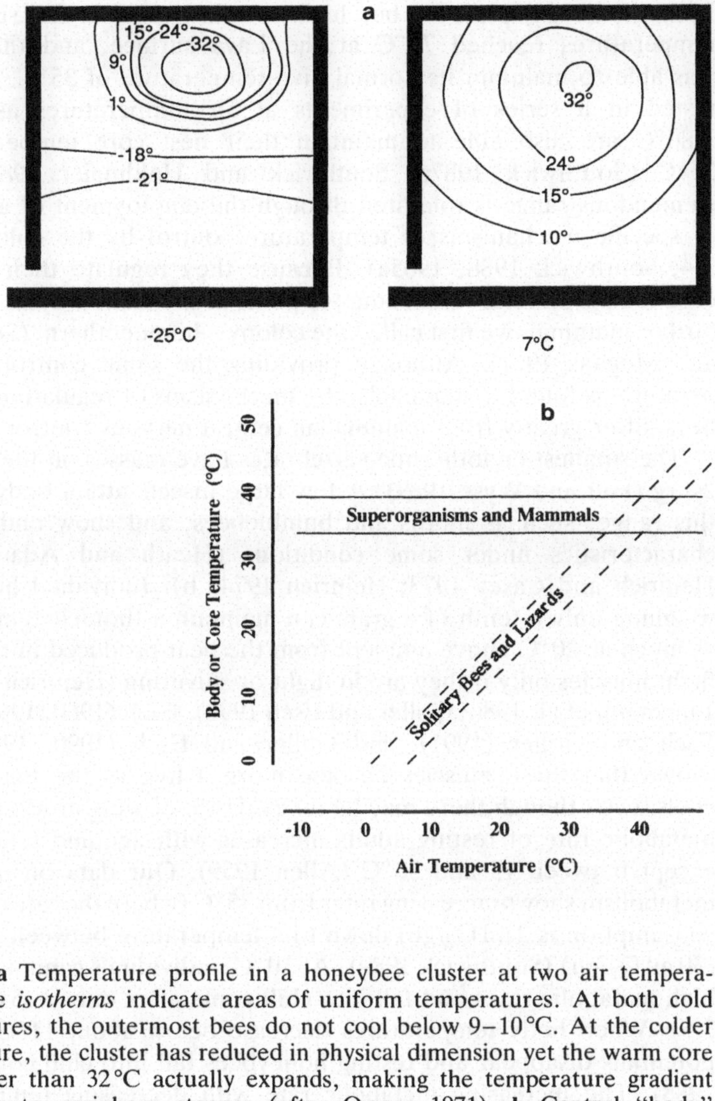

Fig. 3.3. **a** Temperature profile in a honeybee cluster at two air temperatures. The *isotherms* indicate areas of uniform temperatures. At both cold temperatures, the outermost bees do not cool below 9–10 °C. At the colder temperature, the cluster has reduced in physical dimension yet the warm core held higher than 32 °C actually expands, making the temperature gradient from core to periphery steeper (after Owens 1971). **b** Central "body" temperature control in endothermic colonies of bees and in mammals compared with body temperatures in solitary bees and other ectotherms. As environmental temperatures rise or fall, ectotherm temperatures follow the changes, while endotherms maintain constant high temperature over a wide range of air temperatures

Lindauer (1954) placed a bee hive on a lava field in full sun where temperatures reached 70 °C at the Lava surface, and the colony was able to maintain its normal hive temperature of 35 °C. Colonies tested in a series of experiments at air temperatures as low as −80 °C are also able to maintain their nest core temperature at 35 °C (Southwick 1987a; Southwick and Heldmaier 1987). This tremendous range is tolerated through the employment of a number of specific mechanisms of temperature control by the colony (Fig. 3.4; Southwick 1988, 1985a). Because they regulate their colonial core temperature around some set-point in the same way as a typical bird or mammal, we first called the colony a homoeotherm (Southwick and Mugaas 1971). Although providing the same control of temperature as found in mammals, the mechanisms of regulation used by bees differ greatly from mammalian central nervous control.

The smallest endothermic vertebrates have masses on the order of 2–3 g (Hill and Wyse 1989). A few large insects attain body mass in this range, such as moths and bumblebees, and show endothermic characteristics under some conditions (Heath and Adams 1967; Heinrich and Casey 1973; Heinrich 1974a,b). Individual honeybees weighing only a tenth of a gram can maintain a thoracic temperature as much as 20 °C above ambient from the heat produced in the active flight muscles only if they are in flight or shivering (Heinrich 1980a,b; Jungmann et al. 1986; Goller and Esch 1990). Esch (1960, 1964, 1976), Esch and Goller (1991), and Goller and Esch (1990, 1991) have shown that these muscles become more active as the temperature cools, even though there may be no evidence of wing movement. The metabolic rate of resting adults increases with age and temperature except between 15 and 30 °C (Allen 1959). Our data on individual metabolism shows increasing rates from 35 °C (where the rate of oxygen consumption is 2 ml O_2/gh) down to a temperature between 10–15 °C (30 ml O_2/gh) (Southwick 1989). At 10 °C, individuals cannot maintain high metabolic rates and fall into chill coma after only 2–5 min (Fig. 3.5). When body temperatures decrease below about 11 °C, muscle potentials disappear and resting honeybees fall into chill coma (Esch 1988). The increasing metabolic rate with decreasing temperatures above 15 °C is the norm for large vertebrate endotherms but sharply contrasts with other individual insects that do not form superorganismic groups, such as flies (Diptera) and wasps (Hymenoptera).

Respiratory metabolism monitored at several air temperatures by measuring oxygen consumption of the entire group gives data indicating that when the outside air temperature falls, the bees increase their metabolic heat production. In principle, the modus

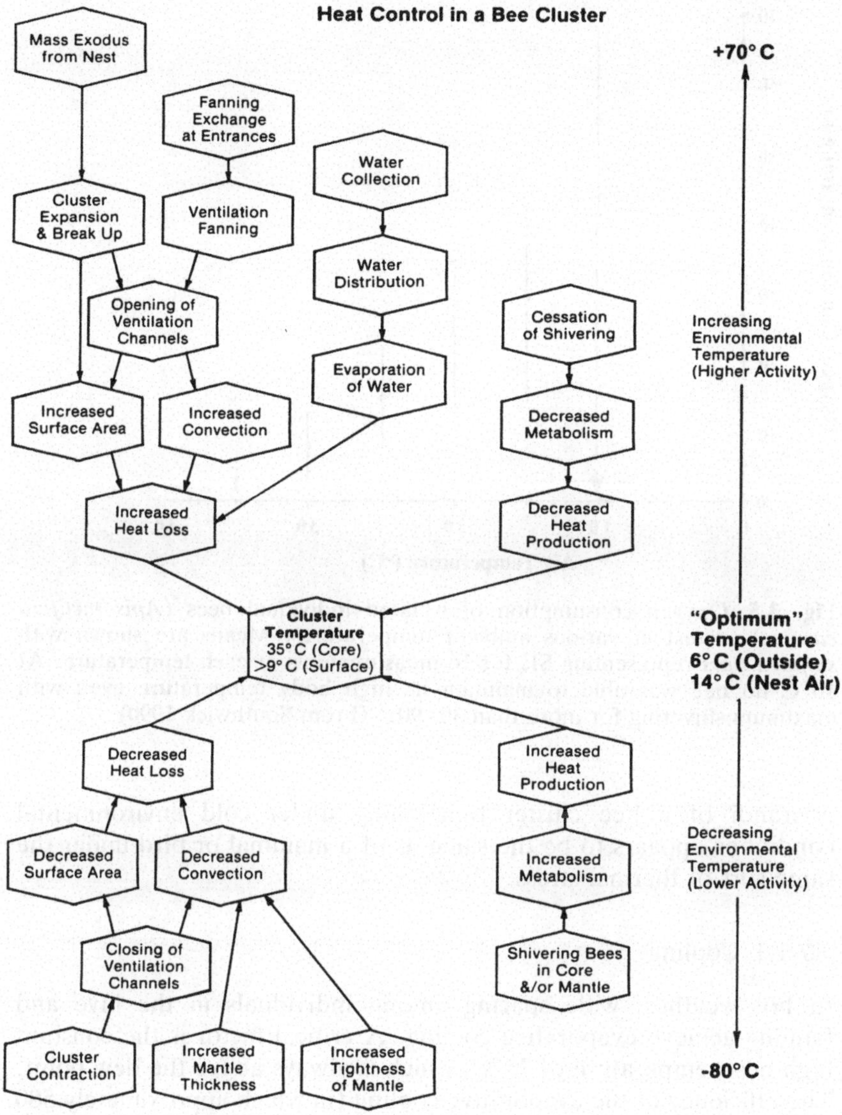

Fig. 3.4. Effective temperature control through modes of adjustment within the social context of a colony of European honeybees. "Optimum" air temperatures require the least metabolic work for maintenance of brood and cluster temperatures. (From Southwick 1988)

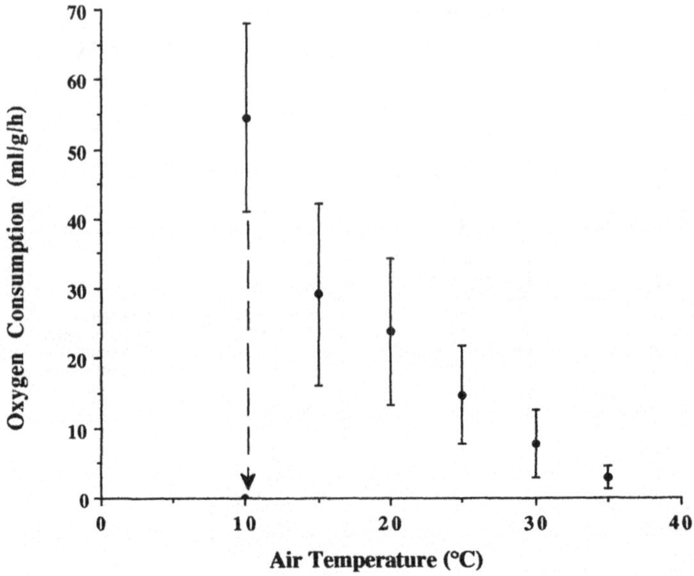

Fig. 3.5. Oxygen consumption of isolated individual bees (*Apis melifera carnica*) at rest at various ambient temperatures. Means are shown with *stopped bars* representing SE for 16 measurements at each temperature. At 10 °C no bee was able to maintain its high body temperature even with maximum shivering for more than 30–90 s. (From Southwick 1990)

operandi of a bee cluster functioning under cold environmental conditions appears to be the same as of a mammal or bird under the same sort of thermal stress.

3.7.1.1 Cooling

In hot weather, wide spacing among individuals in the hive and fanning achieve evaporative cooling. A critical factor is the constant high nest temperature of 35 °C, which is always above the dew point. The efficiency of the evaporative cooling (in which approximately 580 calories of heat are lost for each gram of water evaporated) is directly dependent on the difference between the hive temperature and the dew point temperature. The actual water vapour pressure in the hive is, therefore, always lower than the saturation vapour pressure, resulting in a large saturation deficit. Individual bees can take advantage of this and increase evaporative cooling effective for the whole colony. The worker holds a water droplet in a thin film on its

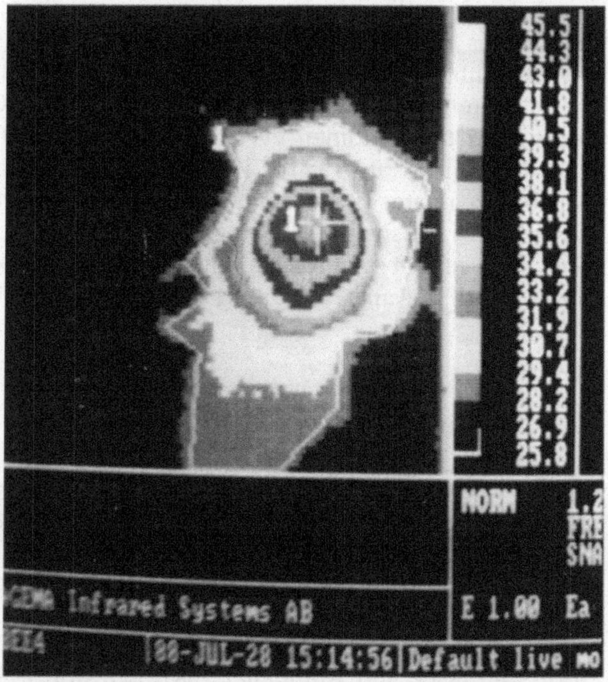

Fig. 3.6. Infrared spectrogram (with colours reduced to shades of grey) of a worker bee at rest and feeding on a droplet of honey. There are 4758 individual points of temperature within the boundary lines on the bee body. The hot spot centrally located in the thorax at the cross hairs is 39.8 °C and indicates the active muscle mass. Cooler points are seen as the distance from this heat source increases. The abdomen is the coolest at 28 °C

mouth parts, and places small droplets on surfaces throughout the hive, thus increasing the water surface area available for evaporation, and then causes drying air currents by fanning behaviour (Lindauer 1952). So even if a colony of bees is in a very hot environment such as a desert, it has no problem cooling as long as there is a sufficient supply of water. Problems could be encountered in humid tropical regions if environmental temperature approached that of the brood (35 °C) and if the air was highly saturated with water vapour.

In flight, individual honeybees can maintain a high thorax temperature over a wide range of air temperatures (Esch 1960) because the flight muscle activity generates excess heat. Heinrich (1980) recorded body temperatures as high as 47 °C in flying honeybees. The high temperatures achieved in flight stimulate modes of reducing heat load, including increased haemolymph flow to the head and mechanisms of evaporative cooling (Esch 1976). The temperature may vary substantially in different regions of the active bee (Fig. 3.6). The head temperature may be by as much as 3 °C cooler than the thorax, and the abdomen even cooler (Esch 1960; Heinrich 1980b, 1987; Cooper et al. 1985). The honeybee under these conditions probably regulates head temperature, with thorax and abdomen temperatures passively following. However, as isolated individuals, honeybees cannot maintain a high enough thorax temperature for flight (>27 °C) at air temperatures of less than about 12 °C (Heinrich 1987).

3.7.1.2 Heating

The colony precisely controls its temperature and does not allow the brood nest temperature to rise above or fall below about 35 °C (Gates 1914; Himmer 1927; Lindauer 1954; Owens 1971; Southwick and Mugaas 1971; Ritter 1982; Southwick 1982, 1985a, 1987a, 1989). In the bees comprising the outside surface of the cluster it rarely falls lower than about 9 °C (Southwick and Mugaas 1971; Heinrich 1981, 1985; Southwick and Heldmaier 1987; Winston 1987). Such ability to regulate temperature is not developed well in bee species with small colonies, but is very conspicuous among species with large complex societies (Tauber et al. 1986). The release of heat energy by the breakdown of carbohydrate food is the primary source of thermogenesis for grouped honeybees. They take in honey which is about 82% hexose sugars (3.4 M) and break these down immediately in their shivering thoracic muscles (Beenakkers et al. 1984; Crabtree and Newsholme 1975; Esch 1960). The metabolic breakdown of simple sugars and the resulting heat production follows the set of general chemical reactions typical for aerobic metabolism. A summary of the more than 20 intermediary reactions is given as Eq. (3.3):

$$C_6H_{12}O_6 + 6O_2 + 6H_2O \xrightarrow{\text{enzymes}} 6CO_2 + 12H_2O + 36\ \text{ATP}. \tag{3.3}$$

About 673 kcal (2814 kjoules) of energy are bound up in the bonds of a mole of glucose molecules. Some energy is trapped in ATP and utilized in the body to power cell functions such as muscle contraction. Eventually all of this energy degrades to heat given off to the environment.

Shivering muscles show only minute movement with amplitudes of contraction and relaxation so small as to be undetectable in movement of the wings. Recently, Surholt and Bertsch (pers. comm.) found that, at least in thoracic musculature of bumblebees, there is warm-up even before actual shivering vibrations are detectable by a sensitive transducer. This implies the presence of a mechanism of non-shivering thermogenesis which is followed by shivering thermogenesis once higher muscle temperatures (ca. 30 °C) are reached. Several earlier reports on bumblebees (e.g. Newsholme et al. 1972) have apparently documented a possible "futile cycle" between fructose-6-phosphate (F6P) and fructose diphosphate (FDP) which releases heat energy by continuous hydrolysis of ATP. There is still some contention about the significance of this heat generation system in bumblebees (Plowright and Laverty 1984; Esch et al. 1991). Esch (pers. comm.) suggests that muscle tetanus may operate in bumblebees during warm-up providing heat without obvious shivering. In any case, there is no evidence of biochemical heat generation by substrate cycling or non-shivering thermogenesis in thoracic muscle of honeybees. The necessary enzymes are found up to 30 times higher in bumble bees than in honeybees, in which they are about the same as in other insects (Blum 1985; Smugor 1991).

The small amount of glycogen stored in flight muscle can be converted to glucose and will support flight or shivering metabolism for only 10–20 min (reviewed in Wigglesworth 1984). After this time, the glycogen stores are depleted and trehalose taken from the haemolymph is the primary fuel, along with sugars from the honey stomach. As isolated individuals, honeybees cannot shiver at temperatures of less than 8 °C (Esch 1988; Goller and Esch 1990), which might explain why isolated honeybees fall into a chill coma when air (and body) temperatures decrease below this level.

3.7.1.3 Maintenance of Nest Temperature

A thermal profile of the bee colony at low ambient temperatures shows high stable temperatures in the central core area (where the brood is located) and concentric isotherms of lower temperatures at greater and greater distances from the centre (Fig. 3.3). The slope of the temperature – distance profile is small near the core, indicating little control of heat loss within this loose portion of the cluster. The slope is steep through the main body of the packed cluster, showing a better insulative structure.

Nest thermoregulation is particularly precise in colonies with brood (Phillips and Demuth 1914; Riter 1978, 1982). Broodless colonies and groups allow their central core temperatures to fall below 30 °C, while colonies containing brood try to maintain the brood area at 34–35 °C. Brood development, like that of other insects, is highly temperature-dependent (May 1985), and larval development of bees is impaired by temperatures below (or above) the narrow range maintained in the cluster core. Under artificial experimental conditions, in which the colony is deprived of a large enough number of worker bees to completely cover all the brood area and protect it from the low temperature imposed, outlying segments of the brood will be abandoned and other more centrally located areas will be kept warm (Southwick, unpubl. data). In another experiment, a small amount of brood was located in one sector of a comb on which workers were gathered, and food located a short distance away on the same comb. The numbers of workers again were insufficient to cover both the areas simultaneously. Under low temperature conditions, the bees were required to choose between the food in one area and the brood in a different area. Invariably they selected the food sector and allowed the brood to become cold. The same results occurred when the choice was between a tethered queen and food at another location. That is, under low temperature stress, the bee cluster moved to the food, not the queen, who was allowed to cool below chill coma temperatures. Although these conditions may not reflect natural conditions, it seems clear that the queen does not control heating behaviour of workers under low temperature conditions. She follows the cluster and moves to the warmest central core. The cluster does not follow the queen, keeping her somehow centrally located.

Clustering. As the ambient air temperatures drop below 15 °C, the bees draw close together forming a tight cluster, thereby conserving heat (Moebus 1980). The reduction in heat loss results from the precipitous reduction in animal surface area for heat exchange (although the number of bees and their living mass remains the same). This cluster expands and contracts in direct response to environmental temperatures at least between about 15 °C and −15 °C. Before the cluster is formed, each bee loses heat through its total body surface of about 1.76 cm^2. This is a body surface area-to-mass (SA/M) ratio of about 14.8 (1.76 cm^2/0.12 g body mass). With such a high SA/M ratio, the heat generated by metabolic processes in the tissues is quickly lost to the surrounding air by convection, radiation and conduction. On the other hand, when the individuals pack themselves in the form of a cluster, the SA/M ratio is greatly reduced. For example, if we take a typical feral colony in northern Europe at about 50° north latitude entering winter with 16000 individuals (cluster mass of 1.9 kg), the sum total of individual surface areas for heat loss would amount to about 28160 cm^2 (1.76 cm^2 × 16000). Of course, not all bees would be exposed to the same air temperatures even when spread out. Those individuals more centrally located would experience a moderated temperature surrounding as compared to those some distance away. When cooled, the clustering phenomenon reduces the total exposed surface for heat exchange to a small fraction of the combined individual surface areas; in this case, to only 1257 cm^2 (assuming the cluster is formed as a sphere with a diameter of 20 cm). Like cells in the core of a vertebrate endotherm, centrally located bees contribute no heat loss surfaces. The clustered group has the same mass but the surface area has shrunk to only 4.5% of the original size, with a SA/M ratio of only 0.65! From a heat transfer standpoint, this is much better than SA/M ratios typical for endothermic animals in the same size range, such as *Marmota monax* at 3 kg with a SA/M of 1.32 (Palmer and Fowler 1975). This reduction in heat exchange surface is reflected in the great savings of metabolic cost and associated savings in consumption of winter food stores. This is, therefore, the key to overwintering success and a perennial nest existence.

Honeybee clusters of different sizes show different responses to falling temperature. The greater the number of bees in the cluster, the smaller is the proportion that is exposed at the surface. The larger the population of bees, the less is the cost of temperature maintenance for each bee during the winter (Free and Racey 1968; Southwick 1988, 1985c). To be a member of a small cluster is very costly (see Sect. 7.2).

Retention of Heat. Heat generated by the bees is apparently further conserved by outer layers of packed bees forming an insulating shell (ca. 25–75 mm in thickness) around the core (Moebus 1980). This shell is composed of tightly compacted layers of bees with their heads pointed inward, and abdomens exposed to the cold. Their inward-directed heads may be important in their precise thermal control since they have been shown to control head temperature as individuals under some circumstances (Heinrich 1980a,b), and their abdomens are cooler than their thoraxes.

Honeybees possess hairs distributed over the body but especially densely distributed over the thorax. These body hairs are plumose, morphologically resembling goose down (Fig. 3.7). The hairs are

Fig. 3.7. Hair on the thorax of honeybees is dense enough to provide good heat retention when individuals pack themselves together in a cluster. The interlacing hairs of adjacent bees comprising the periphery of the cluster form numerous small pockets of trapped still air greatly reducing convective heat loss

small (<1 mm) and cannot trap the same proportion of air in a poorly conducting boundary layer as the feathers and fur of birds and mammals which are as much as a million times larger. However, when the bees form the superorganism as a winter cluster, the individuals in the outer portion or shell have their thoraxes close together and the interlacing of thoracic hairs forms small dead air spaces (as found in fur or feathers of large mammals and birds that have adapted to winter cold, and in effective man-made insulating

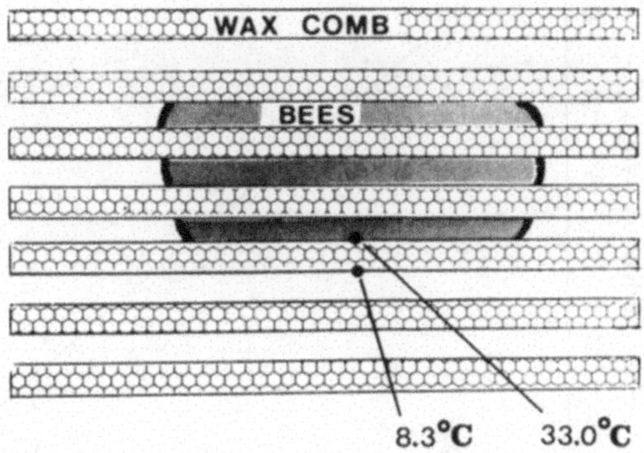

Fig. 3.8. Outer empty comb functioning as insulation against the cluster of bees in the cold. The trapped air in the comb serves to reduce heat loss and the effective surface area for heat exchange of the cluster. (From Southwick 1985c)

materials such as closed-cell expanded polystyrene). Thus, the bees succeed in trapping similarly large relative volumes of air in an effective insulating coat for the colony as a whole; something they could not achieve as individuals. The insulating effectiveness of the hair was revealed in experiments testing normal bees with hair, and their sisters that had little or no thoracic hair (hairless mutant), all offspring from the same queen mother (Southwick 1983, 1985b). All other genetically controlled aspects were the same. Heat retention by the hair of outer bees in groups was reflected in metabolic energy savings at low temperatures. When tested in groups at 2 °C, small clusters of 100 hairless bees consumed 21% more oxygen than their hairy sisters (hairless ($\dot{V}O2 = 10.3 \pm 1.17\,ml\,O_2/gh$, haired ($\dot{V}O2 = 8.5 \pm 1.20\,ml\,O_2/gh$ (Southwick 1985b). The groups of hairless bees also lost a greater number of individuals to chill coma than the hairy ones. As a check, tests at thermoneutral temperatures (30 °C) showed no difference in metabolism between hairy and hairless groups.

Another factor contributes substantially to further heat savings in cold condition, and this is a behavioural attribute of the bees clustering between empty wax combs. The bees pack themselves in a pattern that leaves empty combs serving as outer walls of the cluster (Fig. 3.8). The bees form a compact living layer over the comb effectively trapping air in the empty cells. As much as two-thirds of the surface area of the cluster is insulated by these combs in the nest cavity (Southwick 1985c). The total heat conductance of the cluster [about 0.15 cal/(gh °C)] shows similar values to that of homoeothermic mammals of the same mass, but much lower than ectotherms, like lizards (Herreid and Kessel 1967; Southwick 1985a).

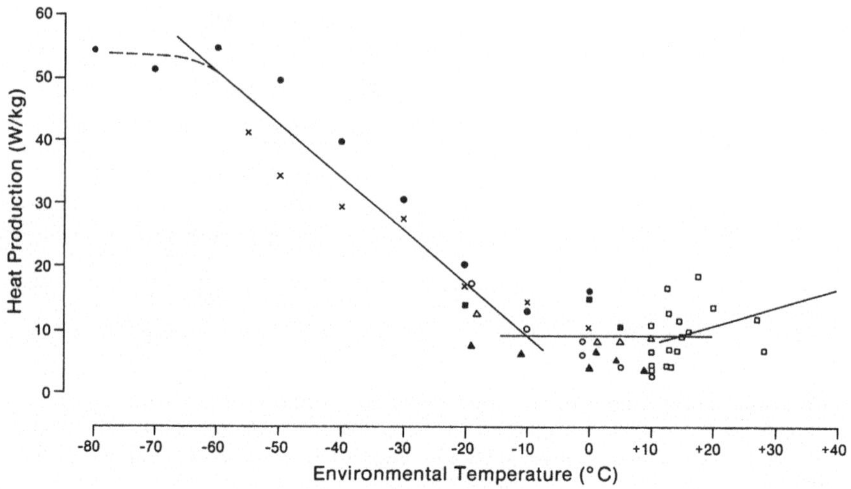

Fig. 3.9. Minimum heat production of full-size bee colonies held under various temperatures for at least 24 h. The shape of the plot indicates a typical endothermic response with increasing heat production below a lower critical temperature (ca. −10 °C) and above an upper critical temperature (ca. 10–15 °C). The temperature range from about −10 to about +15 °C is a "thermal-neutral zone" requiring little physiological control but good physical controls, such as cluster formation and compaction. (From Southwick 1988)

Low Temperature Response. When all means of heat savings are employed and the temperature continues to fall, the honeybee cluster generates more heat metabolically by burning up the honey stores as a fuel. This cluster metabolic heat production increases with decreasing air temperature in full-size colonies (just as it does in humans exposed to cold) following the function plotted in Fig. 3.9. Esch (1960) found that the shivering flight muscles of the thorax are the heat producers in honeybees. The thoracic muscle is one of the most metabolically active tissues known. Although the shivering amplitude is too small to be seen, Esch and Bastian (1968) were able to measure the muscle potentials, demonstrating that rapid low-amplitude contractions were taking place. The amount of oxygen consumed is directly proportional to muscle potential frequency and shivering muscle activity, and is therefore a good measure of heat production. The relationship between ambient temperature and oxygen consumption (heat production) for clusters of honeybees is the same as that shown by mammals and birds, and not at all like in ectotherms (such as lizards or individual insects). Whereas the body temperature of ectotherms tends to follow that of the environment, the core temperature of the bee cluster remains remarkably constant. Muscle tissue in honeybees utilizes predominantly carbohydrate as a fuel, as demonstrated by respiratory quotients, RQ, of 1.0 (Jongbloed and Wiersma 1934; Chippendale 1978; Beenakkers et al. 1984; Storey 1985; Southwick and Heldmaier 1987; Rothe and Nachtigall 1989). Under cold-induced thermogenesis, the total energy utilized for mechanical movement (shivering) and dissipated as heat amounts to about 55 mW/g cluster mass in winter clusters of 16000 bees with brood (Southwick 1987a). The aerobic flight muscles are shivering and using on the average about 14% as much energy as for free flight (400 mW/g body mass; Nachtigall et al. 1989).

In fact, when smaller groups of 1500 workers are exposed to extreme low temperature stress (−50 °C), their maximum maintained heat production reaches 111 mW/g cluster mass, twice the nominal level reached in normal large clusters under similar cold stress, and probably a true cold-induced maximum (Southwick 1987a). The short-term peak heat production recorded for small broodless clusters amounts to 194 mW/g cluster mass, or 1082 mW/g muscle tissue. This is similar to the heat production in brown fat tissue of cold-exposed Djungarian hamsters, *Phodopus sungoras* (Heldmaier 1975), and about one-half the values for flight reported in individual honeybees by many authors and reviewed in Nachtigall et al. (1989).

Caution must be taken here in attempting to compare energy use by bees clustered in groups with that of individual flyers because, within the large cluster, we do not yet know which or even how many bees are actually doing the shivering and generating the heat. The overall metabolic rate is simply averaged over the total mass. One cannot assume that all workers in the cluster are generating heat at the same average rate. It is nevertheless significant that such high values are obtained for cold-induced heat production, especially in the smaller clusters. This indicates that a large fraction of the bees must be actively shivering in the group.

Since individual bees show increasing heat production with decreasing air temperatures from 35 to about 10 °C (Southwick 1989; Goller and Esch 1990), it would seem that if those in the core in an overwintering cluster are acting like individually tested bees, they would be producing virtually no heat, and those on the periphery are too cold to be heat producers. Then how is the temperature in this superorganism regulated? The core rarely rises above 35 °C, and the periphery is not allowed to drop below 7–9 °C (Southwick and

Mugaas 1971; Fahrenholz et al. 1989). It appears that the effort in heat production and regulation is for maintaining a temperature necessary for survival of the most bees. This means that enough heat is required to insure survival of the bees comprising the surface.

Heinrich (1985) recognized the problem of apparent control of temperature within the cluster, but argues against any knowledge of thermal "needs" by bees in one portion of the cluster of bees in another locality. He tested for presence of "messenger" bees between the core and periphery, chemical control, and buzzing communication from cooler bees on the periphery to core bees. None of these routes of possible communication had any support from his data, and he concluded that core bees play no active role is thermal control.

It may be that although most bees in the core are not metabolizing much, a few "turn on" their heat-generating throacic shivering muscle for short time periods. Individual bees tested at several air temperatures by Esch (pers. comm.) show bouts of heat generation during which they shiver until their body temperature reaches about 36 °C. This activity occurs at air temperatures down to about 8 °C, but if the body temperature falls below 20 °C, the bees are no longer able to shiver enough to reach their 36 °C body temperature, probably because the enzyme "machinery" is less efficient. Measurements of metabolic rate using flow-through systems (for example, Southwick 1989) are likely to mask these bouts and instead record averaged bouts, giving the impression of smooth increasing rate with decreasing air temperature. Moving out from the core of the cluster, more and more bees may turn on their heat generators for longer time periods to make up for the heat loss in the decreasing temperature environment and ever increasing volumes. This increasing time and proportion of bees generating the heat as one moves further outward from the core in concentric spheres continues to the point where the bees are generating their maximum (at an air temperature of about 20 °C). At points further out, the contribution to heat generation falls of drastically as these bees are exposed to temperatures too low for shivering maintenance of body temperature to occur. In this way, local thermoregulatory responses could result in overall thermal control in the superorganism.

3.7.1.4 Optimal Environmental Temperature and Nest Choice

Metabolic heat production of large clusters of honeybees is a function of environmental air temperature (Fig. 3.9, Southwick 1988). Below

+10 °C, to about −60 °C the metabolism increases with decreasing air temperature following Eq. (3.4):

$$H = -0.83\,T_a + 1.21 \quad (r^2 = 0.88, n = 20), \tag{3.4}$$

where H = heat production; T_a = ambient air temperature.

Examination of Figure 3.8 reveals that there may be a range of optimum winter air temperatures under which the bee colony will consume the least amount of food stores. These temperatures would lie between about −10 °C and +15 °C, the range in which the energy cost is lowest. With the biological variation found naturally in colonies taken into account, it would be safer to take a smaller median temperature range of about −2 to +8 °C as the optimum. Bees apparently utilize their honey stores most efficiently at outside air temperatures of 4 to 8 °C (Betts 1943). Under these conditions, the air temperature inside the hive, but outside the cluster, will be about 11 to 14 °C, low enough to cause the bee colony to maintain the tight cluster form.

Typically, in summer, scouting honeybees searching for a new homesite for a reproductive swarm have a preference for cavities located high off the ground, with walls that are "tight," (Lindauer 1955; Seeley 1985), and will choose, if possible, a small, low, south-facing entrance in northern temperate climate (Seeley and Morse 1976). They probably cannot locate nest sites with conditions approaching optimum air temperatures for minimal energy use, but these behavioural traits of honeybees work together to conserve energy and to aid in the maintenance of the high constant temperature inside the colony, especially around the developing brood (see Chap. 4).

3.7.2 Mass-Metabolism Relationships and Colony Fitness

Oxidative metabolism is an integrator of general physiological functioning in any organism. Metabolic efficiency and maximum attainable metabolic rate reflect the overall condition of the animal and are likely to be tightly correlated with fitness (Hawkins et al. 1986). Colonial oxygen consumption, then, is expected to be an important index of colony level fitness. Fitness effects of metabolism occur in at least two important ways which determine the success of the colony. First, honeybees capable of higher metabolic rates may be able to carry larger loads of pollen and nectar during their foraging trips to and from the field. This resulting increased foraging efficiency would provide the means for increased rate of colony growth and reproduction. The data of Heinrich (1975) on bumblebee foragers seem to support such a hypothesis. He found a positive correlation between oxygen consumption and nectar load in foraging bumblebees. The more heavily laden bees simply expend more energy to fly. Metabolic scope could, thus, be a limiting factor on foraging loads. The drones of honeybees do not forage, yet a higher maximum metabolic rate would likely lead to increased probability of mating success through increased agility and flight speed (Coelho and Mitton 1988). The second way that metabolism affects fitness of the colony is through thermoregulation, especially under low winter temperature conditions.

When vertebrate animals of different masses are tested for their rates of resting metabolism, it has been found by many investigators that metabolic rate is highly dependent on body mass (Brody 1945; Hemmingsen 1960). Larger animals have higher rates of metabolism but not as high as would be predicted based on size alone. It is more common to look at mass-specific metabolic rates for comparative purposes. Mass-specific rates are expressed as energy expended per unit body mass, for example, oxygen consumed or calories of heat produced per gram body weight. When data points for metabolism of endothermic animals are plotted against body weights of these animals, a curve results that first falls steeply and then has a more and more gradual slope. Thus, the rate of mass- specific oxygen consumption decreases dramatically as the mass-to-surface ratio improves. A gram of tissue in an elephant consumes oxygen at a rate of about 1/100th that of a gram of tissue in a shrew. Many generalizations can be made about the physiology of blood flow, heart function, respiration, food requirements, etc. that apply to the organisms in this "mouse-to-elephant curve" (Schmidt-Nielsen 1985). A regression function [Eq. (3.5)] can be utilized to predict the metabolism for a mammal of any given size (Schmidt-Nielsen 1985):

$$MR = 0.676 \times BW^{-0.25} \quad (r^2 = 0.89), \tag{3.5}$$

where MR = mass-specific metabolic rate (lO_2/kg/h); BW = body weight (kg).

The relationship is easier to use if log-transformed with the equation for the regression of daily energy production on body mass:

$$MR = 70.0 \times BW^{0.75}. \tag{3.6}$$

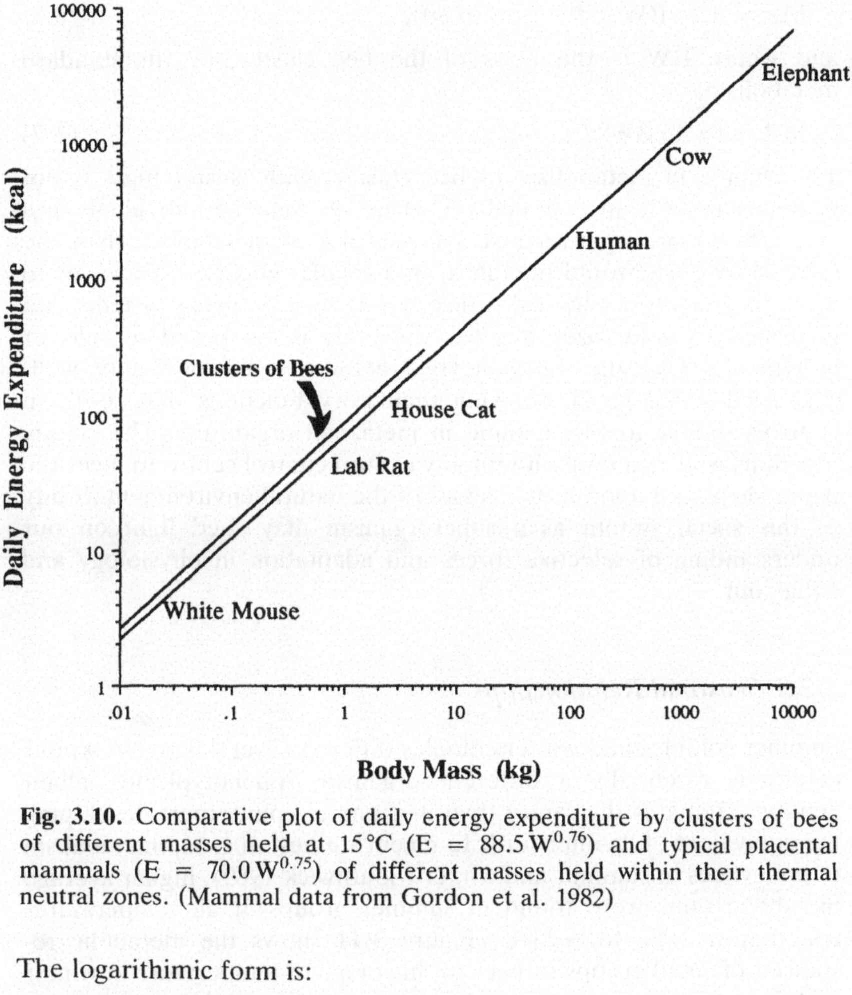

Fig. 3.10. Comparative plot of daily energy expenditure by clusters of bees of different masses held at 15 °C ($E = 88.5\,W^{0.76}$) and typical placental mammals ($E = 70.0\,W^{0.75}$) of different masses held within their thermal neutral zones. (Mammal data from Gordon et al. 1982)

The logarithimic form is:

$$\log MR = \log 70.0 + 0.75 \log BW. \quad (3.7)$$

The log-log plot of oxygen consumption against body mass shows linear regressions for a wide variety of organisms with slopes of about 0.75 (Fig. 3.10). We wanted to test the hypothesis that bee clusters of different sizes would show this same relationship. This seemed highly unlikely as large clusters are simply composed of more individuals than the smaller clusters. We tested clusters ranging in size from 50 bees to 12 000 bees (at 15 °C, the temperature above which bees do not stay in a cluster). This results in the regression of mass-specific metabolic rate for honeybee colonies (Southwick 1985):

$$\mathrm{MR} = 4.23\ \mathrm{BW}^{-0.27} \quad (r^2 = 0.89), \tag{3.8}$$

and where BW is the mass of the bee cluster, or for standard metabolism:

$$\mathrm{MR} = 88.46\ \mathrm{BW}^{0.76}. \tag{3.7}$$

The change in metabolism of bee clusters with cluster mass is not distinguishable from mammals. That means that the individuals in a large cluster are operating at a lower rate of metabolism than the same individuals would operate if in a smaller cluster. This seems to point to an overall physical principle followed by living systems that is related to body size, whether the body is composed of cells or individuals. Overall, the honeybee colony is a remarkably well-organized social group showing regulatory functions that result in controls similar to those found in metazoan organisms. The colony functions well as a unit without any central control centre to meet the demands placed upon it by stresses of the natural environment. Study of this social system as a superorganism may shed light on our understanding of selective forces and adaptation in physiology and behaviour.

3.7.3 Seasonal Relationships

Summer colonies and winter colonies differ in several ways. A winter colony is essentially a different organism (phenotypically, albeit genotype remains the same) than its summer counterpart developing the equivalent of "winter fur". In recent studies on group metabolism of honeybees in summer and winter (Southwick 1989), higher average metabolic rates were found in summer groups at all temperatures tested from +25 to −25 °C. Figure 3.11 shows the metabolic responses of small groups of bees to this range of temperature in winter and summer. Group responses in summer show a moderate metabolic rate (ca. 35 ml O_2/g/h) at 25 °C, and increasing metabolism directly with decreasing temperature from +25 to +15 °C. At 15 °C the group shows a maximum sustainable level of heat production of 43.6 ml O_2/g/h which matches the level later attained at −25 °C (45.5 ml O_2/g/h). At 15 °C, allowing sufficient time to pass (1 to 3 h), even summer bees will cluster, thereby reducing their cost of thermal maintenance (to 14.0 ml O_2/gh). Winter bees form their clusters immediately and show no similar phenomenon at 15 °C. It can also be noted at air temperatures of less than 15 °C that the slope of metabolism vs temperature is greater for summer bees than winter bees. These results imply that not only are the winter bees clustering earlier

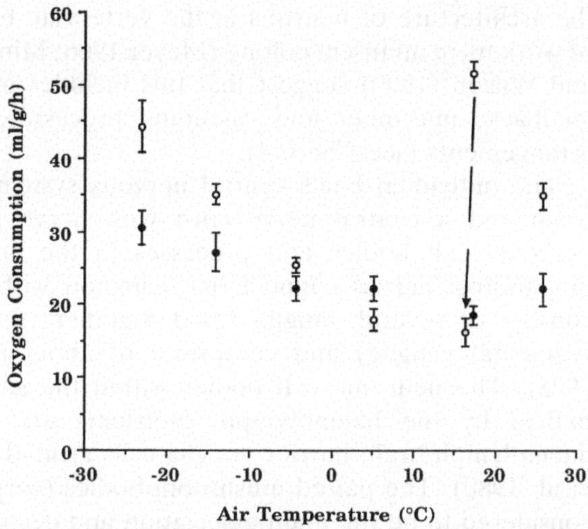

Fig. 3.11. Oxygen consumption in groups of 300 honeybees (*Apis mellifera carnica*) in summer (○) and winter (●). Each group was allowed 30 min to accommodate to stepped temperatures (except at 15 °C at which 120 min was needed). Means and stopped bars representing SE are shown for 15 groups at each temperature. The drastic drop in oxygen consumption at 15 °C in summer resulted from the clustering effect of the bees over about 100 min. (Data from Southwick 1990)

(i.e. at higher temperatures) but also that they are clustering better. They form a cluster that is better at heat retention and therefore they can spend less metabolic energy to stay warm. This sort of organismal seasonal acclimatization has obvious selective advantages.

3.8 Neurophysiology

In the broad context of this volume dealing with the colony as a superorganism, there is no intact nervous system to delineate, and the intention in this section is merely to introduce but not to treat in detail the subject of neurobiology of bees as it pertains to colonial biology. The neuroanatomy and signal processing in the brain of individual honeybees is an exciting and active field, has been well studied, and is reviewed in several articles and texts (for example, Menzel and Erber 1978; Erber 1981; Menzel 1983; Gould 1984, 1991; Menzel and Mercer 1987; Brandes et al. 1988).

Even though there is no intact anatomical nervous system in any superorganism, some authors have articulated similarities between the architecture of neurons in the vertebrate brain and organization of workers in an insect colony (Meyer 1966; Minsky 1986). Hölldobler and Wilson (1990) suggest that this includes principles of biological feedback, multiplier and cascading processes, and optimal spatial arrangements (see Chap. 4).

The individual bee's central nervous system (CNS) consists of a brain and a central nerve cord with nerve ganglia comprised of neuronal cell bodies and processes in the body segments. Paired longitudinal nerves connect one ganglion with the next. The brain consists of several ganglia fused together (three supraoesophageal segmental ganglia) and comprised of about 860 000 cells (Pickard 1988). The neuronal cell bodies within the ganglia are not directly bathed by the haemolymph. Nutrients are conveyed across the haemolymph/brain barrier via glia-cells around the neurons (Mordue et al. 1980). The paired mushroom bodies (corpora pedunculata) are considered to be the main association and decision-making centres of the honeybee brain (Schurmann and Elekes 1987; Pickard 1988). The cellular basis of decision making is associated with chemical transmitters operating in complex local synaptic circuits (Schürmann and Elekes 1987). Visual perception of movement and direction of stimuli have been localized in the lobula region of the optic neuropile of the bee brain, and spatial and colour contrast are detected in the medulla (Hertel and Maronde 1987). Intracellular recording using microelectrodes has shown that single neuronal cells respond to both specific odour molecules and specific visual signals (Masson and Arnold 1987; Milde 1987).

3.8.1 Vision

The one dorsal and two lateral ocelli in the head of each worker bee contain light-sensitive receptor cells (i.e. sensitive to light intensity) as do the ommatidia of the compound eyes. Each of the 6900 ommatidia in the eye of a worker registers an average light intensity from a small area of the visual field with the whole eye forming a mosaic pattern (apposition). These compound eyes also respond to polarized light and colours including ultra-violet light in a wavelength band of 0.30 to 0.65 μm (von Frisch 1965, 1967; Menzel 1990; Rossel and Wehner 1984). Their colour vision is important in visitation to floral resources and includes optimization of colour contrasts between the flower and its background under varying light conditions. The importance of ultraviolet reception is brought out in investigations by several authors, showing floral patterns used by bees in these wavelengths (e.g. von Frisch 1965; Jones and Buchmann 1974; Menzel 1990; Werner et al. 1988). The neural strategy in colour coding is based on spectal antagonism, and discrimination of colour is likely to occur as a result of spectrally opponent visual interneurons in the bee brain (Menzel 1990). The visual image formation is adequate for efficient discrimination of shapes and contrast, and detection of movement (Mordue et al. 1980). Honeybees can learn and distinguish flowers and landmarks. Some authors (Gould 1986, 1987a) suggest that they can store this information in a visual map for later recall, while others convincingly argue that no cognitive map is required for memorizing resource positions or flight patterns (Wehner and Rossel 1985; Dyer 1991).

Karl von Frisch disconvered in the 1940s that bees can detect and orientate to the celestial e-vectors in polarized skylight (von Frisch 1949). It has been found since that only a small part of each eye at the dorsal rim (POL region) is used in polarization vision (Helversen and Edrich 1974; Labhart 1980; Rossel and Wehner 1984). In this region only the ultraviolet receptors are sensitive to polarized light. Rossel and Wehner (1984) show that bees scan the sky and detect compass direction from e-vectors even in small patches of polarized light (Fig. 3.12). In chosing compass course for flight, the bee is thereby able to orientate relative to the celestial symmetry plane. Individual vision capabilities have obvious connections with successful colonial life including forager focussing in on home sites and floral resource patch locations provided in the dance language communication.

3.8.2 *Chemical Sense*

Chemoreceptors are found mostly on the antennae as pits, hairs and pores in neuronal plate endings (Fig. 3.13). Only the pores are sensitive to odours, but there are 3000 of them on each worker antenna, all located on the eight distal segments (Slifer and Sekhon 1961; Lacher 1964). Bees can use their paired antennae to determine direction of an odour source by differential intensity of odour molecules between antennae (Lindauer and Martin 1963). The differential originates during angular flight across and into a plume of odour molecules. The upwind antenna would be in a more dense portion of the plume as the downwind antenna would be emerging into the more dilute air stream outside the plume. The bee, detecting the

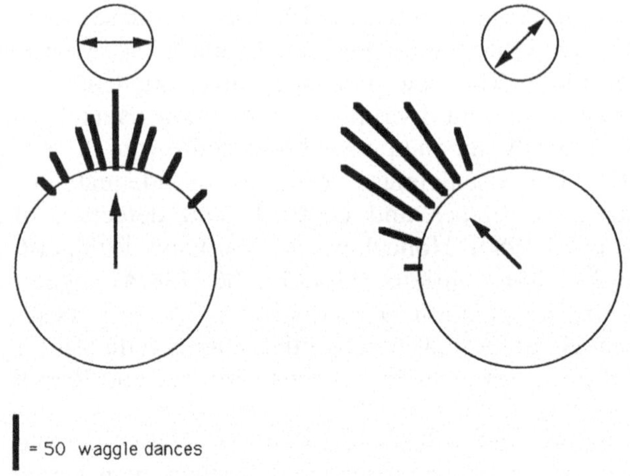

Fig. 3.12. Dance orientation to polarized light showing two different e-vector directions. Direct sunlight is unpolarized with increasing polarization (e-vector directions) as the angle from the sun approaches 90°. Bees can detect and use the e-vector components for orientation. The mean directions of the waggle dance runs are indicated by *solid arrows*. (Frisch 1949, redrawn from Rossel and Wehner 1987)

odour differential turns toward the more concentrated side, thus back into the plume, and follows a zig-zag pattern up the concentration gradient toward the source (Wells and Wenner 1971; Kramer 1976).

The threshold for perception of taste of sugars dissolved in nectar is between 1% and 2% sugar (0.03–0.06 M, von Frisch 1967). Carbon dioxide detection by bees insures that concentrations within the colony are controlled by fanning at less than 4.3% (Hazelhoff 1941; Seeley 1974). Chemosense has importance in collective colony co-ordination and other behaviours, particularly related with the queen and worker chemical pheromones (Free 1987; Naumann et al. 1991).

3.8.3 Acoustic and Mechanical Reception

Mechanoreceptors responding to mechanical stimuli usually consist of socketed hairs on the bee's body which when deformed cause tactile stimulation. Innervated hairs occur on various parts of the body and appendages. In the compound eyes, for example, sensory hairs are arranged at the junctions of the facets. Deflection of these hairs allows perception of air flow during flight (Neese 1965). At the base

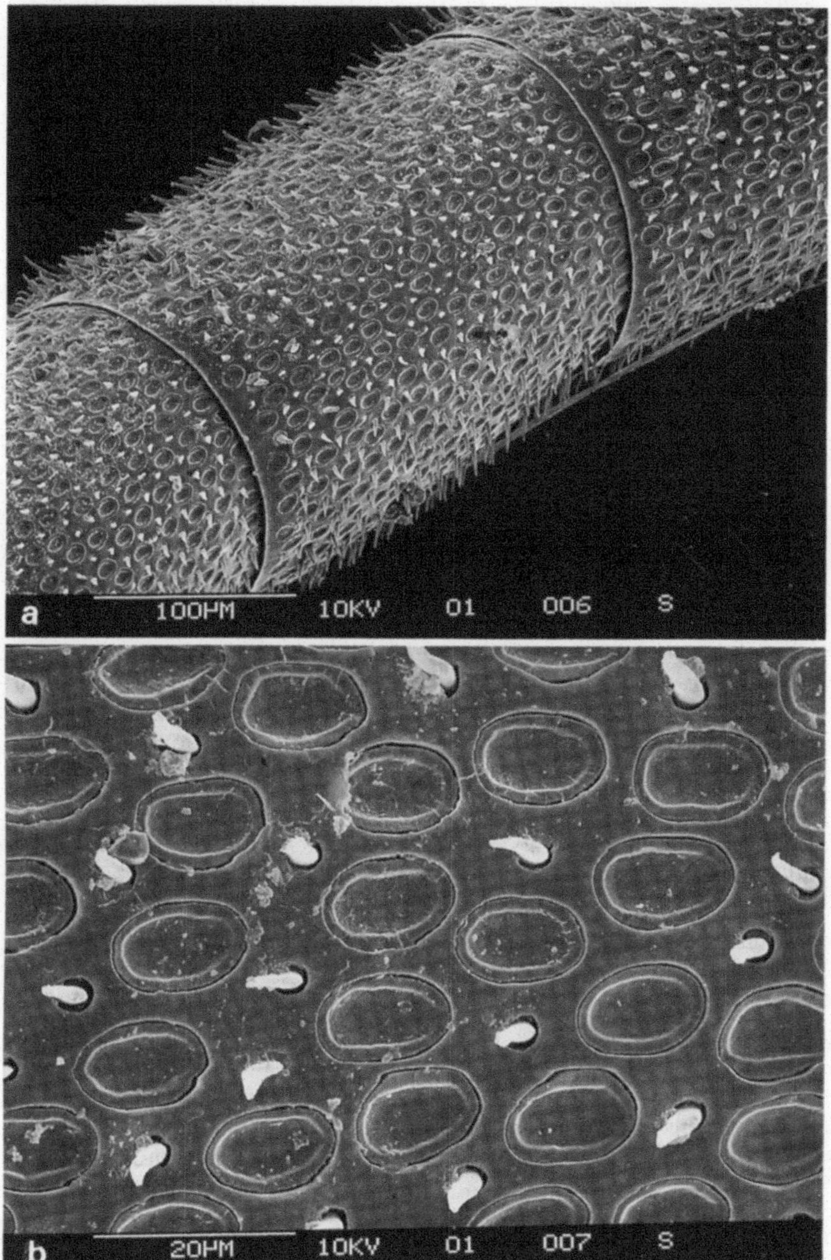

Fig. 3.13a,b. Antennal receptors of a worker honeybee (*Apis mellifera carnica*). **a** Three segments of the flagellum at ×280; **b** plate organs and *sensilla trichodea* at ×1400

of the antennae (in the pedicel) are groups of sensory cells comprising the Johnston's organs. They are sensitive to small changes in antennal position, and they function by bending in air flow during flight. They are thought to be used in measuring flight speed (Heran 1959).

The bees produce a variety of sounds (Esch 1961, 1967a; Esch and Willson 1969) some of which have been shown to be used for communication. Sound production and perception by honeybees occurs during the waggle dance and at other times (Esch et al. 1965; Wenner et al. 1967; Michelsen et al. 1986; Towne and Kirchner 1989; Kirchner 1990; Kirchner et al. 1991). A mechanical model of a dancing bee used by Michelsen et al. (1989) indicated that sound perception and airborne sounds are necessary components of the dance information (Kirchner 1990) (see Chap. 4). Wing beat frequency is tied in with most sound production including queen piping and high frequency sounds of workers within the hearing range of wax moths (*Galleria mellonella*) (Simpson 1964; Spangler 1986, 1991).

3.8.4 Special Sense, Learning and Integration

Neuronal inputs are integrated for complex behaviours such as foraging and search, decision making, learning and memory (Menzel and Mercer 1987). Foragers performing the waggle dance compensate for the rotation of the earth relative to the sun's position using an internal "clock" (von Frisch 1967; Dyer 1987). If bees are placed on a day–night schedule in a flight room with dawn and dusk shifted by several hours, they enter the incorrect time into their internal clocks and orient with a compass deviation corresponding to the artificial phase shift in the day–night cycle (Lindauer and Martin 1963).

Bees can detect and orient to the earth's magnetic field (Lindauer and Martin 1968; Martin and Lindauer 1973). Drones and queens may be able to orient to even small anomalies in local magnetic fields in location of mating areas (Loper 1985; Loper et al. 1987, 1988). Just how this magnetic field is detected is still unanswered. Honeybees and a few other animals, including some birds, mollusks, and bacteria, contain magnetite, an iron-containing compound in specific areas within the body (Lindauer and Martin 1968; Martin and Lindauer 1977; Gould 1980). Markl (1962) demonstrated that bees orientate to gravity which is important for the colonial existence with its comb construction and dance communication in the dark nest cavity.

Extensive behavioural experiments have shown conclusively that honeybees exhibit learning (for example, von Frisch 1965; Menzel and Mercer 1987). They can learn to distinguish odours, learn and remember shapes and colours and a vairty of other tasks (von Frisch 1967; Gould 1985, 1986; Pickard 1988). Learning and memory play key roles in the foraging behaviour of individual honeybees (Gould 1985; Menzel and Mercer 1987; Menzel et al. 1973). Foragers specialize on flower shape, odour and colour, and tend to visit a single species of flower (Aristotle 1949) on any trip or even on successive trips. This sort of floral constancy reduces search time and retrieval time of the floral reward. In the right behavioural context, the important cues are learned in a very few visits (von Frisch 1967; Menzel et al. 1973; Menzel and Erber 1978). A specific odour associated with a sugar reward may be learned in a single experience. The result can be a constancy to that odour in 90% of the workers. A food reward associated with colour will induce that level of constancy after just five exposures. The memory of orientation cues to floral resources may last for weeks. Trained foragers will return to the same feeding station area searching for food even following winter nest confinement for 5 months or more (Menzel et al. 1973). Only through the learning capability can a large worker force of a colony forage a given nectar flow efficiently. Learning is a key for the functioning of the superorganism.

3.9 Muscle Function

Also in this section, we only briefly review the physiology of individual bees, since there is no apparent correlate to a muscle in the superorganism. When the whole colonial superorganism flies (in absconding or swarming), each of the individuals must fly on its own and all food materials must be carried by individuals.

3.9.1 Locomotion and Flight

The muscles involved in walking honeybees respond to single nerve spikes (Symth 1985). Central control in the brain is probable and has been found in other insect species. Sensory feedback is likely to be important at each step. The indirect flight muscles of bees show no relation between the timing of the individual contractions of the muscles and the timing of the motor nerve impulses during flight (Bastian and Esch 1970; Esch et al. 1975; Eckert and Randall 1983;

Symth 1985). Bee flight muscles are asynchronous (because they do not contract in synchrony with each motor impluse). The muscle maintains an active state of fibrillation from a train of motor nerve impulses, but the contraction rate is much higher than the rate of nervous impulses. The stiff thorax changes shape and effects flight patterns of the wings. Since the muscle force is applied close to the fulcrum, it is large and acts through a short distance. The force transmitted to the wings moves them through a greater distance in the air. The wings pivot through a circuit and assume an airfoil shape during flight, and the efficiency of flight and hovering is directly related to this. Wing thickness varies from 2 to 6 μm with a mass of only 10 g/m^2 (McMasters 1989). Instead of being flat, their cross-section is rather irregular with structural stiffening characteristics that provide amazing aerodynamic properties in an air stream. The roughness promotes a boundary layer of air that results in a smooth envelope contour resembling the wing of an airplane. Bees can fly as fast as 29 km/h (von Frisch 1967; Casey 1981). The mechanics and energy consumption of bees during free flight are discussed in detail by Casey et al. (1984) and Nachtigall et al. (1989). Nachtigall et al. (1989) estimate the cost of free-foraging flight of workers at about 400 mW/μg body mass.

3.9.2 Stinging and Biting

Other activities of muscle include stinging and biting and are not only associated with defensive behaviour. Individual workers use their mandibles for eating pollen, to manipulate wax and propolis, and to carry debris from the nest. The workers of *Apis cerana* use their mandibles to pick off and cut external Varroa mites (see Chap. 5 and Peng 1988). Mandibles of the queen are larger than those of the workers, and those of the drone are smaller (Snodgrass 1956). The sting apparatus is similar in structure to an ovipositor. It is housed in the sting chamber at the posterior end of the abdomen. It is made up of a stylet and moving barbed lancets which form the poison canal through which the venom travels (from the poison sac). Once the stinging bee has penetrated its stinger in a victim with fleshy tissue (vertebrates), the barbs prevent its withdrawal causing autotomy (not so in insects). The sting apparatus remains intact and works itself further into the flesh of the victim. This is achieved by the muscles on one side of the sting contracting pulling one lancet against the other with the barbs holding the fixed side in place. Then the other side contracts pulling the sting into the flesh further, and each subsequent contraction causes the sting to penetrate deeper. Alarm pheromones released by the sting apparatus attract other attacking workers, which is important for the colony defending against large predators.

3.10 Circadian Rhythms and Sleep

The ability to transmit photoperiodic information among individuals comprising the colony is unique to highly social insects (Tauber et al. 1986). Many honeybee activities follow a daily cycle during the summer season (for example, Moore and Rankin 1985; Frisch and Aschoff 1987). The rhythms are apparent in activity (Frisch 1984), colonial oxygen consumption (Stussi 1972; Southwick 1982, 1987b; Southwick and Moritz 1987a), and colonial breathing (Southwick and Moritz 1987b). Bees are diurnal animals, being highly active during the day, with little activity at night. Although it was assumed that they rested at night, it was only recently shown by Kaiser (1988) and Kaiser et al. (1987) that they have all the characteristics of actual sleep. This includes a reduced metabolism, even when isolated (Esch, pers. comm.; Frisch and Koeniger 1991). Their periodicity is unique among the insects in showing sleep as part of the daily cycle (Beck 1980). Their time sense (Beeling 1929, Renner 1957) is shown in foragers with a daily rhythm in foraging bouts. They can be trained to feed at a feeding station at a specific hour of the day for several days. They even show anticipation, arriving at the feeding station minutes before the daily filling of sugar water. They continue to come to the feeding station at the same hour for several days, even if there is no food reward (Schreffler 1981).

3.10.1 Cyclic Metabolism

At moderate temperatures (10–20 °C) lowest metabolic rates in colonies of honeybees are found to occur at about 0400 h (Southwick 1982). When air temperature is held constant, a cyclic colony metabolism is keyed to photoperiod. The change from high daytime metabolic rates to low night-time levels is large, as much as eight times higher in the day (Southwick 1982). Even under constant low light or dark conditions, the rhythmic metabolism persists, indicating a truly endogenous clock. This free-running circadian rhythm of group metabolism is apparent in colonies with or without brood.

Groups of bees from indoor flight rooms entrained to an inversed light cycle (D:L) show oscillations in metabolic activity which are shifted nearly 180° out of phase from those of outdoor controls (Southwick and Moritz 1987c). Their high metabolic activity occurs during the subjective "day" and their low metabolism is at "night". Somehow, in each group, each bee's metabolic rhythm is "in step" with her sister's, resulting in a coordinated group oscillation. It seems that tactile interactions among bees, and the temperature generated by worker groups and the queen, function as social Zeitgebers that synchronize the social coordination and reset the internal clock of the group rhythm (Southwick and Moritz 1987a; Moritz et al. 1991b; Moritz and Sakofsky 1991).

3.10.2 Cyclic Locomotion

As honeybees are diurnal (day-active), during the daylight hours they forage for nectar and pollen, and carry out many other activities. At night, when nearly all the foragers are back in the colony, their activities are greatly reduced. If they are kept under continuous illumination (LL) and constant temperature (25 °C) and humidity (65% RH) with sugar syrup available ad libitum, the free-running activity rhythm of colonies of honeybees is found to be about 23.5 h (Frisch 1984). Frisch monitored activity using the signals recorded from an interrupted light-beam of a photocell located at the hive entrance. The activity and rest phases are dependent on the intensity of illumination, with greater intensity decreasing the period.

3.10.3 Cyclic Ventilation

The social control of respiratory activity of the bee colony shows a conspicuous daily pattern like that of metabolism, even under constant dark conditions. Respiratory frequencies of small colonies measured in the laboratory are described in Section 7.4. The rhythm of ventilation in honeybee colonies has many typical characteristics of vertebrate breathing, including the sleep–awake daily rhythm. The daytime respiratory frequency is seven times (7×) higher than at night (RFnight = 0.4 "breaths"/min cp. RFday = 3/min), reflecting the daily cycle of oxidative metabolism (which is eight times higher). Though there are periods of "apnea," the tidal volume remains about the same. The change in respiratory frequency found in honeybee colonies is in the range of that found for humans (active RF is 3× the resting RF) and marmots (active RF is 11× the hibernating RF).

3.10.4 Sleep

"*. . . how very remarkable the psychic energy and the activity of these agile insects whose "manual" abilities and "intellectual" capacities are a never-ending source of amazement for us. It view of these intense expressions of existence, the industrious workers have earned a few hours of rest, and we should be able to understand that they, too, have a more intense requirement for rest, that insects also "become weary and sleep"*. (Fieberg 1912)

This quote dating from the beginning of this century has recently received good experimental support. The behaviour, reaction time, muscle tone and electrophysiological wave patterns of honeybees indicate profound rest periods fulfilling the criteria for sleep (Kaiser and Steiner-Kaiser 1983; Kaiser 1985, 1988). Lindauer (1952) observed a forager for 2 days and nights and found it to be idle at rest 78% of the time at night. Kaiser (1988) notes that the bodies of some bees resting on horizontal surfaces "sink down" so low to the wooden surface on which they are resting that their mandibles contact the wood. It seems obvious that the muscle tone in their legs and neck is reduced. Their antennae also hang down, unmoving, and their body temperatures are indistinguishable from the substrate temperature. These bees show relative insensitivity to strong light stimuli, seem to adopt a particular resting location within the hive, and have a preferred resting posture. It appears that they are in a sleep-like state. Using position and least motility as the criteria, bees experience their deepest sleep late in the night. Electromyograms (EMG) from neck muscles of bees also show reduced amplitude corresponding with the deep sleep period. When bees rest, their body temperatures fall to that prevailing in their ambient surroundings, and only when their flight muscles are activated does body temoperature rise (Esch 1960; Bastian and Esch 1970; Heinrich 1987). Since these sleep signs are present in honeybee workers at night, they provide strong evidence that the bees and the colony as a whole undergo a sleeping phase.

Chapter 4 Communication Network of the Superorganism

No life lives unto itself alone; each is an intrinsic part of all; each administering to its own needs while perpetuating the entirety. (G. Frostic 1981)

4.1 The Analysis of Communication Networks

Clearly, communication is essential in a superorganism. It is achieved via chemical, acoustic, and other sensory signals (Lindauer 1961; Prosser 1986). Yet, understanding all modes of individual communication resulting in a coordinated system is complex because of the large number of individuals involved. The analysis of complex communication systems has a history in the sociometric era of the 1930s to the 1950s (Moreno 1934; Simmel 1964). Using general systems theory and dynamic communication links, modern researchers in the discipline of the social sciences show how network analysis aids in understanding processes of mutual exchange of information in complex populations. They also show that any simple individualistic concept must fail if one wants to understand the more complex group. The term, communication, is defined as a process in which participants create and share information with one another and reach a consensus (Rogers and Kincaid 1981). The key in the network analysis is derived from systems analysis and cybernetics and does not distinguish between source and receiver but focuses on communication among "transceivers" as carrying on mutual exchange.

Communication network analysis identified the communication structure in a system including relational links. Even though network analysis developed as the favoured approach in social science research, communication links were too often treated as static. Even the early network analyses involved too many practical problems to be workable and this contributed to the reduction in their use in the field of social communication analyses until the formalization and computerization of sociograms. The immensity of the task of analyzing even a small group of 100 individuals is realized if one considers that each could have a communication link with 99 others. Since, in communication networks, pairs of individuals are taken as single relationships in which each is a transceiver with direction of information flow being unimportant, there are half the total number, or 4450, relational pairs possible (each relationship is a communication link between two individual transceivers or a "dyad"). The general formula for the number of dyads in any communication system of N individuals is:

$$\text{number of dyads} = N(N - 1)/2. \qquad (4.1)$$

A group of 200 individuals would have 19 900 dyads, and a colony of 16 000 bees would have 128 million possible communication links! Study of communication within such a system is made manageable by invoking network analysis. In fact, the sociologists have reached the conclusion that individuals cannot serve as the analysis units of variations in the social structure and as factors determining this, but that the groups themselves, whose social structure is under study, must be made the units of analysis (Blau 1969). A central issue in the theoretical study in the social sciences is why some systems in human culture, such as political or economic ones, for example, have better performance and are more effective than others. Zaltman et al. (1973) give an example in the human realm which shows that organizations with greater decentralization and less bureaucracy are more innovative and flexible. A similar approach could be used by neurophysiologists in attempting to understand more clearly the complex anatomy and function of the 10 billion single neurons interconnected by a meshwork of fibre nets acting in concert as a united whole in the brain (Beck 1971). The theoretical framework originally designed for examination of cummunication pathways in human societies, and now used in multiple-path computer networks with parallel processing and neurological pathways in nervous tissue, can easily be adapted to communication systems in superorganisms.

The systems of information transfer in some social insect groups require special modes of communication and these have drawn more attention from behavioural biologists. For example, pheromonal regulation of colony activities through semiochemicals released by the queen are of fundamental significance for the colony. Yet, the flexibility of colonial responses to environmental changes are unlikely to be affected by these overall colony regulators. The queen pheromones, which comprise a very hierarchical communication medium, usually control stereotypic situations which require permanent information for all colony members. Generally, pheromonal communication leaves very little room for context-dependent responses. For most tasks the workers themselves must generate decisions within their local environment without any knowledge of the whole, and it is the sum of all these small individual decisions that must accumulate into an overall adaptive global response with no central control of any kind implied or required. In such a system, colonial behaviour can be controlled by social feedback loops in a self-organizing way. Social feedback loops are used in many functions and behaviours of superorganisms, such as in their responses to food and water needs, thermal stimuli from the environment, hygienic behaviour, and reaction to enemy intrusion.

4.2 Division of Labour

The many tasks within the colony are generally handled by workers of certain castes and specific ages, thus, caste or age polyethism. There are three requisites for social communication regarding division of labour in bee colonies (Lindauer 1961). These include:

1. a large repertoire of instinctive behaviour available,
2. a "socially directed drive" which results from the degeneration of ovaries of worker bees, and
3. individuals available for community tasks at any moment.

There are three general task categories associated with age distribution for hive workers (Wilson 1985): the cell-cleaning tasks, broodnest tasks, and food-handling tasks. These general task categories involve many different behavioural duties. Three dozen or more different duties are performed by workers in the hive (Rösch 1927; King 1933; Lindauer 1952; Seeley 1982a; Robinson 1991). The duties include bees removing debris (including dead bees), concentrating nectar, capping honey, guarding, and other social behaviours. Bees 2–11 days old show strong probabilities of feeding brood, capping brood, cleaning and polishing cells, and attending the queen, but they practically never receive nectar from foragers, nor pack pollen, nor store nectar. Workers 11 to 20 days old routinely receive nectar, pack pollen, and store nectar, but almost never feed brood,

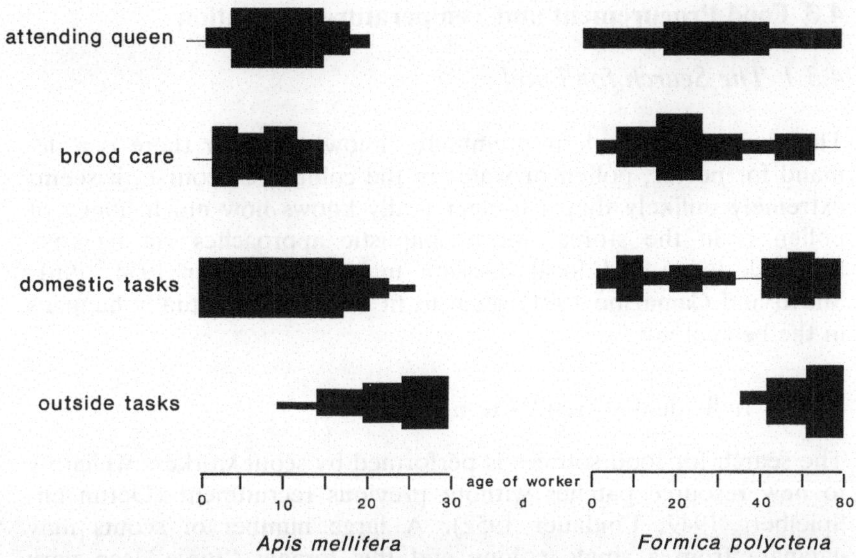

Fig. 4.1. Age profile of activities carried out be worker honeybees in a queen-right colony, and in ants. (After Wilson 1971)

cap brood, clean and polish cells, or attend the queen. The typical age profile duties most commonly performed by workers is represented in Fig. 4.1.

The age polyethism is not a rigid structure, however. There are usually some bees outside these age categories that also perform the duties, as represented by the low-lying portions at the extremes of Fig. 4.1. If something happens to the colony so that a number of workers of one or several particular ages are lost, workers of other ages will pick up the duties even though they are not of the "right age" (Rösch 1930). This fulfils the third requisite that individuals are in a constant state of alertness to do jobs needed at the moment. Division of labour among nestmate workers is central to the colonial organization and ecological success of the superorganism. Recently, in has been found that also genetic variability affects division of labour (see Chap. 8).

4.3 Food Procurement and Temperature Regulation

4.3.1 The Search for Food

The workers of a colony seemingly "know" whether there is a demand for nectar, pollen or water in the colony. Of course, it seems extremely unlikely that a forager really knows how much honey or pollen is in the stores. More simplistic approaches via negative feedback loops and local decision making (Lindauer 1954, 1961; Sneyd and Camazine 1991) seem to fit better what actually happens in the bee colony.

4.3.1.1 Individual Search Pattern of a Scout

The search for food sources is performed by scout workers which fly to new resource patches without previous recruitment (Oettingen-Spielberg 1949; Lindauer 1952). A large number of scouts may emanate from a single colony and this forager "force" can vary considerably depending on environmental and hive conditions. During the peak of a good nectar flow, as few as 5% of the population in a single nest may search for new food sources (Seeley 1983). In periods of dearth, one-third or more of the worker population may be involved in scout activities. The costs and the rewards to locate and harvest a food source influence the bee's decision whether or not to continue foraging at a particular resource or to abandon it and move on to another (von Frisch 1967; Southwick et al. 1981; Real et al. 1982; Southwick and Schreffler 1984; Schmid-Hempel 1984; Seeley 1983, 1989; Seeley et al. 1991). Honeybees assess and collect sugar solutions preferentially in the concentration range of 30–50% rather than higher or lower (Waller 1972) and prefer flowers yielding higher reward per visit (Waddington 1987). The scouts recruit other workers to the floral patches they are working via a dance communication system (described below). The efficacy of these dances strongly depends on the quality of the nectar (Nunez 1970).

Floral nectar production varies over the day with some flowers presenting nectar or pollen early in the morning, and others later. Honeybee foragers visiting these floral resources learn their times of presentation and visit them at the correct times (Beeling 1929). In fact, using artificial feeding stations, it has been shown that the bees actually anticipate the availability of food by arriving at feeding sites slightly before nectar is available (Forel 1910; Renner 1957;

Schreffler 1983). Kalmus (1939) suggested that foragers stop collecting nectar after reaching a critical load weight or volume and then return back to the colony. In fact, bees weigh costs in deciding when to stop feeding and this suggests that energy efficiency is increased when the crop is not fully loaded (Schmid-Hempel et al. 1985; Schmid-Hempel 1987; Waddington 1987).

4.3.1.2 Foraging a Patch of Flowers

When isolated individual foragers are observed in the field, they seem to follow specified patterns in their searching behaviour for nectar at floral resources. Their behaviour depends on the memory of their previous foraging patch location and on the temporal presentation of nectar rewards by the floral species. The combination of efficiency in feeding and efficiency in nectar production has led many investigators to hypothesize that this co-evolutionary pattern has adapted over time into an optimal scheme of foraging (Pyke et al. 1977; Pyke 1978). Invoking the term "optimal" means that there is not a better method by which the animal could possibly obtain food. Jander (1982) finds no support for such foraging as a selective scheme, but instead sees a "random walk" as sufficient to explain the observed behaviour in bee species such as bumblebees (*Bombus* spp.). Indeed, it seems very unlikely that "optimal" organisms have evolved through natural selection since there are stringent constraints by the bauplan of the species. Yet, clearly one might expect some optimization within these constraints, which in fact have been observed in bees (e.g. Waddington and Holden 1979). Real et al. (1982) and Southwick and Schreffler (1984) showed that honeybees and bumblebees prefer to forage in nectar patches that provide small consistent rewards rather than in patches that provide less sure bonanza rewards, even if the average amount of nectar available in the patches is the same. In other words, the foragers choose the more stable supply system over the more variable one. The amounts of nectar available in any standing crop are tiny, amounting to only a few microlitres in most blossoms (Southwick et al. 1981). Most blossoms are empty, so considerable exploring and probing are required in any search procedure. With these facts in mind, the overwhelming advantages of recruitment in the social honeybee colony can be seen, as a single colony can quickly harvest new nectar resources before depletion by competitors.

4.3.2 *Communication Mechanisms*

4.3.2.1 Round Dance

The communication dance of the forager honeybee was described most eloquently by von Frisch (1967). Our current understanding of the dance is that workers which have been foraging for nectar in flowers return to the hive and perform a dance on the vertical honeycomb that contains information related to both the distance and the richness of the food source which can be perceived by other potential foragers. There is recent evidence that direction is indicated in the round dance as well, and sound communication may be important in this component (Kirchner 1990). As long as the feeding source is near the hive, the returning foragers first deliver some of the contents of their honey stomachs to house bees and then perform what von Frisch called a round dance. On the vertical wax comb the dancer moves around first to the right and then to the left, in repeating sequences of clockwise and counterclockwise circling around the perimeter of the described circle at the same spot. The dance is done with vigour and continues at the same location for half a minute or more. The dancer may then move to another location on the honeycomb and repeat the round dance, after which she leaves on another foraging trip to the feeding place. During the dance some of the bees near the dancer are stimulated by her and follow her closely, keeping their antennae in close contact with her body as she circles. At intervals the dancer may stop and offer a droplet of her nectar load for the potential recruits to sample. The followers may pay close attention to the pollen loads as well, if the dancer has any. By placing small coloured marks on the bees following the dancer in the hive, von Frisch found that they are able to obtain information adequate to guide them to the same feeding place and can, in fact, be found there. The new recruits on returning to the hive from their feeding bout also dance, and the more the bees that dance in the hive, the more bees appear at the feeding place. It is clear that a message concerning an attractive floral patch is transmitted by the dance.

4.3.2.2 Waggle Dance

When the floral patch is at a great distance from the hive, von Frisch noticed that the returning forager performs a different type of a dance on the honeycomb. In this case, the forager performs a tail-wagging dance. This dance is more complex than the simple round dance. The dancer runs a short distance in a straight line across the centre of the described circle while "wagging" its body rapidly from side to side with simultaneous wing vibration, and then it circles around to the left and reruns the same straight segment again wagging itself, and follows by circling around to the right and rerunning the straight line segment again, repeating the tight figure-eight pattern time and again. When a feeding station is moved progressively further and further from the hive, the dance gradually changes from a round dance into a waggle dance, and between 25 and 100 m the dance takes on an intermediate pattern, looking more and more like the tight figure-eight waggle dance as the distance approaches 100 m. It is clear that these dances are transmitting different bits of information about the food source to the followers.

Bee recruits responding to the round dance seem to perceive only that there is food in the vicinity of the nest, and they fly out in all directions searching for food nearby. The recruits responding to the waggle dance derive much more information. By careful observations and measurements of dancing bees, von Frisch showed that the distance to the feeding place is indicated as well in the waggle dance. By counting the number of turns made by the dancer in 15 s, he was able to show a relationship of dance tempo with distance, the greater distance being indicated by the slower tempo. His data from 6267 measured dances cover distances from 100 m to 10 km from the hive to the feeding source. At 100 m there are 9 to 10 complete turns in the 15 s time, at 500 m about 6 turns, at 2000 m 3.3 turns, and at 10 000 m 1.25 turns. The conclusion is that the frequency of the linear wagging run in the dance figure is the indicator of distance. The distance indicated is that of the route out to the food source and reflects the costs of transport, whether uphill, with a tail wind, or against a head wind. There are racial differences in dance language that are evident especially in the transition from the round dance to the waggle dance. The transition occurs at about 85 m in *A.m. carnica*, at about 65 m in *A.m. intermissa*, and at 35 m in *A.m. ligustica* (Lindauer 1961).

Another piece of information about the food source is revealed in the waggle dance. The direction of the floral patch is indicated by the angle of the straight part of the dance on the vertical honeycomb. This angle, as measured from the vertical, is similar to the angle of the feeding place, as measured horizontally with respect to the sun. More detailed observations showed that the angle of the straight segment of the dance gradually varies with the movement of the sun during the day. The mode of orientation by using the sun has been shown in several species of bees and ants (Santschi 1911; Wolf 1926, 1927). With the vertical combs serving as the base on which the dances are performed (in the dark), it is likely that gravity serves as the cue to which the bees are sensitive. Workers aroused by the waggle dances thus recognize elements of information related to distance and direction of the feeding patch.

Communication through dancing is not unique to *A. mellifera*, but is found also in *A. florea*, *A. dorsata* and *A. cerana* (Lindauer 1956; Dyer 1985a,b). The early descriptions of the Asian species claiming more primitive forms of dance language are being questioned as new data show that these species are no less sophisticated in information processing through means of their dances (Dyer 1987). *A. dorsata* is even capable of foraging on moonlit nights and performing waggle dances on return to the nest. *A. florea*, once thought to dance only on a horizontal plane under celestial cues, demonstrates dances with the sun and blue sky occluded utilizing visible landmarks and will dance on a steep slope (but not transfer the celestial orientation to a gravity projection like the other species).

4.3.2.3 Acoustic Signals

When the returning honeybee foragers from distant floral patches dance, they emit short bursts of sound produced by their wings during the wagging movements. These sounds consist of a series of bursts of 200 to about 330 Hertz tone (Spangler 1991). Bees following the wagging bee closely can detect the vibration with their sensitive antennae. This sound is an important cue for recruits corresponding to the rate of wagging and the distance to the food source (Esch 1967b; Spangler 1991). Towne and Kirchner (1989) and Michelsen et al. (1987) have studied sound detection of honeybees and found them to be stimulated by airborne sounds in the range produced by waggle dancers.

Michelsen et al. (1989) constructed a mechanical model of a dancing honeybee that mimicked the movements of the round and waggle dances including the vibrational motions and sound. The model dancer with vibrating wings transmitted the oscillating air current in close proximity to its wings to the antennae of followers just as a real dancer would. Furthermore, the model bee could do the waggle dance including the straight line run and circling back portions by stiff wire connections to small remote servomotors and an X-Y chart recorder mechanism. The model is also equipped with a small tube connected to a motor-controlled syringe which released tiny droplets of sugar water at selected intervals. The model did not respond to begging signals of the follower bees, but offered the scented "nectar" samples at one-minute intervals from its "head" end. The same scent was added to filter paper baits placed at specific locations in the field. The model was allowed 12 h inside the colony before experiments were conducted so that it would absorb the odours of the group into its beeswax covering, ensuring that the workers would not perceive it as some foreign intruder. The dance of the model indicated one specific location in the field. Observers counted the number of bees approaching the "correct" bait location where they could get no food. This prevented dancing by returning bees.

Information on both direction and distance was transferred from the dancing model to recruits, indicated by the numbers of bees appearing at correct sites as compared to those approaching control sites (Michelsen et al. 1989). Directional cues were lost when sound or wagging movements were left out of the model's repertoire. This is further verification of the importance of sound emitted by the dancer as an important cue to recruits. The mechanical dancing bee demonstrates that the dance language really does function, as found by Karl von Frisch, though better results are obtained when real bees recruit.

4.3.2.4 Odour Signals

The bees following a waggle dancer pick up both direction and distance from cues given by the dancer. In addition to the dance movements and sounds, two other components are important in transmitting complete information to the recruits: the taste as a measure of richness, and the odour from the flowers. Also foragers too are thought to leave a chemical odour on the surface of flowers they have visited (Nunez 1967; Free and Williams 1983). This odour may help the returning forager as well as new recruits to home in on the floral resource increasing the efficiency of foraging (Wells and Wenner 1971, 1973), Wenner and Wells (1990) argue that the floral fragrances and perhaps Nasanov gland and/or footprint pheromones are sufficient to explain most of the information transfer from the dancer to recruits. Wenner and Wells (1990) hypothesize from their data that neophyte foragers fly out of the hive in the general direction of a food resource but their final homing in process is an upwind flight toward the source of the floral fragrance in a criss-crossing pattern. They argue that the bees are in fact flying into an ever increasing concentration of fragrance molecules in an odour plume emanating downwind from the feeding station. The recruit flies through the odour plume at some distance downwind from the source and then follows a zig-zag searching pattern upwind into the ever increasing concentration until it locates the feeding site. This certainly may be a realistic scenario but does not answer the question of whether or not there is any communication in the dance of the honeybee. A way to determine conclusively that dance provides directional information is to provide foragers dancing in a direction indicating a food patch where they had not been to pick up any odours. This was performed by Gould (see Gould and Gould 1988). Bee dances usually occur on the vertical combs inside dark hives with the up direction away from gravity corresponding to the direction of the sun. When the sun is visible, the dancing bees orientate toward it instead and ignore gravity. Experimentally, one can use a bright light in place of the sun and if it is moved the dancing bees shift their indicated direction along with it. Recruits also align their dance interpretations with the light (instead of gravity) just like the dancers. If the three small ocelli, the simple eyes on top of the head, are coated with an opaque paint, such paint-treated bees require very bright light in order to orientate to it, and a level of brightness can be found that affects untreated bees but not treated ones. Just such a light was used to orient untreated recruits to treated foragers. This then caused the foragers to dance on the vertical comb indicating direction to a feeding station with the straight line run angled according to gravity, while the recruits following them ignored gravity and orientated to the displaced light. If the dance language contains information used by recruits, most of them should arrive at the false feeding station indicated by the light (but where the dancing forager had never been). If odour played the major role, most should arrive at the feeding station actually visited by the dancing forager. The results are clear as the recruits fly where the dance tells them to go. If the light is shifted, the recruits fly to the newly misdirected feeding station. The conclusion is that the dance language really does communicate direction and distance.

Generally, it seems that the discoveries by the many investigators have shown that bees utilize a remarkable variety of cues to orientate and find food sources. They certainly use odours, the sun, gravity, sounds, patterns of polarized light, familiar landmarks, presence of other bees feeding, and even magnetism. They have a dance communication system that indicates distance and direction, but it is used along with odours in determining where food sources are located. It seems most likely that recruits fly out in dance-indicated directions and finally home in on the floral resource using local odour cues. Once the patch location is learned, bees fly directly to and from it and their nest, using landmarks for orientation.

4.3.2.5 Dorsoventral Abdominal Vibration (DVAV) Dance

The dance involving rapid up and down motion (vibration) of the abdomen transmits information related to the state of queen emergence and reproductive swarming as well as general stimulation of foraging activities (Milum 1955; Fletcher 1975; Schneider et al. 1986a,b). The vibrations stimulate potential recruits to move into the dance arena where they may have contact with returning foragers. Fletcher (1975) describes how workers vibrating the queen reduce her fecundity in a preparatory period prior to swarming. During the days before swarming the frequency of vibrations of workers on the queen becomes greater in an ever increasing crescendo from zero to nearly 300 times per hour. After swarming, vibration may cease in that context altogether or it may fluctuate depending on the number of maturing queen cells which are themselves vibrated by the workers. This vibration behaviour then serves at least two functions under different conditions: the swarming preparedness function, and the foraging stimulation function.

4.3.3 Central Food Handling and Social Integration

In essence, the honeybee superorganism satiates itself according to its perceived nutrient or water needs. The "hungrier" it is, the more food it seeks. Under hot and dry conditions, it attempts to procure water in lieu of food to satisfy its "thirst" and to maintain the required supply for its machinery involved in evaporative cooling. But, what rules "hunger" and "thirst" in a colony?

Through experimental regimes designed for analysis of potential feedback loops that regulate the foraging efficiency of a colony, it was revealed that food collecting and temperature regulation of the brood under conditions of heat stress are two colonial behaviour patterns that are tightly coupled together. Phenotypically, it seems that the honeybee colony acts as a unified whole, "pooling" the information from its many foragers and uses this information to reap the maximum benefits from the available flower patches. Yet it is the individual recruitment of other bees by the foragers to their flower patches that ultimately focusses the colony's forager force (Lindauer 1961; von Frisch 1967; Camazine and Sneyd 1990; Seeley and Camazine 1991). Foragers are preferentially directed to high-quality food sources. This happens because returning foragers from rich floral patches recruit more successfully in the colony than those from poorer food sources. By the global effects of steadily focussing its foragers on rich patches of flowers, a colony achieves high efficiency in food collection throughout the day, certainly an adaptive colonial behaviour.

4.3.3.1 Amoeboid Foraging Pattern

It is useful to form the mental picture of looking down on the superorganismic honeybee colony from high above, and considering the feeding arms of foragers as pseudopods of a giant feeding Amoeba extending outward toward food particles (Fig. 4.2). Wilson (1984) first used the expression of "amoeboid" pattern of foraging for colonies of leafcutter ants (*Atta cephalotes*) in tropical South America. He visualized the foraging columns of ants as they "snake out like pseudopods to engulf and shred plants" pulling the green pieces back down holes into their fungus gardens. This is descriptive for the central honeybee colony as well, which sends out pseudopods of foragers in search of the best food sources available at the time. The pseudopods of foragers extending out to forage patches from the centrally located colony are dynamic and fluctuate and vacillate

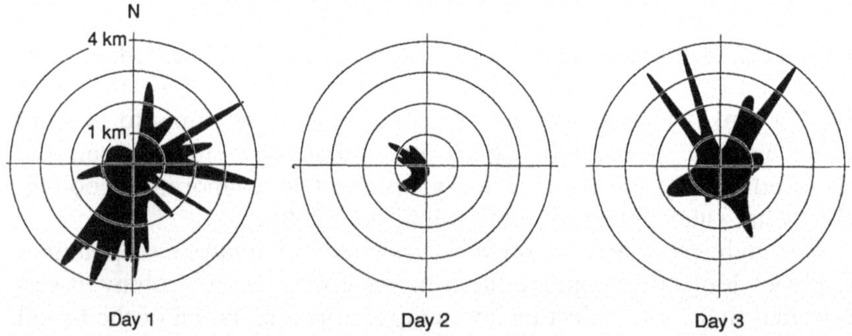

Fig. 4.2. Amoeboid foraging pattern shown on three consecutive days in June. The colony nest site is located at the centre of concentric circles indicating radial distance. The amoeboid arms are rough outlines of foraging locations inferred from honeybee recruitment dances. Day 2 was cool and rainy. (Map data from Visscher and Seeley 1982)

throughout the day and from day to day. Some of the foragers recruit more workers than others, resulting in some pseudopods being much larger than others; the larger ones extending to the richer floral patches. On an average day in the temperate deciduous forest, a typical colony will have pseudopods extending out to feeding patches in an area covering as much as 100 km^2 (Seeley 1985). Levin (1959) observed that young bees foraging during the first day were found to be 300–450 m from the nest. As the foragers age, they forage farther from the nest, but Visscher (1982) found that most foraging actually takes place at 600–800 m from the nest, although under conditions of dearth a single pod may extend to a maximum of 13 km from the central colony in the procurement of food or water (Eckert 1933; von Frisch 1950). The mean length in most habitats is likely to be considerably less than that, and even less than the average 2.3 km foraging distance (and median of 1.7 km) found by Visscher and Seeley (1982). When foraging patches are rich and plentiful in the vicinity of the nest, there is no need to fly farther to other patches, and one would expect the foraging distances to be less. If the case of the honeybees is given in a human scale, the median foraging pseudopods extend out more than 215 km (with the individual foragers carrying back their 40 kg loads of nectar), sampling over an immense foraging area that would be twice the size of the greater Los Angeles-San Bernardino metropolitan region. The complexities of the communication system and travel routes are no less than those in the human social realm. Though some individual humans claim to understand the functioning mechanics of their system through political and hierarchical means, that individual foraging workers have any knowledge of the structure of their complex system is inconceivable.

The pseudopods always retract or are resorbed back into the central body of the colony at night or earlier when the food sources are deemed of low quality or when weather conditions warrant. They may not extend at all if the air temperature falls below about 10 °C or under other adverse conditions, such as rainfall or high wind. The following day there may be some of the same pods as the previous day extended again, as well as new ones with some pods of the previous day no longer functional. The individual workers comprising these feeding pseudopods may stick with their own floral resource and thus continue in the same pod or change pods (and floral patches) depending on the recruitment intensity of other foragers. When there is little food left at one feeding patch, the foragers may be recruited to and enlarge other feeding arms. Pseudopod size, direction, and timing are all dependent on summed local decisions by workers comprising those foraging pods, and not by any central colonial decision-making powers.

4.3.3.2 Nectar Storage

Returning nectar foragers do not unload their cargo into empty cells on the honeycomb. The house bees acting as receivers accept the foragers' regurgitated loads of nectar and deposit them into cells near the periphery of the brood area. Initially, the nectar contains about 40% dissolved sugars as harvested from the floral resources and brought back to the hive (Southwick et al. 1981). The stored nectar ripens by invertase enzyme activity and drying to about 81% sugar before the cell is capped with wax by house bees. Even in unmanaged hives of feral honeybees, much more honey is usually stored during the foraging season than can be utilized over the rest of the year. There appears to be a margin of safety built in that encourages hoarding the excess. Hoarding behaviour is enhanced by empty comb, brood rearing and availability of nectar resources. When cells are full and there is nowhere to store incoming nectar loads, receivers do not unload their own honey stomachs and do not return to the receiving area in the colony, and therefore cannot unload foragers. The returning foragers form a queue, wait, and refrain from dancing, so there is no recruitment. Seeley (1989) suggests that the returning foragers try to activate unloaders by the DVAV dance.

4.3.3.3 Queueing

When returning foragers are not serviced quickly enough, they may form a resting entourage waiting to be relieved of their nectar loads. It takes some time for the storers, once they have secured a nectar load, to move to the food storage area from the food delivery area, deposit their load, and return to the delivery area to receive another load. In small colonies of a few thousand bees, food storers comprise about 20% of the workers (Seeley 1989). When nectar flow is great and foragers bring in nectar loads at a high rate, the returning foragers can overpower the capacity of the food storers to relieve them of their loads. The waiting time before unloading is a sensitive indicator of the colony's nutritional status. The length of waiting depends on the rate of arrival of empty food storers, and the rate of arrival of nectar foragers in the nectar delivery area. These two rates are determined by a set of colony and field factors including availability of nectar and availability of empty comb. Colony integration is provided automatically by the interaction of these two rates as determined by local activities of individual workers. A nectar forager does not know about her colony's nutritional needs but just does not get rid of her nectar load and simply does not fly out again.

4.3.3.4 Comb Utilization

Camazine (1991) and Camazine et al. (1991) have developed a deterministic model based on experimental results, which shows that the pattern formation in the honeybee comb is likely to be the result of a self-organizing process of activities at the local level by individual bees. Using simple behavioural rules based on local cues, their model generates the characteristic concentric pattern of brood, pollen and honey seen on the honeycomb in a typical colony. In essence, the pattern develops when the following rules are followed:

1. the queen lays eggs next to cells with eggs,
2. pollen is placed in cells next to those containing eggs and pollen,
3. nectar is placed in cells next to those containing pollen and nectar,
4. nectar and pollen are relocated if surrounded by eggs.

The pollen placement and active movement of nectar and pollen from one place to another on the comb to clear cells for use in egg laying are a dynamic and global result of individual decisions.

4.3.3.5 Water Handling

The ultimate example of this local effect is illustrated by the nectar forager that returns to the hive with a full load of rich nectar under emergency conditions requiring water for cooling. This fully loaded nectar forager may sit for tens of minutes or even hours without being relieved of its nectar load because the house bees only take nectar of low concentration or water loads from returning foragers. The loaded forager does not place its nectar in wax cells on its own, and then return to the floral patch. Obviously, then, the forager holding the rich nectar will not be returning to the field for another load nor will it be involved in recruiting efforts as long as it is sitting unattended by receiver bees. Temperature regulation demands precise cooperation between water-collecting foragers and water spreaders in the nest. Foragers collecting water or dilute nectar are recruited from the nectar foragers according to the needs of the colony. Under high heat loads, when water is needed to cool the hive, nectar foragers are needed to secure adequate supplies for evaporative cooling of the nest. As in nectar collecting, the delivery time is an accurate gauge of water demand (Lindauer 1954, 1961). In hot weather, more and more individuals in the colony require water for evaporative cooling.

Food and water foragers are usually the older experienced bees in the colony. They have become familiar with the landscape of the area in the vicinity of the nest site and have located flower patches containing good nectar sources, or places to obtain water. Although there is monitoring and adjusting of the forager force in the hive, its control is not by any central administration. The heart of the working system lies in the behaviour of the individual foragers when they return to the hive bringing back their food load. In 1948, Lindauer (1954) deciphered that the colony global control is brought about by the summed effect of how receptive the house bees are at the local level to returning foragers. When water collecting is in progress, each forager returns to the nest with its load carried in the honey stomach and transmits it to younger house bees by trophallaxis. How quickly these receiver bees take up the water loads determines, in part, how encouraged the foragers are to go back out and bring back more, and how diligent they are in their recruitment efforts. In other words, their acceptance by the house bees, on the average, gauges the water demand of the colony. The span of time between arrival at the nest and giving up its water load determines the enthusiasm the forager shows in its dance. The faster the load is received, the more lively is the dance. Times of delivery between 20 and 40 s result in an active recruiting dance. As long as the delivery time is less than about 60 s, water collection will be continued. When delivery times are long, taking perhaps 2–3 min or more, the individual foraging bee's "eagerness" for collecting, dancing and recruitment dramatically decreases. An important point here is that foragers exhibit individual "eccentricities". Some bees collect water even when it is not needed and is actively refused by unloaders. Yet, by persisting in the face of audience apathy (or antipathy) they can effect a remarkably rapid response to a cooling emergency.

4.4 Search for Housing

4.4.1 Nest Site Selection

Swarming is a natural part of the annual cycle of the colony, and is the means of reproduction and migration of colonies (see Chap. 6). Several days before swarming takes place, some retinue workers may perform the DVAV dance on the queen for several seconds, which seems to deter her from laying eggs and helps her to prepare physiologically for flight (Winston 1987). Before swarming a search for a new nest site may already begin (Lindauer 1955). The scouts, the only ones likely to know the location of the new nest site and comprising about 5% of the swarm, excite the swarm to take off in flight with buzzing runs over the surface (Esch 1967a). The workers carry honey in their honey stomachs. providing sufficient food for several days. When the swarm bivouacks on a tree branch, it settles into two distinct parts (Meyer 1956). Bees forming the outer layers provide a protective shell to those hanging in chains inside. There is also a division of labour in the swarm with bees older than 21 days being the searchers, bees in the shell about 18 to 21 days old, and the younger bees being inside forming the chains.

Choosing the new nest site is a process involving the whole organism, which is highly dependent on the accumulated information from returning scout bees who have individually gathered information about single possible nest sites. Lindauer (1951, 1955) and von Frisch (1967) investigated the nest site selection process of swarms which is reviewed in Wilson (1971). The global choice is determined by the liveliness of the dances of individual scouts and their effectiveness in recruiting other scouts. The scouts may dance for minutes or hours (compared to seconds or a minute or two for dances performed by food gatherers). Preference seems to be for the greater distance from the mother hive (Lindauer 1961). Gould (1982) found that the optimal distance is race specific. When a possible site is discovered by a scout, it spends as much as an hour in examination of the inside and outside (Seeley 1977). The scout walks rapidly over the outside of the edifice and carries out hovering flight around the nest site inspecting the surroundings. The inside of the hollow is measured by the scout by walking over its inside surfaces. The scout walks over all its inner surfaces covering more than 50 m and amassing a perception of the cavitiy's volume (Seeley 1977). Once a measure of the suitability is achieved, the scout returns to the swarm. All the colony's scouts report back via waggle dances which indicate the direction, distance and desirability of about two dozen potential homesites (Lindauer 1955). The global swarm decision involved in nest site selection may take less than a day to several days.

4.4.2 Swarm Orientation

The global decision of just one homesite among the many potential sites is made by the summed local responses of the recruits to the dancing scouts. Homesite location is determined then by the liveliness of the dance. Some of the initial scouts will dance with more vigour than others and thereby recruit other scouts which abandon their initial sites and join the first scout. The recruited scouts visit the new site and assess its suitability. Only if the new site is superior to the old will the recruited scout return and dance for the new site. In this fashion, scouts change their allegience among the numerous choices (Fig. 4.3). If several potential sites are of similar value, it may take hours or days to formulate the global decision among the scouts. When most of the scouts (numbering 3 to 600 in a colony 10000 to 12000 bees) agree on one or two sites, they are the only members of the colony that know the exact locations. They have communicated with the other scouts through their waggle dances but do not attempt

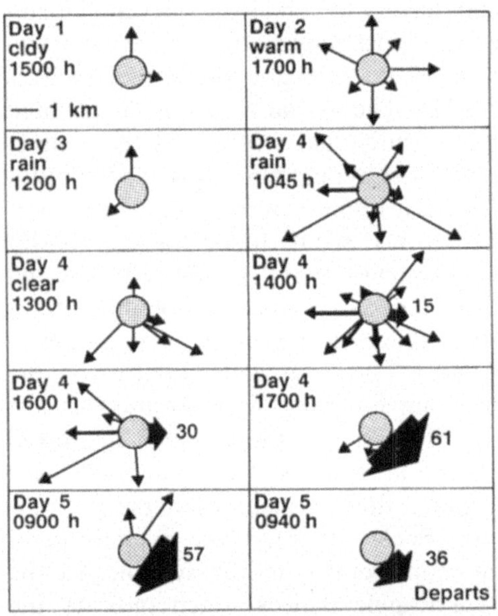

Fig. 4.3. Swarm decision-making process in home site choice over 5 days in June. The swarm of honeybees hangs on a branch with scout bees returning to it after having found potential new nest cavities. The scouts dance on the surface of the swarm announcing distance and direction to the sites. *Length* of *arrows* shows distance and *thickness* indicates relative number of dancers promoting a particular site over a 1–2h time interval prior to the times indicated. In this case, the swarm arrived at a decision at 0940h on 30 June after hanging from the branch since 1335h on 26 June. (Modified from Lindauer 1961 and Wilson 1971)

to carry out such a means of communication to the remainder of the colony. Instead, the scouts utilize buzzing runs through and over the group, exciting the bees to launch into flight all at once. The scouts then pilot the large cloud (ca. 5 to 10 m across) of flying bees toward the chosen homesite (Esch 1967b; Seeley et al. 1979). This piloting apparently involves their streaking through the swarm cloud in the compass direction of the new homesite (Lindauer 1955). Once the cloud has reached the near vicinity of the home site, the scouts land at the entrance and release Nasanov pheromone from a gland at the tip of the abdomen while fanning away from the entrance. Members of the circling swarm are induced to follow the chemical attractant toward and into the entrance, and soon the entire cloud flows into the new home.

Ratnieks (1991) noted that waggle dance indication of distance and direction to homesites was remembered overnight. This was demonstrated by continued dances early in the morning before any scouting flights after a careful forced move of the swarm many kilometres from its original location during the night. Dancers continued to dance initially as if nothing had happened. This relatively long-term carry-over or memory of location could serve a useful function in the normal context in the field where the relative location of the homesite would not change.

Sometimes, a single global decision cannot be reached before the swarm vacates its bivouack and the swarm cloud attempts to fly en masse in two different directions (Lindauer 1961). If the swarm cloud actually splits, the portion without the queen will again rejoin the queen-containing cloud, and the group may again bivouac on a tree branch. If no decision is made, by default they build comb in the branches of the bivouac site and resume normal homesite activities including foraging, storing food and raising brood. In regions of long cold winters, the lack of locating a protective cavity results in death of the colonial organism.

4.4.3 Migration

Migratory swarms result from absconding by the entire colony which abandons the nest site because of infestation by disease-causing organisms, mites, wax moth, predator disturbance or excessive moisture or other causes discouraging to the bees. Migratory swarms must locate a new nestsite just as reproductive swarms do. Honeybees that nest in the open such as *Apis dorsata* and *A. florea* must contend with a variety of predators not encountered by cavity nesting species. As they nest by building combs hanging from tree limbs, they are easily reached by visual predators and crawling intruders such as ants. To prevent predation, the little *A. florea* colonies nest low in dense vegetation and if the cover is lost as might occur when plants lose their leaves during the dry seson, the bees move to a new nest site (Seeley 1985). Specifically for protection against ants, they coat the branch on which their nest is located with sticky plant resins which effectively block ant invasions. The much larger *A. dorsata* is highly effective in its sudden defensive attacks and is less likely to be forced to migrate. *A. dorsata* does, however show long distance migrations (Koeniger and Koeniger 1980), whereas *A. cerana* and *A. florea* migrate over only short distances.

Another example is the annual migration of the cliff-dwelling honeybee, *A. laboriosa*, in the rugged climate of the Himalayas. This species builds large combs like *A. dorsata*, which hang from the undersides of high cliffs (Valli and Summers 1988; Underwood 1990). Nests are located in the high country as much as 4000 m above sea level. Good sites have as many as two dozen cliff bee nests exposed as hanging combs 1 m or more across and as much as 120 m above the ground. *A. laboriosa* does not build perennial nests, but in the summers migrates to the high cliffs so as to gather the flower nectar from the mountain flowers. In the autumn after the summer rains and with falling temperatures at these high altitudes, the bees retreat into the valleys. In winter the bees simply remain clustered tightly near the ground under the cover of logs, plants, overhanging rocks, or tree limbs and do not build comb until their return up to the cliffs in early spring. Underwood (1990) observed that the bees could not effectively defend themselves against birds when the air temperatures were low because their low body temperatures did not permit immediate flight. Many of the swarms in the valley, if they are not sufficiency hidden, are exposed to bird predation.

The contrast in the cliff dwelling places in dramatic. These colonies are not attacked by birds, at least not in warm weather. Any attacker is quickly met with an efficient defence behaviour. Thus, Underwood reached the conclusion that the bees leave the cliffs when the weather cools and hide in the dense vegetation because of their defenceless position when cold.

They show two extraordinary adaptations to survive the cold witners without combs and a nest shelter. First, these bees are capable of engorging honey, holding in their honey stomachs as much as 135% of their body weight. Only a few guard bees remain unladen with these large stores. Second, there is no temperature regulation in the wintercluster. The significance of this finding is that these swarms do not have to consume much honey to maintain a warm central temperature as does *A. mellifera.* By reducing its metabolism and thus lowering its food requirements, colonies of *A. laboriosa* remain essentially dormant through the winter phase.

4.5 Pheromones

4.5.1 Superorganismal Control

Without an anatomical equivalent to the central nervous control of an individual organism, the colony is regulated chemically in a molecular environment composed of a constellation of pheromones and semiochemicals (both volatile and non-volatile) released primarily from the queen but also from workers, brood, and possibly drones (Boch and Shearer 1962; Butler and Fairey 1964; Pain 1973; Breed 1981; Blum 1985; Prestwich and Blomquist 1987). Pheromones are chemical substances, usually highly volatile, that are secreted externally and elicit behavioural or physiological responses in other members of the same species (Karlson and Butenandt 1959). A specific pheromone may be a primer which alters the physiology of the receiver individual and prepares it for new behavioural assemblage, or a releaser which directly evokes a behavioural response in the receiver (Free 1987). In honeybees, primer pheromones play a role in the organization of the social group, such as inhibiting reproduction in a worker caste. Releaser pheromones function in sexual attraction, trail marking, recognition and alarm. A number of pheromones have been identified that are important for the social organization of bees (Avitable et al. 1975; Pickett et al. 1980; Winston 1987; Table 4.1). A bouquet of 30 or more chemicals comprises the endocrine system of the honeybee superorganism, and parallels the hormonal environment in individual organisms, causing changes which effect global responses of the system.

The specificity is in the blend of pheromonal components (Blum 1974). Besides being additive with synergistic interactions, pheromone action may be context dependent (Silverstein et al. 1966; Velthuis 1970). Under some conditions a particular concert of pheromones may be active in controlling colony behaviour and individually separated component effects can be seen. In other situations, no individual effects are seen, yet additive effects are very apparent. In

Table 4.1. Some of the best-known pheromones identified in the honeybee. Over 30 pheromones have effects in honeybee colony behaviour (Free 1987; Winston 1987)

Name	Key chemical	Source	Effect
		Queen pheromones	
Queen substance	9-Oxo-2-decenoic acid 9-Hydroxy-2-decenoic acid	Mandibular glands	Drone attraction Worker attraction Colony and swarm Stabilization Worker ovary inhibition Queen rearing inhibition
Footprint	??	Tarsal glands	Queen rearing inhibition
Tergal secretions	??	Tergal glands	Worker ovary inhibition Queen recognition Drone attraction
		Worker pheromones	
Alarm pheromone	Isopentyl-acetate 2-Heptanone	Sting chamber Mandibular glands	Defensive behaviour Defensive behaviour
Scent	Geraniol, citral	Nasanov gland	Recruitment, orientation
Footprint	??	Tarsal glands	Orientation at flowers?

his review of pheromones in social bees, Free (1987) concludes that virtually all bee pheromones are multicomponent with communication being more complex than earlier supposed. We will confine our discussion to the few pheromones known to play important roles in honeybee colonies.

4.5.2 Queen Pheromones Inside the Nest

4.5.2.1 Queen Retinue Behaviour

The classical queen substance, a pheromone blend synthesized in the queen's mandibular glands, is known to be the key control chemical in the honeybee colony (Butler 1959). It is likely that the queen deposits the substance on her own body surfaces which is spread by grooming. This attractive queen substance has a composition of long-chain fatty acids including 8-hydroxy-octanoic acid (8-HDA), 9-oxo-2-decenoic acid (9-ODA), 10-hydroxy-2-decenoic-acid (10-HDA), 9-hydroxy-2-decenoic acid and 10-hydroxy-decanoic acid (Barbier and Lederer 1960; Barbier et al. 1960; Callow and Johnston 1960; Barbier and Hügel 1961). The blend of 9-ODA and 9-HDA is especially potent in stimulating some dozen workers to be continually attending the queen (Allen 1957). These workers feed the queen royal jelly, a highly proteinaceous and nutritious substance produced and secreted by their food glands. Positions of workers around a stationary queen are such that they are able to move their heads forward and reach the queen with their antennae and tongues. The antennal contact and licking of the body of the queen, followed by self grooming, moves the active molecules all over the body (Naumann et al. 1991). Subsequent moving about in the colony and sharing surface carried and imbibed queen-secreted substances with other bees are extremely important activities in maintenance of colony order and normal function.

Workers that have been a part of the queen retinue and have contacted her increase their antennal contact with other workers (Free 1987; Naumann et al. 1991). The contact is mostly initiated by the other workers. This suggests that they can perceive the presence of the attractant on the body of the retinue workers. They are even stimulated to make contacts in experiments where excised heads alone of bees from the retinue are presented (Free 1987). The longer the bee has been in contact with the queen in the retinue, the greater is its attractiveness to other workers (Naumann et al. 1991). On leaving the retinue, the bees move about the brood area widely, but within a short period of 10 to 30 min or less, the workers' behaviour and attractiveness return to the levels of workers outside the queen's immediate vicinity (Velthuis 1972; Juska 1978; Juska et al. 1981) (Fig. 4.4). This probably comes about because of the diminished amount of queen pheromone on their bodies and the lack of stimulation from the queen. Some of the loss of the queen substance from the workers may be through volatilization and enzymatic breakdown into inactive substances (Pain and Barbier 1981).

4.5.2.2 Colony Stabilization and Ovary Suppression in Workers

Queen substance is a primary superorganismic hormone as it inhibits the growth and development of the ovaries in all the female workers and prevents the establishment of new queen cells. Within the superorganism this chemical acts like a vertebrate hormone, being produced in one place (queen mandibular glands) and released on her surface, where it is taken up by physical contact of the attending workers (Velthuis 1970; Simpson 1979; Free 1987). It is spread throughout the colony by body contacts to affect its target tissue, the inhibition of ovaries in workers (Voogd 1955; Butler 1959; Jay 1970; Velthuis 1970; Slessor et al. 1988). In addition, the queen substance stabilizes the colony, stimulates brood rearing, comb building, hoard-

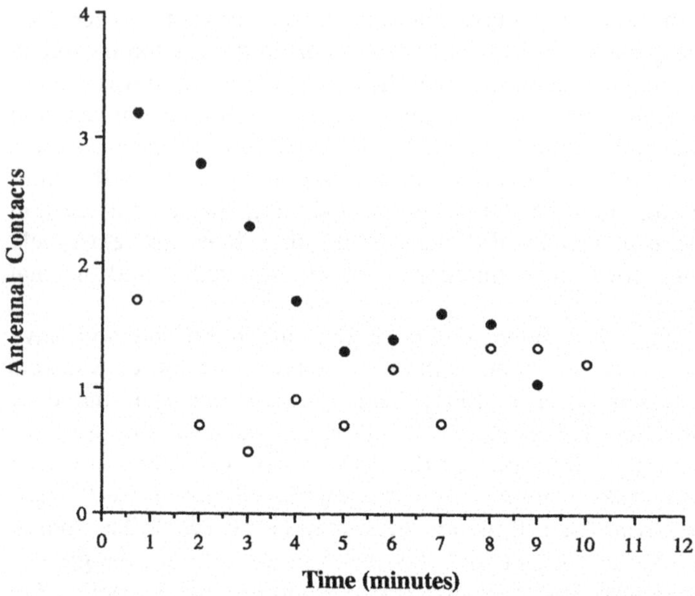

Fig. 4.4. Number of reciprocal antennations between a worker from the queen retinue (●) and other workers during a 10-min period after leaving the queen, and control workers (○) not closer than 40 mm to the queen showing a decline in the effect of queen pheromone over time. (Data from Ferguson and Free 1980)

ing, and foraging (Butler 1954, 1959; Butler et al. 1961). Queens with mandibular glands removed are much less attractive to workers (Gary 1961; Velthuis and Es 1964), yet body parts from recently killed queens are attractive (Verheijen-Voogd 1959; Velthuis 1972).

The instability in a colony resulting when the concentration of queen substance circulating throughout the group is reduced could occur in two ways; either the queen has left or been lost to the hive, or she has died or slowed or stopped her production of queen substance. The queenless workers will begin to construct queen cells in an effort to raise up a new queen. This activity occurs well before ovarial changes are manifested in workers.

The spread of queen substance via messenger workers is now believed to be understood. Naumann et al. (1991) established that the important molecules are imbibed and distributed over the body of retinue bees. Self-grooming and chemical translocation and migration spread the molecules over much of the body surface.

The queen also leaves pheromone trails on the combs which are sufficiently persistent to be detectable by workers (Juska 1978). Butler (1973) showed that workers that had been in contact with the queen in her retinue and lost her will examine the comb where she had been with their antennae. Also, the rate of loss of attractiveness of a beeswax-coated cage after a queen's removal (Juska 1978) is consistent with the amount of time for a colony to exhibit the restlessness of queenlessness. Such a trail chemical is likely to be produced in tarsal glands on the fifth tarsomere of her legs (Free 1987). The rate of secretion of these glands in the queen is up to 13 times greater than the rate seen in workers. The queens footpads deposit the secretion on the comb surface as she walks over it. The combination of mandibular gland secretions and tarsal secretions are effective in inhibiting queen cell building (Lensky and Slabezki 1981; Winston 1987). A schematic summary of the modes of communication of queen presence are represented in Fig. 4.5.

4.5.3 Queen Pheromones Outside the Nest

4.5.3.1 Swarm Attraction

During swarming the queen's odours insure the consolidation of the swarm (Morse 1963; Butler et al. 1964; Butler and Simpson 1967; Morse and Boch 1971; Avitable et al. 1975; Winston et al. 1982). Her odour attracts flying workers to the swarm, and if a swarm leaves a bivouac or homesite without its queen, or if she is artificially restrained, the bees will not fly far before breaking up and returning to the point where she was last with the group.

4.5.3.2 Sex Attractant

At particular locations which are determined by landscape, magnetic anomalies, and geophysical features, mature drones from many nearby nest sites congregate on warm and sunny summer afternoons. Virgin queens are attracted also to these areas where mating ultimately takes place. When queens fly through drone congregation areas they leave a trail of scent that is attractive to the drones which form a comet tail in pursuit (Gary 1962, 1963; Renner and Vierling 1977). Experiments using models dabbed with queen substance show remarkable attractiveness to drones when the models are moved through the air high above the ground in simulated flight. The models,

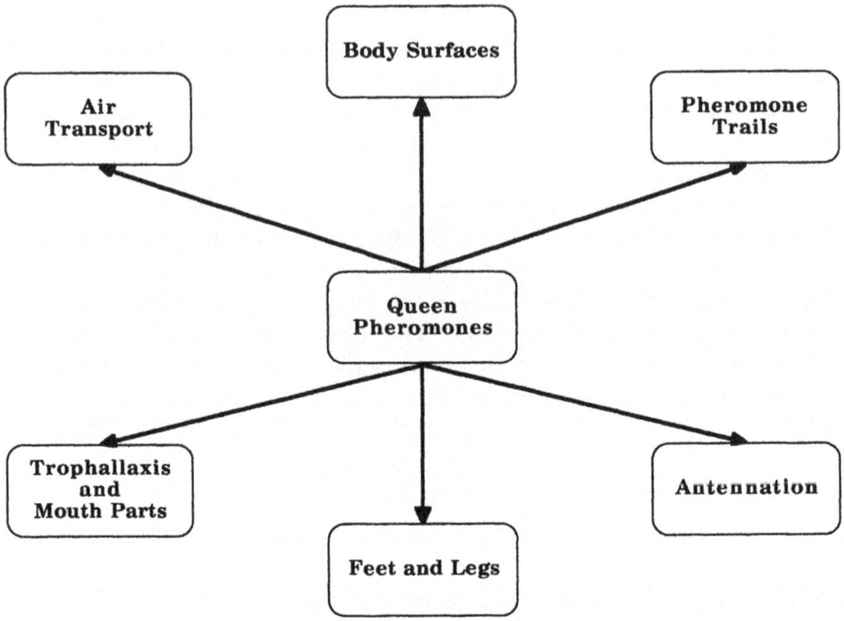

Fig. 4.5. A schematic summary of the modes of queen pheromone transfer and communication of queen presence in a honeybee colony. Although volatile pheromones move readily through the air, non-volatiles must move throughout the colony by transfer of molecules on the surfaces of the head, thorax and abdomen, legs and feet, and by trophallaxis. Some semiochemicals are left on the comb surface. Detection is via sensitive receptors in the antennae. (From Free 1987; Naumann et al. 1991)

even though only pieces of wood or hollowed dowels, are attractive enough to cause attempted copulation by the pursuing males (G. Koeniger 1987, 1988b). The pheromone mediating the main attraction is 9-ODA, although 9-HDA and tergal gland secretions have synergistic effects.

4.5.4 Worker Pheromones

Worker honeybees also produce several important pheromones. Winston (1987) lists 16 identified and three unidentified chemicals as worker-produced pheromones. Geraniol and citral are released from the abdominal Nasanov gland when workers are at the entrance to their nest or on the surface of a swarm and the scent plume carried in the air helps returning foragers and scouts find home. A different chemical set is released from mandibular and sting chamber sources when there is danger and initiates a group alarm behaviour described below. Other chemicals are released which aid in the recognition of members of the particular social group.

4.5.4.1 Orientation

Queen-produced pheromones act together with worker-produced substances to effect nest-site identification and swarm formation and movement. One of the queen pheromone components, 9-HDA, stimulates workers to release Nasonov pheromones at the homesite (Ferguson and Free 1980). The Nasanov scent gland is present only in workers (not in drones or queens) and is located under the dorsal surface of the abdomen where it releases its components (seven volatile chemicals, the most active components being geraniol, citral and geranic acid) through minute ducts into a canal between the sixth and seventh tergite (Snodgrass 1956; Boch and Shearer 1962; Pickett et al. 1981; Williams et al. 1981, 1982). The scenting worker exposes the gland by raising and flexing its abdomen and usually fans its wings which disperse the pheromone. That the odour attracts other bees was first recognized by Sladen (1901) and subsequently experimentally verified by Free and Butler (1955) and others.

Nasanov pheromone performs many functions involving orientation and aggregation of workers. These include:

1. indicating the entrance to the home,
2. bringing together a swarm, and
3. marking a water source.

Nest entrance fanning of Nasanov scent is exhibited by experienced workers as well as young bees returning from orientation flights (Hazelhoff 1941; Ribbands and Spiers 1953). Any activity that causes disorientation by flying bees initiates Nasanov release. When one or a few bees are scenting, others are stimulated to do so. The fanning at the entrance also carries out in the odour plume a mix of nest odours emanating from within the nest cavity. These include comb, brood and perhaps queen odours, and undoubtedly help in identifying not only the entrance, but the correct homesite as well.

Nasanov scent is released when a swarm is moving or first settles on a tree branch (Sladen 1905; Jacobs 1924). Once the new nest site is determined and while the swarm is airborne, scouts release Nasanov scent at the site, and upon arrival many bees release scent at the entrance as they enter the new cavity (Morse and Boch 1971; Free 1987).

A. cerana and *A. florea* foragers have been observed releasing Nasanov pheromone in Sri Lanka (Butler 1954), but few observations have been made of *A. mellifera* foragers exposing their Nasanov glands while foraging on flowers, and these observations occurred only in glass houses (von Frisch and Rösch 1926; Free and Racey 1966). Nasanov exposure is more observable on sugar feeding stations (Free 1968; Free and Williams 1972). The lack of strong floral fragrences seems to stimulate bees to add their own scent at the food source. Nasanov pheromone is released at water collection sites by foraging *A. mellifera* (Free and Williams 1970). Free (1987) suggests that this may help focus water foragers to the more discrete and odourless water sources.

4.5.4.2 Colony Defence

Honeybee colonies are well known for their defensive posture, with some races, or even colonies within the same race, being more highly defensive than others when they are attacked by a predator or otherwise disturbed. Such disturbance results in an organized colonial defence demonstrated by the colony that follows a specific pattern of detection, alarm, recruitment, and attack, ultimately with stinging (Maschwitz 1963, 1964a,b; von Frisch 1967; Morse et al. 1967; Collins et al. 1980). The pattern of defensive behaviour commences with the alerting of the colony by the few guard bees standing at the entrance. This is followed by numerous bees taking flight in defence of the colony. The defensive response is elicited by alerted or stinging workers which release volatile alarm pheromones from their Koshevnikov's glands and the setose membrane near the sting organs (isopentyl acetate) along with mandibular gland secretions (2-heptanone) (Boch and Shearer 1966; Boch and Rothenbuhler 1974; Boch et al. 1962; Koeniger et al. 1979; Southwick and Moritz 1985). The synthesis and release of the alarm pheromones is related to the behavioural polyethism and age of the bee (Collins 1980). Electroantennogram responses (Winston 1987) and metabolic responses (Moritz et al. 1985) to isopentyl acetate increase with age from emergence to about 16 days and remain high throughout the rest of the worker's life (to 60 days).

When groups of honeybees are exposed to physiological doses of the alarm pheromone, the defensive behaviour is highly dependent on the number of bees in the group (Southwick and Moritz 1985).

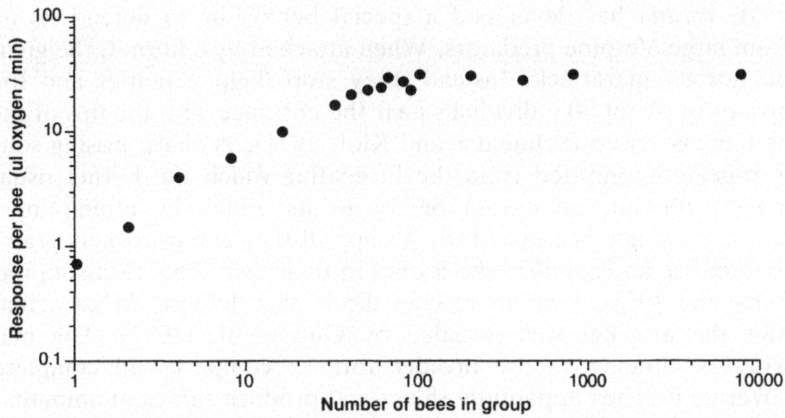

Fig. 4.6. Group effect of defensive response to the alarm component, isopentyl acetate. (Data from Southwick and Moritz 1987)

There is very little response from one isolated bee or a few bees. The response per bee increases as the number of individuals increases in the group, giving an ever larger group response. Not until the group size is about 100 bees or greater do we see a large consistent response. Thus, individuals respond differently when they are a part of a group than they do when solitary (Fig. 4.6). The defense response we find with the larger groups is highly correlated with actual stinging behaviour of the colonies in the field. Yet, when one encounters solitary honeybees foraging in the field, there is rarely any defensive behaviour exhibited. Here the bees demonstrate a colonial pattern of temperament not observable in individuals. The intensity of defensive behaviour of honeybees is highly dependent on external environmental factors as well as the genetic make-up of the colony (Schua 1952; Crewe 1976; Collins 1981; Southwick and Moritz 1987b).

All species of honeybees release alarm pheromones at the nest site when disturbed (Free 1987). *A. mellifera* and *A. cerana* colonies nest in protected cavities and have restricted entrances. Their release of alarm pheromone is perceived quickly in the interior of their nests. *A. dorsata* and *A. florea*, being open nesters, produce more alarm pheromone when disturbed (Morse et al. 1967; Koeniger et al. 1979), perhaps as more is required to alert nestmates in the exposed open-air sites. The fact that *A. dorsata* nests in aggregations with many nests within a few metres of each other provides an additional mechanism of cooperative defense against large or persistent predators.

A. cerana has developed a special behaviour to defend its nest from large Vespine predators. When attacked by a hornet, the guards do not counterattack. Instead, they stop flight activities and form groups of about 30 individuals near the entrance with the tips of their abdomens raised (Schneider and Kloft 1971). A sharp hissing sound is repeatedly emitted from the hive after which the hornet usually leaves. Should the hornet persist in its attack by attempting to capture a single bee out of the group, all the *cerana* workers grab its extremities and smother the hornet in their own mass (Matsuura and Sakagami 1973). Few or no bees die in this defense. What actually kills the attacker was revealed by Ono et al. (1987). The many workers clinging to the invader form a compact ball completely covering it. They apparently shiver and produce sufficient amounts of metabolic heat to raise the internal temperature of the ball, where the hornet is located, to lethal levels for the predator. Temperatures in excess of 43 °C kill the hornet but are tolerated by the bees.

4.5.5 Other Pheromones and Odours

Chemical odours have now been identified with the queen, workers, comb, brood, mating, nest and colony, swarm, trail and scent marking of food and water sources, alarm, and possibly drone and drone comb. Many of these odours are true pheromones, while others have not yet been tested sufficiently to be classified as such. Odours are also significant cues for workers to discriminate between nestmates and individuals from other colonies (Butler and Free 1952). The queen, too, seems to have a set of nest-specific and perhaps genetically determined odours which identify and distinguish her from other queens (Boch and Morse 1982; Moritz and Crewe 1988a). Tergal glands secrete potential components for kin recognition. Queen recognition by swarms shows that they cluster about their own rather than a foreign queen when given the dichotomous choice (Boch and Morse 1974, 1979, 1981; Ambrose et al. 1979). These behavioural repertoires result in colonial individualism in which there is a single queen with a consort of workers, and outsiders are not allowed to mingle (see Chap. 8).

Drones produce compounds in their mandibular glands that are attractive to other drones (Lensky et al. 1985), but their function is unknown. Brood itself releases chemicals that identify it as brood and cause the grouping and incubating behaviour of workers (Koeniger and Veith 1983). In addition, brood odour stimulates foraging activity by workers for both pollen and nectar (Free 1967; Scott 1986). The odour of empty, but previously used combs, increase hoarding behaviour as well (Rinderer and Baxter 1978).

Other chemicals such as those picked up at flowers by foraging bees may be important externally derived odours. These floral odours are chemically close to pheromones produced by flower-visting insects, and it would be unusual if they did not possess some of the characteristics to elicit responses in members of the animal species. Certainly, some plant-produced chemicals have quantitatively more attractiveness than others to bees. Anise oil (from *Pimpinella anisum*) for example, is highly attractive to honeybees, whereas some of the pungent odours emitted by nectar-rich flowers on plants of the hawthorn family (*Crataegus* spp.) are repulsive to bees (Southwick et al. 1981).

4.6 Learning

The colony ascertains through the amassed learning of the individuals of which it is comprised. When, for example, a large pseudopod of foraging workers is active at a floral patch as the evening becomes darker, they must return to the hive while still enough light remains. Yet, the next day, this arm will extend again to the same floral patch, having learned that it is a good source of nectar. This global learning, of course, is only the result of the behaviour patterns exhibited by individual workers as described above. Extensive behavioural experiments have shown conclusively that honeybees exhibit learning (for example, von Frisch 1965; Gould 1984, 1986, 1987a,b, 1988, 1991; Gould and Towne 1987; Menzel and Mercer 1987; Menzel et al. 1991). They can learn to distinguish odours, learn and remember shapes, tell time of day, formulate maps or landmark locations in their brains, and a large variety of other tasks. The exciting neurophysiological basis for this learning is becoming clarified.

Learning at the individual level has been the focus of study for years by Menzel and co-workers (Menzel et al. 1973, 1991; Menzel and Erber 1978; Menzel 1983; Menzel and Mercer 1987; Brandes et al. 1988). Among other tests, they use a simple motor activity of proboscis extension response to demonstrate learning of odours. He found that bees associate the olfactory stimulus quickly with a sugar-water reward. Even a single associative learning trial lasting only a few seconds establishes a stable memory. The one cubic millimetre brain contains only a few hundred thousand neurons (not several billions like the human brain). The mushroom bodies and antennal lobes are considered to be the major association- and decision-making centres, possibly because they are the sites of synthesis of cyclic AMP. However, no precise relationship has been established between stimulation locus and behavioural pattern evoked (Pickard 1988). Low levels of biogenic amines can reduce the proboscis extension response. Dopamine inhibits information retrieval from memory but does not interfere with recognition of chemosensory stimuli or the motor responses. Epinephrine and octopamine improve classical conditioning, facilitate habituation and restore the proboscis extension respoonse. The cellular studies show that the formation of long-lasting and stable storage of olfactory information can take place even without protein synthesis.

4.7 Social Intelligence and Network Analysis

4.7.1 Social Intelligence

Social intelligence is defined as the apparent colony-level behavioural responses to environmental stimuli. It is likely to be possible only through multiple-path communication networking of individual local behaviours resulting in global responses that resemble information centre strategies in problem solving. Application of communication network analysis may be a key to understanding the communication architecture, flexible problem solving and efficiency in superorganisms. Applying the formula Eq. (4.1) for determining the number of possible communication links in a thriving honeybee colony of 30 000 workers gives a total of 449 985 000 dyads. Adding 1000 drones and the single queen ups the total dyads to 480 million in a typical honeybee colony during the summer. These 450 to 480 million communication links are certainly not all used and there must be plenty of redundancy in the system. It is implausible to attribute any knowledge of this complex system to the individual worker. Instead we must search for simple behavioural explanations at the individual level. Linear communication links exist within the system with multiple tasks being performed through multiple communication routes being utilized simultaneously. An efficient multitask approach to solving colonial problems with multiple parallel paths followed simultaneously by dyads of individual linkages rather than the more direct and faster serial pathway with sequential linkages is most reasonable. Like the supercomputers now in use, parallel processing seems to be central in the division of labour (Robinson 1991) and to global decision making. The credo of systems theory, that the whole is more than the sum of its parts (Rogers and Kincaid 1981), seems to fit nicely the behaviour patterns seen in the honeybee colony. The colony's global response in feeding at a number of floral patches with varying richness of nectar is to focus its forager force on those patches yielding the highest return. This can be achieved by individual foragers following three simple rules:

1. in the hive, deliver food and dance, fly back to patch,
2. when delivery time is long or food is not accepted, do not dance and stay in hive,
3. follow other dances if not working a rewarding patch.

The simple rules are illustrated in Fig. 4.7. Following these rules at the individual level will result in the global responses that are actually seen in the honeybee colony.

Even though the honeybee colony has no defined central information processing leader, there is no absence of any cohering principle, which in the political world of humans would lead to anarchy. Instead, the superorganism is a self-organizing system using feedback control, and we conclude that the coherence is achieved through implementation of the following three elements:

1. Self-organizing local factors at the individual level form the global decisions (that is, local decision making yields a global response).
2. Heirarchies within the organism only rarely contribute to decision making (for example, pheromones may contribute to heirarchical decision making).
3. Parallel and network operations comprise the elements of communication, not series or hierarchical routes.

Coherence is also mediated through pheromones and is not completely based on local decisions.

4.7.2 Network Analysis

Using the approach of network analysis, the communication system of a colony can be quantitized for comparative studies of certain behaviour patterns within or between colonies, populations, or even different species. This method is a means of identifying the communication structure or arrangement followed in the superorganism. The elements in the system are the individual bees, and their dyads

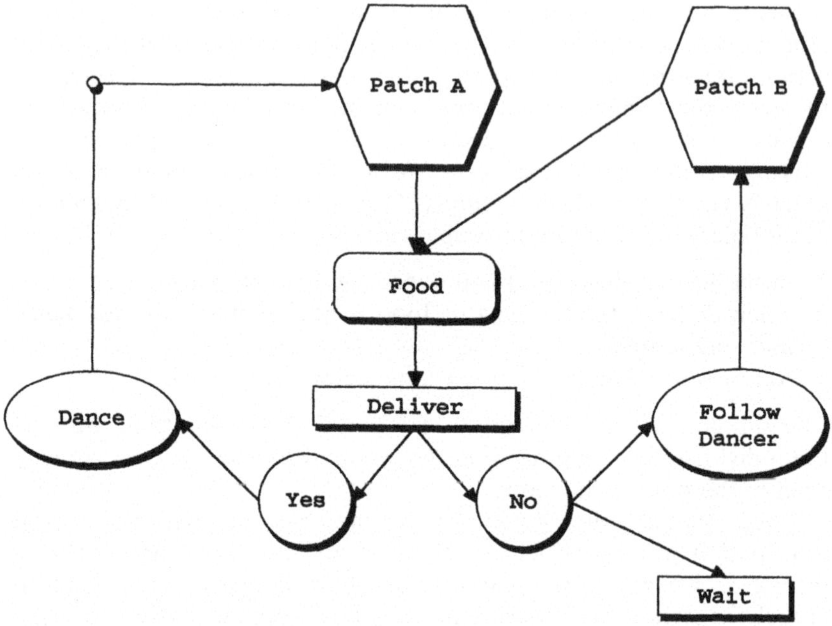

Fig. 4.7. Simple local rules for foraging by individual workers from a colony. A forager returning from Flower *Patch A* with food to the colony. If she delivers the food to a receiver bee, she dances and returns to the same patch. If no receiver takes her food, she does not dance, does not recruit, and does not return to the patch. If her honey stomach is filled, she has no choice but to wait. If she is only partially filled or when she eventually is relieved of her load, she follows a dancer and may be recruited to a new Flower *Patch B*

form the necessary communication linkages. Social scientists have found that physical distance separating individuals is a major determinant of whether or not a dyadic communication link occurs (Festinger 1949; Allen 1977; Blau 1977). The shorter the distance, the greater is the probability of a communication link. In the highly social entity of a superorganism then, the communication pattern would seem to be highly affected by spatial location. As a returning nectar forager, for example, moves to a specified dance floor in the hive, those workers located elsewhere in the hive structure will certainly be unaware of the dancers' efforts to communicate information regarding the food resource in the field. Only those workers close at hand will be able to share in that information directly. Others in the vicinity can obtain some information, but only second hand, as in the exchange of the food resource samples via trophallaxis. A

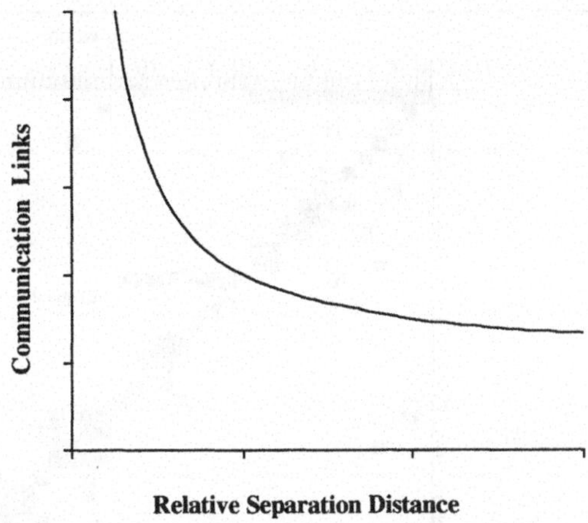

Fig. 4.8. A plot of proportion of communication linkages between individuals as affected by the distance separating them shows that very many links are found when they are in close proximity, followed by a sharp decrease in number of linkages at middle distances and a gradual decrease in links over greater distances

plot of the proportion of communication linkages as affected by distance separating individuals shows that very many links are found in close proximity, followed by a sharp decrease at middle distances, and a gradual decrease over greater distances (Fig. 4.8). What remains to be done here is an actual determination of just what communication links and distances are found in the honeybee colony.

Another feature of network analysis is the recognition of subsystems of communication, or cliques, in which the elements interact more frequently with each other than with other members of the communication system. In context, the dozen or so honeybee workers comprising the queen retinue certainly communicate more within their group (including the queen) than they do with the one or two hundred house bees receiving nectar loads from foragers returning from the field. The queen retinue forms one communication clique while the forager-receiver bees form another clique. The comunication within the cliques would be extensive, but between these two cliques would be very limited. Several cliques are operating simultaneously in a bee colony, with their temporal character depending on the length of time of participation of specific workers. The worker composition of any clique changes with workers moving into a specific clique and participating for a time before moving on to

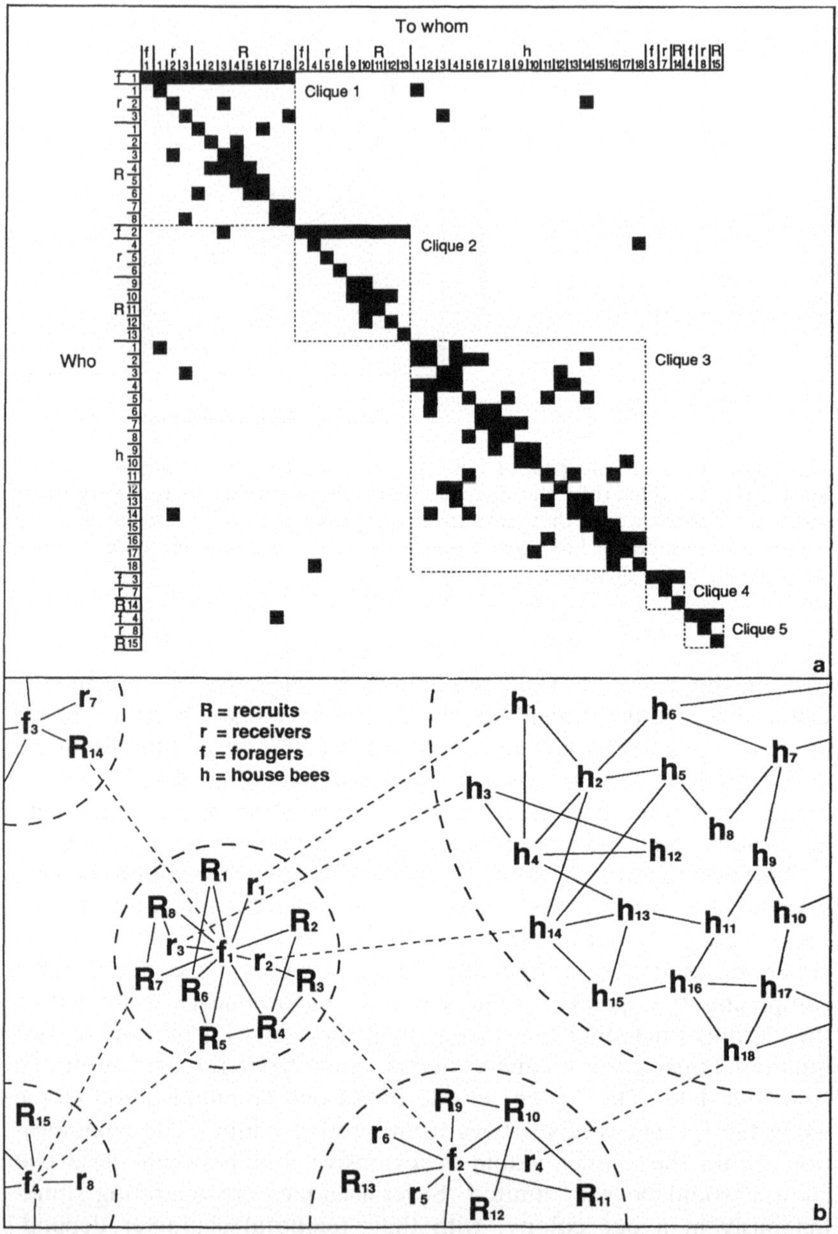

Fig. 4.9a,b. Matrix (**a**) and sociogram (**b**) of estimated communication linkages involved when a honeybee forager returns to the hive under warm temperature conditions with its water load. While the matrix differentiates ▶

another clique. Any particular worker may serve in a number of cliques during its life as a house bee. Their clique membership is determined by age, genotype, and individual perception.

Network analysis provides a way of examining communication within the colonial system by using matrix analysis and the sociogram. Illustrated in Fig. 4.9 is a matrix and a sociogram of probable communication linkages involved when a honeybee forager returns to the hive under hot conditions with its water load. Using the cliques and dyads shown in the sociogram, an index of average system connectedness (or more simply, the ASC index) can be determined. This index is defined as the degree to which the average member of a social system is communicatively linked to other individuals in the system, and it is calculated as the ratio of the number of actual communication links found in the system to the number of possible communication links as given Eq. (4.2):

$$\text{ASC Index} = \text{Actual dyads/Possible dyads.} \quad (4.2)$$

The possible communication links are determined from Eq. (4.1). The number of actual links must be determined experimentally, but the particular behaviour is unimportant. It is the number of contacts that is needed; it is not important what the behaviour is or whether some of the contacts were antennal and some trophallaxic. Values for ASC indices must fall in the range of 0 to 1, with those values approaching 1 demonstrating a network in which nearly all the individual members are connected with each other via direct communication links. The system with an index of 1 would have complete connectedness and there would be no communication structure per se because there would be no pattern of information flow in a system of perfect and direct communication linkage. This is probably true for most volatile pheromonal communication. At the other end of the scale, an ASC index of 0 implies that no individual is linked to any other individual in the system, and again no communication structure

between the giver and receiver of information, the sociogram only indicates the communication linkage between the two. The sociogram serves as the base for quantitative assessment of the tightness of communication within the group. For example, the forager, f_1, delivers its load to receiver bees r_1, r_2, and r_3, and provides samples and/or communicates to recruits following its dance, R_1, R_2, R_4, R_5, R_6, R_7 and R_8, and possibly to R_{14}, a recruit from another clique. Some house bees communicate with each other and with the receiver bees as they move from the dance floor to the honey storage sites but almost never with the forager or recruits on the dance floor exists. The ASC index could be considered also as the fraction of all possible comunication linkages that are actually utilized under any set of circumstances, and then expressed as a percent usage. A communication system in which the index = 0.112 would be utilizing 11.2% of the possible linkages.

Using values for linkages in the communication network of honeybees accepting water or dilute nectar from returning foragers under the heat stress conditions (estimates from Lindauer 1961), we calculate an ASC index for water handling to be 0.031 or 3.1% of all possible linkages utilized (see sample calculation below). When compared to values obtained from other systems or other behaviour patterns, its significance becomes apparent as a quantitative measure. The ASC indices calculated for 39 humans communicating about family planning in each of two Korean villages, for example, are 0.116 (11.6%) and 0.137 (13.7%) respectively, showing that the second village is somewhat more closely linked (or more tightly connected) than the first (Rogers and Kincaid 1981). Both villages of humans, however, have better communication links than those shown in our water-handling bees. The index is highly dependent on the total number of members possibly involved in the particular communication. The larger the number of individuals, the smaller the index is likely to be because of the proximity problem where distance alone separates out those not in the immediate vicinity during any particular time period.

In short-term colonial defence, which may last only a few seconds to minutes, the ASC index should be high because the defence response results almost entirely from the release of alarm pheromone that permeates the air and solves the problem of one-to-one communication linkages by nearly instantaneous communication among all workers in the immediate vicinity (that is, complete connectedness). Workers not likely to be involved in colonial defence (and therefore unlikely to be recruited and unlikely to participate in this communication pattern) are not included in the index calculation. These would include all foragers, workers located anywhere except in the entrance area, and all bees younger than about 12 days including all nurse bees, most house-keeping workers and the queen retinue, and all drones. This leaves three groups of workers comprising two communication cliques under normal (non-defensive) conditions: the six to eight guards and the dozen or so workers feeding the guards (feeders) forming one clique, and perhaps 30 house workers in the immediate vicinity of the entrance forming the other clique, but both of which might rapidly respond to a cloud of alarm pheromone released by the guards for the initial defensive response. Very little communication is apparent between guards at the entrance (we estimates 4 links); each of the feeders probably operates rather independently from other feeders but there is likely to be a little contact (24 links); the house bees have some interaction ($30 \times 3 = 90$ links); the communication between guards and feeders is likely due to begging by the guards and would be limited to short bouts of trophallaxis ($8 \times 4 = 32$ links); there is no interaction between guards and house bees (0 links), but some between feeder and house bees ($12 \times 5 = 60$). Then using the Eq. (4.2) for communication among the fifty bees (a) in defence "posture" prior to defensive response, and (b) and (c) during initial defensive response, we get:

a) communication before defensive behaviour (defense posture):

$$\text{ASC Index} = \frac{\text{guard} + \text{feeder} + \text{house} + \text{guard} \times \text{feeder} + \text{guard} \times \text{house} + \text{feeder} \times \text{house}}{\text{all possible guard} \times \text{feeder} \times \text{worker dyads}}$$

$$= \frac{4 + 24 + 90 + 32 + 0 + 60}{50(50-1)/2} = 0.171\ (17.1\%)$$

b) communication during defensive behaviour if the total response is due completely to alarm pheromone released from the guard bees alone:

$$\text{ASC Index} = \frac{8(8-1)/2 + 0 + 8(12) + 0 + 8(30) + 30(12)}{50(50-1)/2}$$

$$= 0.552\ (55.2\%)$$

or,

c) communication during defensive behaviour if the total response is due to alarm pheromone released from all the bees active in the vicinity:

$$\text{ASC Index} = \frac{8(8-1)/2 + 12(12-1)/2 + 30(30-1)/2 + 8(12) + 8(30) + 30(12)}{50(50-1)/2}$$

$$= 1.0\ (100\%).$$

We find that the defence behaviour is a very tightly connected communication pattern in a relatively small group of individuals (with 55% to 100% participation), and understand better the resulting rapid transitory nature of the response brought about by the local effect of the volatile alarm signal. The ASC Index is expected to be large because of communication mediated by the pheromone rather than via localized individual dyad channels. On the other hand, for water handling by a full size colony over a critical hour or two in midday, many more individuals are involved and we therefore have a much larger system with which to deal. The matrix and sociogram are shown in Figure 4.9. Each of the 1000 or so returning foragers in this time slot has contact with perhaps two or three receiver bees and eight possible Recruits on each of five foraging forays. The 1000 receiver bees may have contact with another dozen workers each, on their way to and from the cell storage or water placement area. There is likely to be only a little contact between foragers when a forager from a rich nectar source becomes a recruit to a water source. These, then, are included in the recruit groups. The 200 or so recruits may contact two or three other foragers and perhaps two dozen house bees. The 1000 house bees each have contact with probably 50 others during this time. Although house bees have contact with the queen retinue of a dozen workers, none contribute to water handling. Neither do the 1200 drones, which are contacted by feeder bees, participate in water handling, and both of these groups are therefore left out of the calculation. In this case there are four worker groups (forager, receiver, recruit, and house bees) which fall into two communication cliques (the forager-receiver-recruit clique and the house bee clique) (Fig. 4.9). Again using Eq. (4.1) for communication in water handling by 2280 bees,

$$\text{ASC Index} = \frac{\text{forager} + \text{receiver} + \text{Recruit} + \text{house} + \text{fr} + \text{fR} + \text{fh} + \text{rR} + \text{rh} + \text{Rh}}{\text{all possible worker dyads}}$$

$$= 0.031\ (3.1\%).$$

The result in this case reveals that the communication system has much less connectedness (only 3.1% participate) among all those participating workers than does defence behaviour, or even defence posture before a defensive response is provoked. The global water-handling behaviour is expected to result in a small ASC Index, however, because it is based on local decision making. It is mediated by numerous individual activities with relatively few of the many possible dyads actively being employed. This does not imply that water handling is not effective or efficiently done or is a poor system, but instead reflects the great number of possible communication links in any behaviour pattern that involves a large number of bees. The numerical results in these examples may be off to some degree because of error in our estimates, but it would require substantial changes to alter the index much, and the relative values as calculated are important in contributing to our understanding of the communication patterns involved.

It is clear from this analysis that the superorganism is a self-organizing system incorporating two very distinct pathways of communication. One mode is via localized individual worker interactions with low connectedness (for example, dance language, unloading foragers, etc.), and the other one via volatile or semiochemical pheromones with high connectedness. If we now examine the communication modes at the functional level, we see that the pheromones reach the entire superorganism, more or less: a global message with a global reaction (for example, the queen pheromones simultaneously and continuously signal to every worker in the colony that it is queenright). In contrast, dance communication is a local message with a local reaction but a global consequence (if the reactions of all bees involved are added together). The problem bees have in order to make a global decision on the basis of dance communication with a low connectedness becomes very clear if we look at the swarm orientation. It can take the scouts weeks of recruitment dancing before a global decision can be made (all the bees have to decide the same thing at the same time) (Lindauer 1961). The low connectedness is the very reason for this poor communication result. Give the scouts a "direction pheromone" and the swarm would move to the new nest site within minutes!

With this introduction to methods of network analysis, we open the door to aquiring observational data of actual linkages for precise calculations, making possible quantitative comparisons of communication links in a variety of behavioural repertoires within and between superorganismic bees and other social insects. The principal weakness in the determination of accurate values for the ASC Index lies in being able to make accurate observation of links, and in making estimates of the sizes of the subgroups and communication group as a whole (knowing whom to leave out and whom to include) when the system is large. Sociograms of small groups are easy to construct because all links can be tracked, and index determinations would be expected to be accurate. The analysis of large groups may require some computing facility in order to handle the large sized matrices.

Chapter 5 Ecology

. . . natural selection is continually trying to economize every part of the organisation. Darwin (1859)

5.1 Demography of Natural Populations

5.1.1 Life History Strategies

In some species of eusocial insects the perennial colony is the only scheme allowing ecological and evolutionary success. Contrary to most social insects and all solitary insects that have adapted means of survival of the genome carried in solitary individuals in a constellation of habitats (for example, see Roubik 1989), individuals of these specialized eusocial groups, apart from the group, are incapable of long-term survival under the natural conditions found in any ecosystem. They spend all of their lives as functional parts of the colonial super-organism with only brief excursions away from the central unit. The following discussion pertains primarily to the true honeybees of the genus *Apis*. For comparisons of the biology and life history of honey-bees and stingless bees, and a thorough discussion of the natural history of social and semi-social bees, the reader is referred to several fine texts in the area, especially Michener (1974), Seeley (1985), Ruttner (1988) and Roubik (1989).

5.1.1.1 Sessile Phase

Once established, the honeybee colony stays at the nest site as long as favourable biological and abiotic niche characteristics, the con-ditions inside the home, pollen and nectar flow, and water availability are met. When the colony stays in place, it is said to be sessile. In many ways, the sessile nature of the superorganismic colony is analogous to other sessile organisms, especially some of those in marine environments, such as sea anemones. Neither moves from its established site, both feed by extending tentacles or pseudopods, and expand their territories only through reproductive means by casting out reproductive units which may end up some distance from the parent, depending on environment factors such as currents and available habitats.

The nest itself is a critical element in the proper functioning of the colonial organism. The importance of the nest for the functioning of the honeybee colony is borne out in an examination of the amount of time the individuals spend there (as compared to being absent on foraging forays). Workers develop in the nest and only 15 days after emergence will they leave for their first flights (Lindauer 1961). Since even experienced foragers spend only 4 to 7 h foraging in the field, an adult normally spends 90 to 95% of its total life-span within the nest. Nest architecture (combs and cavity) itself is important for the survival of the superorganism because of the need for efficient brood production, food storage, protection from predation and thermal control.

Niche Characteristics. The niche is more than a mere description of where a colony lives. It also includes its behaviour, how it transforms energy, how it responds to and modifies its environment, and what its relationships are with other species (Odum 1971). The most crucial ecological constraint in honeybee populations is probably the lack of appropriate nest sites (Ruttner 1988; Roubik 1989). Homesite selection for cavity-dwelling species such as *A. mellifera* is based on a set of characteristics that are assessed for quality. The colony does not alter the cavity, except to plug holes and fissures with plant resins, and smooth inside surfaces, but must make use of cavities that already exist in nature (Seeley 1977, 1982b). Site characteristics important in the selection of nest locations in temperate regions include at least the following (Avitabile et al. 1978; Gambino et al. 1990; Seeley 1985; Seeley and Morse 1976): distance from the parent colony, tree (or other plant) species and whether alive or dead, dimensions, cavity volume, height off the ground of the entrance, entrance area, moisture and wetness of the cavity, and the compass direction the entrance faces. Nests of feral colonies of honeybees in North America typically fill cavities in living hardwood trees of 15 to 80 l (prefer 40 l, Seeley and Morse 1976) in which the preferred small entrance ($<65\,cm^2$) is about 2 m above the ground, faces south, and is positioned near the floor of the dwelling. Even hot, arid and treeless regions are not devoid of wild colonies that may occupy cavities in large cactus plants (*Saguaro spp.*), fissures in cliff faces, or even in the ground (pers. observ.). Colonies thrive in these habitats under the proper conditions of exposure, shade, cavity, and food and water availability (Gambino et al. 1990).

Benefits of most homesite properties are clear. The large cavity volume is important because of the necessity to store adequate amounts of honey and pollen to survive long periods of dearth or cold winters. Typically, honey and pollen stores in natural colonies found in trees amount to 15 to 20 kg (with about 20% being pollen). These stores require around 20 to 25 l of storage space (including food, comb and bee space). Preferences extend to small entrances minimizing area of defense and reducing convective air currents inside the occupied nest cavity. The small entrance and the preferred height reduces chance of robbing and predator disturbance. The southerly direction of the entrance (northern hemisphere) enhances solar radiation heat gain during the winter and may allow for more frequent cleansing flights during this period. Should the colony be able to move into a site already stocked with comb (i.e. abandoned nest), the cost savings are substantial. Depending on the conversion efficiency, each gram of wax saved allows for 4 to 26 g of honey sugar to be stored or utilized in other ways (Hepburn 1986).

Nest Architecture. The architecture of the dwelling places of bees is highly dependent on long-term environmental conditions. In the tropics, a number of species such as *Apis dorsata* form clusters on exposed sites where the bees cling to themselves in vertical sheets and their parallel bodies may act as water-repellent "shingles" (Ruttner 1988). *A. florea* builds open nests on small branches which are further protected from ant predators by sticky rings of plant resins around the branch on each side of the nest (Lindauer 1956; Fig. 5.1).

The architecture of honeybee colony homes in more northern regions is limited by the dwelling places that already exist in nature. To finish home construction, the workers of *A. mellifera* use propolis to fill gaps and smooth the cavity interior. The wax for comb production is produced by the bees themselves, however the propolis must be collected as plant resins, saps, or gums outside the nest. Normally, collection of propolis occurs in the vicinity of the nest. Foragers for resin gnaw it from sticky buds and bark lesions on trees using their mandibles, and place it on their hind leg corbiculae for the trip home.

Plant resins are basically unchanged, retaining their original composition when utilized as propolis by bees, and they consist of organic solvents including flavonoid pigments which have antifungal

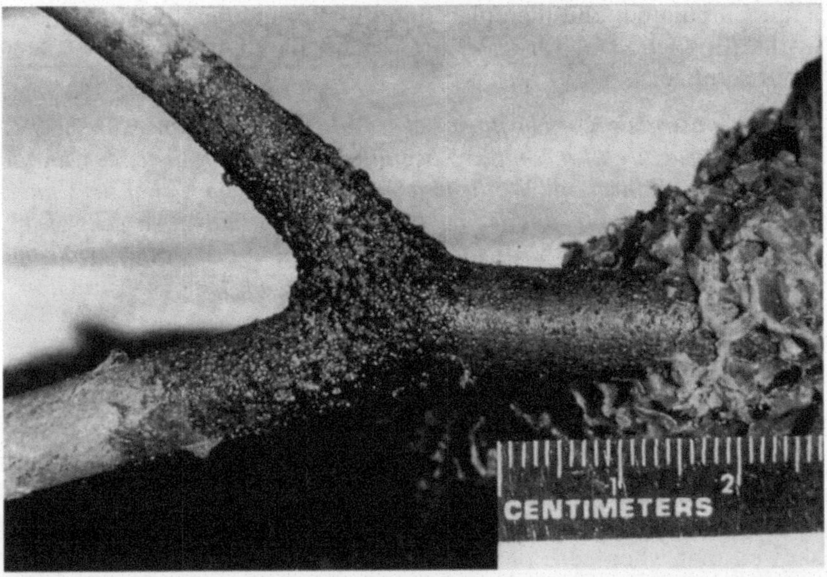

Fig. 5.1. *Apis florea* builds open nests on small branches which are further protected from ant predators by "sticky ends" of propolis mixed with glandular secretions wrapped around the branch on each side of the nest. (Courtesy of T.D. Seeley)

and antibacterial properties (Meyer 1956; Hancocks 1973; Ghisalberti 1979). These properties hinder the growth of molds and bacteria, and should a mouse or other animal be killed or die within the nest cavity, it will be completely covered with a layer of propolis which inhibits its deterioration within the nest. On arrival back at the home nest the propolis collector moves to that portion of the structure where workers are building and have need of propolis. She may wait for minutes or hours until the builders come and gnaw off the propolis for use in construction. In some species that occupy cold climate regions, propolis may be used to reduce the entrance size in winter (Ruttner 1968). In hot climates where melting or softening of wax could destroy portions of the nest, the workers blend propolis with wax used in construction, effectively raising the melting point.

The greatest physical change effected within the chosen honeybee nest cavity is the extensive construction of honeycomb. Wax production requires the input of a great amount of resources by individual workers. Those with mature wax glands secrete wax in small scales (Snodgrass 1956). These are removed with the mandibles, chewed and manipulated, and then placed into the proper spot where needed in the production of vertically orientated, parallel sheets of double-sided comb.

Comb Construction. "*He must be a dull man who can examine the exquisite structure of a comb, so beautifully adapted to its end, without enthusiastic admiration. We hear from mathematicians that bees have practically solved a recondite problem, and have made their cells of the proper shape to hold the greatest possible amount of honey, with the least possible consumption of precious wax in their construction*" (Darwin 1859).

A complete detailed analysis of the geometric structure of the comb was accomplished by mathematicians only after the invention of the calculus by Newton 300 years ago. However, 16 centuries ago, the Greek mathematician and astronomer, Pappus of Alexandria, proved the advantage of the hexagonal structure (Huber 1814; Hancocks 1973; Moody Inst Sci; Anon 1985; Hepburn 1986). He showed that when cells of unit size are packed together, the regular hexagon cross-section uses the least material because of the sharing of every wall by adjacent cells (Fig. 5.2). No other cell geometry is as economical in its use of space and material. Only three regular polygons would result in all walls shared between adjacent cells: triangles, squares and hexagons. Other shapes such as circles, pentagons or octagons leave spaces in-between adjoining cells where walls are not shared.

Stress patterns monitored with a polariscope show that the hexagonal pattern is best for lateral stress, which is then distributed evenly along the walls (when applied from point to point) (Moody Inst Sci). This feature is an adaptation which is important in honey storage in the vertical stacking of the nearly horizontal cells. The combined observations, insight and mathematical abilities of Maraldi (1712), Reaumur (1759) and Koenig (1739) revealed the economy of not only the hexagonal shape with its 109°28′ angles, but also the optimum angle (about 70°) for the rhombic structure of the intersecting bases of the honeycomb cells. Utilizing Newton's calculus, Kearney (75 years later) provided the necessary rigor to calculate the exact rhombic angle (70°32′) for using the least material for the special shape of a hexagonal cell with a trihedral base. The comb is constructed so as to utilize these optima by offseting the bases shared by three cells on one side of the comb with a single cell on the opposite side of the comb. Of course, the bee-constructed cells are not perfect nor identical but plainly variable and irregular. The cells

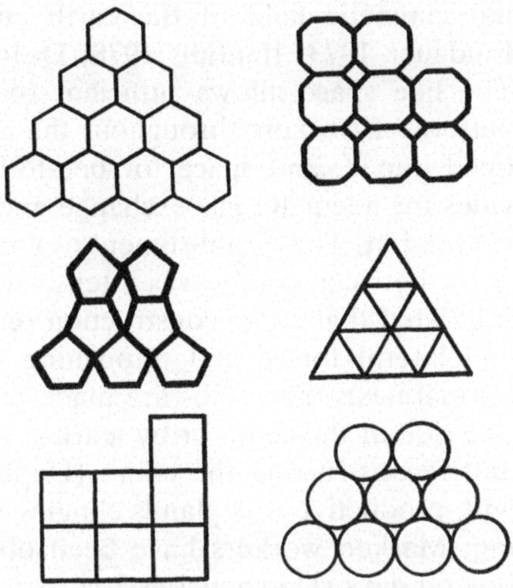

Fig. 5.2. Wax comb cells with the hexagonal cross-section use the least material and provide the most efficient use of space because they share all walls with adjacent cells. Circles, pentagons and octagons leave large gaps and do not share all wall surfaces. The hexagon shape is stronger than squares or triangles for the vertical forces encountered by the weight of honey stored in them

are not precisely uniform in wall thickness, shape, angles or orientation. They possess natural variation found in any biological system and are the result of natural selection and physical forces. The cells in honeybee comb are not, however, simply constructed as round tubes packed together, the pressure applied resulting in the hexagonal shapes as found in some tropical meliponine species (Michener 1974; Roubik 1989). Instead, *Apis* comb builders start with the base laying down three rhombuses and pull out the six cell walls. Maraldi carefully observed, "*They first build one of these rhombuses, and trace two new planes on the two edges of the rhombus; they add a second rhombus to the first at a specific angle, as we shall see later, and trace two new planes on the two edges of this rhombus; finally they add a third rhombus to the earlier pair, and raise on the two exterior edges of this rhombus two more planes, which with the four others form the cell, which because of this shape of the base necessarily has a hexagonal shape*" (Anon 1985, p. 76).

Within the nest cavity the workers construct wax combs in parallel vertical planes separated by the "bee space" (8–10 mm) first reported by Langstroth in 1851 (Dadant 1975), and orientated with respect to the magnetic field of the earth and the parent nest (Martin and Lindauer 1973; Ifantidis 1978; DeJong 1982; Gambino et al. 1990). The bee space allows sufficient room for free passage of a traffic pattern of workers throughout the nest, emergence of mature developed pupae, and space for bee-to-bee communication. It also provides for adequate gas exchange and humidity control through active ventilation. The establishment of parallel combs and regulation of the space between combs was intensively studied by Darchen (1968). He suggested that comb construction results from an interplay of vertical and lateral forces and retouching by the constructing workers. In normal nests the combs are made parallel by lengthening the cells of one side of the comb or by tearing apart what they have already built and reconstructing the comb (Hepburn 1986). Festoons of workers with productive wax glands clinging to each other do the comb building. Marked workers have been observed to remain in a chain for several days (Darchen 1962).

Biotic Factors. The homesite becomes unattractive as soon as there is insufficient forage within a reasonable range (Roubik 1989). This is the same consequence seen in human hunter gatherers such as the kung bushmen who find their home location unattractive when the food supply requires more than a half-day's foray (Pimentel and Pimentel 1979). In both cases the home site is abandoned and a new one sought in an area of more promising food supply. In the case of the bee colony, there must be a continuing floral supply of nectar and pollen as no single plant species can provide these resources in sufficient amounts for colonial maintenance over the long-term. The effect on wild honeybee populations of changing food resource availability is seen clearly in studies of Africanized honeybee swarms and colonies in Central America by Roubik and colleagues (Roubik and Buchmann 1982, 1984; Boreham and Roubik 1987; Roubik 1989). The population of honeybees within a large well-defined geographic area of Panama shows rises and declines in numbers found throughout the year. The peak population corresponds with a decrease in precipitation, and the population trough occurs with increasing rainfall. A closer examination of phenological changes in floral resources revealed that the bee colony population changes were reflecting those in the flowering plants, which themselves were responding to the wet season and dry season. The mid-dry to early wet season brought on plant growth and finally flowering. This resulted in progressively increasing supplies of floral resources (nectar and pollen pulses) available for exploitation by nectar and pollen feeders. The honeybee colonies rise to the occasion by rapidly responding to the increased food resources, and show a growth in population size delayed only by the period of incubation of the brood.

Abiotic Factors. When environmental factors prevent successful foraging for food, the colony no longer can tolerate the conditions and absconds in search of a new site (moves from the sessile into the mobile phase, see below). However, as long as they are mobile, they are more vulnerable to predation and direct environmental effects (rain, high or low environmental temperature, wind, direct solar radiation load, etc). Direct effects of the microclimate are deterministic for successful foraging. Considerable evidence exists showing that potential foragers rarely leave the confines of the dwelling under conditions of low air temperature under about 10 °C, or when it is raining or windy, or when light conditions are very poor (Lundie 1925; Ribbands 1953; Burrill and Dietz 1981; Villa 1987; Southwick et al. 1990). Continuously poor foraging conditions override food availability in determining whether or not the colony becomes mobile. In both cases the effect is the same; a lack in build-up of food stores resulting in decreased brood rearing and general demise of the sessile organism.

5.1.1.2 Mobile Phase

Colonial homesite stability is especially dependent on the rate of population growth and size of the nest cavity, and on environmental nectar flow predictability, water availability and direct environmental influences as outlined above. The mobile phase of colonial existence results from a global decision and takes place under two very different circumstances: the reproductive swarm, and the absconding or migratory swarm. Reproductive swarms only issue under favourable external conditions and are mobile always as intact superorganisms, even with sufficient honey stores to survive short periods of unfavourable weather. The reproduction results in a primary swarm that contains the old queen, and facultatively in one or more afterswarms with young virgin queens (Chap. 6).

While in temperate regions the bees rarely leave the stores and nest site, typically in the tropics there is frequent migration from one site to another (Michener 1974; Roubik 1989). In tropical and subtropical regions where wet and dry seasons alternate causing a floral bloom followed by a dearth of floral resources, absconding and migratory swarming occur frequently and predictably. *A. florea* may abscond when too many ant invaders attack the nest site (Seeley 1985). For tropical species, nest selection is less precise, food stores are small, and populations occupying single sites are not as large as in temperate regions. Under poor environmental conditions, the colony may abscond in search of a better homesite (Fletcher 1978). Only one swarm would emanate from an absconding colony. *A. dorsata* and *A. laboriosa* show seasonal migration of individual colonies from spring dwelling places in the valleys to summer dwelling places at high altitude in the mountains of India and Sri Lanka (Koeniger and Koeniger 1980; Koeniger 1982; Underwood 1990, see Chap. 4).

5.1.1.3 Life Cycle

The life-span of a honeybee colony continues as long as it is an intact unit, at least in terms of genotypic survival. The specific colonial genotype is particularly limited by the length of life of the queen, although other factors may decrease it as well. When the queen dies or her fecundity deteriorates, her colony moves into a queen replacement phase which results in the death of the original genotypic colony and its replacement by a new colony genotype, comprised first of sisters and later of the offspring of the new queen (see Chap. 3).

A key characteristic recently documented as an important parameter, which must be considered in evaluating life history strategies, is the foraging effort expended by individual workers and its effect on the fitness at the colony level. In studying the foraging ecology of honeybees, Schmidt-Hempel (1987) showed that individual honeybee workers invest less energy in foraging than required to maximize the net energy delivery to the colony (Fig. 5.3). How can there be selective advantage in performing at less than optimum levels? Normally, it would be assumed that increased efficiency to maximize net energy gain by the colony would be a measure of fitness (Stephens and Krebs 1986). Then, is foraging performance a special case of lower fitness? Foraging is the only means of procuring the required resources for colonial survival and growth. Furthermore, foraging is metabolically expensive for honeybees which have high rates of oxygen consumption when flying (Jongbloed and Wiersma 1934; Nachtigall et al. 1989; Rothe and Nachtigall 1989). In addition, there is a positive correlation between metabolic rate and body mass of the forager. A load of nectar weighing about 50 mg would nearly double the energetic demands (Wolf and Schmidt-Hempel 1989). Houston et al. (1988) argue that during the ergonomic growth stage of the colony

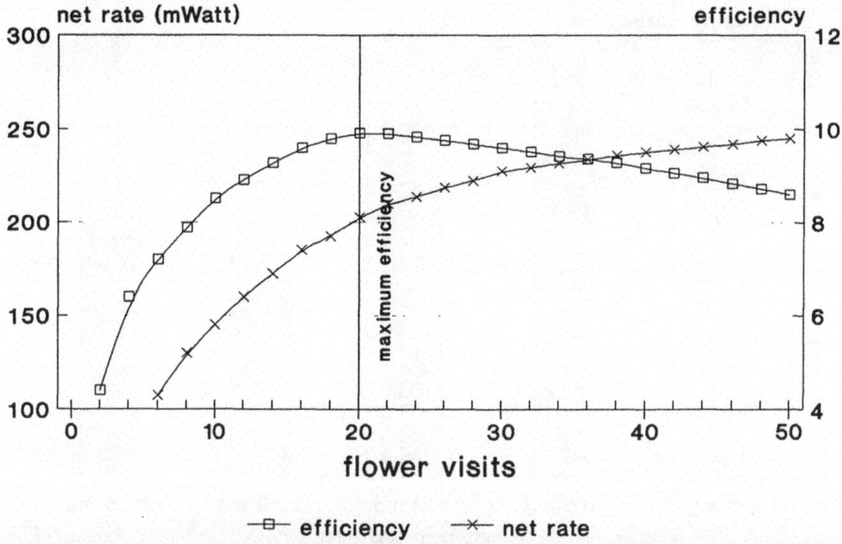

Fig. 5.3. If a forager maximizes the net rate of total energy then it should visit more than 50 flowers. If the foraging efficiency is maximized, only 20 visits must be made. (Data from Schmid-Hempel et al. 1985)

the submaximal foraging strategy could be the best *modus operandi* because the activity of foraging itself entails a survival cost, mainly energetically but also due to predators and other misfortunes. The heavier returning forager would be more vulnerable (and a richer reward) to predators because of decreased manoeuvrability in flight. Whatever the proximate cause, experimentally greater work loads required of foragers by forcing them to carry extra lead weights glued onto their thoraxes while foraging brought about reduced life-spans (Fig. 5.4) (Wolf and Schmidt-Hempel 1989). Clearly there is no simple work-load – life-span relationship. Foraging efforts of honeybees involve non-trivial survival costs in a threshold-like dependency. Extra loads in excess of about 20 mg resulted in a stepwise 30% decrease in life-span. The harder working individual foragers die sooner and thus require that much more of the incoming resources to be allocated to worker replacement rather than colony growth. The extra resources brought in by harder work would be insufficient to compensate for the drain of worker replacement. "*In this scenario, the colony is regarded as a unit where maximum growth is related to maximum fitness*" (Schmidt-Hempel and Wolf 1988). This

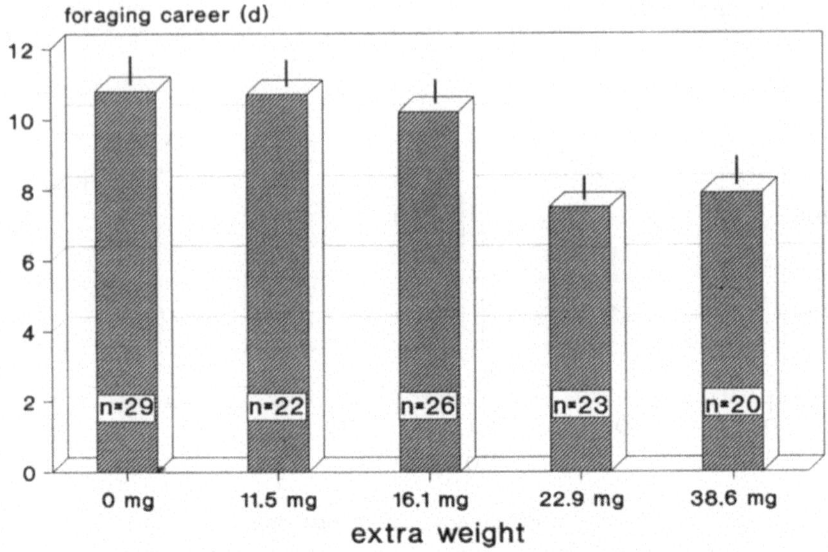

Fig. 5.4. Life span (foraging career in days) is reduced if the bees are loaded with extra weights. (Data from Wolf and Schmid-Hempel 1989)

is colony level selection, and the superorganism phenotype is placed in a central position of a natural selection scenario.

5.1.2 Spatial and Temporal Distribution

Colonial existence is not confined to the physical enclosure of the nest site. Instead, the superorganism spreads itself out over the local habitat in food and water procurement, defecation and debris elimination, defense, and mating. When the activities of an animal or group of animals are restricted to a certain area, this is classically defined as the home range, but when an animal in an area employs mechanisms for spacing the individuals apart from each other or actively repulses other animals, it is a territory (Burt 1943; Odum 1974). Applying to individual animals, a territory is actively defended from other individual intruders. Territoriality and home range at the colonial level do not differ at all from those concepts originating with the behaviour of mammals (Baroni Urbani 1979). In fact, such territoriality is a recognized characteristic of many eusocial arthropods with complex reproductive patterns including nest construction and care and protection of the young (Odum 1971). When compared to defensive behaviour of individual animals, fatalities resulting from territorial defence have quantitatively and qualitatively different significance for superorganisms and other social insects. Division of labour results in valuable reproductives and sterile workers which are relatively expendable. The loss of a worker is quantitatively only a reduction in energy and labour and not destruction of a reproductive operative. The gain in colony security offsets the cost of death of the worker (Oster and Wilson 1978). Hölldobler and Wilson (1990) suggest that natural selection theory implies that an animal should establish a territory only when the size and design of the territory make it defensible in economic terms. Most bees confine their defence to colonial preservation in the vicinity of the nest and are not defensive at their foraging patches.

5.1.2.1 Colony Defence

Eusocial insects regularly defend their nests and are well equipped with coordinated defensive specializations. Examples among the highly eusocial bees are the orchestrated defensive postures exhibited by the honeybee with its barbed sting, potent venom, and alarm pheromones, and the stingless bees with their biting behaviour, sticky resin depositing, and alarm pheromones. In contrast, bumblebees which store little food include few highly defensive species in the tropics, but none exhibits the highly organized massive colony defence effected by alarm pheromone (Sakagami 1976; Morse 1982; Plowright and Laverty 1984). Whereas the area constantly defended by most vertebrates extends well beyond the dimensions of the nest itself, in the case of most honeybee races it is reduced to that area surrounding the entrance to and actually occupied by the colony. Depending on the frequency of disturbance and species or race of honeybee, this defended zone around the entrance to the home may be expanded or contracted, but is rarely more than a few meters. For the highly defensive Africanized honeybee in South and Central America, the zone normally extends with patrols going outside the nest 10 or 20 m (without cause for release of or in response to alarm pheromone, unpublished personal painful experience). Disturbance causes a sequence of programmed alarm and defensive responses which carry attacking workers into the defence range of the nest. The numbers of potential defenders is usually less than 0.5% of the workers in the colony during any interval of time (Gary 1975). Although normally standing alert only at the hive entrance and checking incoming foragers or returning workers from orientation flights, the guards respond to disturbance by increased alertness, flight movement and alarm pheromone release (Free and Simpson 1968; Ghent and Gary 1962; Maschwitz 1964a,b). The defending workers are attracted to dark colours, movement, rough textures, animal scents, and attack the intruder by buzzing, crawling into and pulling the hair, biting and ultimately stinging. Once a sting is implanted, the alarm pheromone given off by the sting glands quickly recruits other workers for attack.

Comparing defensive behaviour of three Asian honeybee species, Seeley (1985) points out that *A. dorsata* is particularly highly defensive. This results from the two obvious characteristics; its large size (weighing one-and-a-half to two times as much as *A. mellifera* and three–five times the weight of other Asian species, for example *A. florea* and *A. cerana*), and the easily visible and unprotected form of its nest (Baroni Urbani 1979). *A. dorsata* often constructs aggregations of nests in particularly favourable nesting sites, usually on large-diameter smooth trees, rock overhangs or man-made structures (Seeley et al. 1982). Their nests are constructed in the open as large vertical combs often more than a meter across over which they cluster (Fig. 5.5).

Fig. 5.5. *Apis dorsata* open nest. (Courtesy of Dadant and Sons)

The defending arm of a bee colony operates remarkably similar to that of the stinging cnidoblasts received by intruders who make contact with members of the marine animal phylum, Cnidaria. The honeybee colony and colonial Portegese man-of-war have similar modes of defence against those who violate their close zones of defence. Both are colonial organisms. In the case of the man-of-war, actual contact with the organism or sufficient disturbance of the nearby environmental milieu is usually necessary to elicit the defensive response. This is similar to the honeybee colony, which requires only close approach to the entrance of the homesite to elicit a defence response.

5.1.2.2 Removal of Debris and the Dead

The colony in residence maintains a clean house through continuous cleaning activities. Much of the debris brought in with foragers and created within the nest simply falls down to the bottom of the cavity which may be quite deep in hollows of trees. Cell cleaning done by the young workers in preparation for egg laying includes the removal of larval excreta, remains of pupal cocoons, cleaning cell walls and removing cappings. Nest sanitation requires other debris to be purposely removed from the confines of the nest cavity. A portion of the older house bees (11 to 15 days old) assume the tasks associated with removing debris such as old cell cappings, mouldy pollen, and dead larvae or adults, although 90% of the old adults die in the field while foraging (Gary 1960). Some of the workers occupied with cleaning activities specialize for a few days in undertaking activities, removing dead bees from the nest. The undertaking crew numbers about 1% of the workers in the colony during any interval of time (Winston 1987). They recognize dead bees from the chemical odours released soon after death of a bee (Visscher 1983). The cleaners carry small items to be discarded to the entrance and then fly off with them to drop them 20–30 m from the hive (Morse 1972). If objects are heavy, such as pieces of wood or propolis or the drones destined for removal in the fall of the year, the cleaners take them to the entrance and struggle with them trying to fly and walk at least a metre or so away from the hive entrance. The cleanliness extends to all adult members of the colony which do not normally defecate within the nest space. This normally takes place during foraging on the outward-bound flight. For house bees or workers which have been confined for days or weeks, considerable waste accumulates in their rectums. The elimination of this waste is accomplished by moving to the nest entrance and leaving on a cleansing flight. Depending on the weather, these cleansing flights carry the frass 2 to 10 m or more from the nest. The cleansing flights are especially important after the long winter confinement in northern habitats, or after extended periods of confinement during the wet season in tropical regions. A few timely days of some warmth and sun are needed for initiation of these flights.

5.1.2.3 Home Range

Most sessile animals obtain their food via mechanisms such as filter feeding (e.g. in marine corals and sea anemones). No terrestrial animals can filter feed because the atmospheric medium is not rich enough in food particles. As a result, most terrestrial animals must move about in search of food. The sessile nature of the colony prevents movement as a whole for food on a regular daily basis, yet specialized individuals within the colony are designated to travel in search of food and to return to the mother organism with it. The model utilized in Chapter 4 of an amoeba with moving pseudopods extended out after food particles is a useful illustration of the dynamics of colonial feeding. The home range of most terrestrial animals as defined above is that range over which they generally restrict their foraging for food and other activity. Using this definition for the bee colony, the home range can extend out in all directions to a distance of 10 km or more (Eckert 1933; von Frisch 1967; Visscher 1982). However, the feeding arms are usually extended to only a fraction of this distance in normal feeding and the mean distance to which the forager arms extend is about 2 kilometres, and the most common feeding distance is actually 600 to 800 m (see Fig. 4.2). There are some data showing that pollen foragers travel to floral resource patches at greater distances from the nest than nectar foragers travel to procure their commodity. One report by Gary et al. (1972) shows that bees foraging for pollen over a field of carrots travelled an average of 1663 m, while those collecting nectar from onions foraged 557 m. Winston (pers. comm.) suggests two factors that may be involved in the farther distance travelled to obtain pollen. The most important factor is probably the stimulation of the constant requirement for pollen as nest reserves are not extensive. He also notes that less time is required to procure a pollen load than a nectar load, and pollen loads are somewhat lighter, allowing for less energy to be expended on flight. However, Southwick and Pimentel (1981) calculated that the energetic return for nectar retrieval is much higher than that for pollen retrieval, making pollen the more expensive commodity to collect. Another factor promoting greater distances for pollen procurement might be the spread of pollen producing flowers. Nectar may be more readily available in nature than pollen that is suitable for collection by bees. Some other foraging statistics that are likely to influence the distribution of workers in the local habitat include (summarized from compiled data of many authors by Winston 1990): number of trips per day (average 10–15 for pollen and nectar collectors); number of flowers visited per trip (widely variable, but typically hundreds to thousands for pollen, and tens to a few hundred for nectar); time spent collecting in a single trip (commonly 10 min or so for pollen foragers, 30–80 min for nectar foragers); weight of a load on return (pollen weights of 10–30 mg are common, and nectar loads weigh 25–40 mg).

The colony needs to obtain other commodities besides pollen and nectar to survive. Probably the most important of these is water as it is used in the cooling of the nest, diluting of brood food, and for individual consumption by adults. Water is used by nurse bees to dilute honey in the preparation of larval food. The colony may go to extremes to obtain water under circumstances of evaporative cooling requirements brought on by hot weather. Colonies in the desert have been monitored that reach out as far as 16 km in their search for water (J. Gould, pers. comm.). Again, the normal circumstance involves a water source within a short distance of the nest site.

5.1.2.4 Mating Distance

For the organism to mate successfully, it must extend itself out again in order to contact unrelated mating partners. This outcrossing involves the extension of mating arms in the form of fertile drones and virgin queens (details see Chap. 6) meeting on drone congregation areas (DCA) (Ruttner and Ruttner 1963, 1965, 1966, 1972). Queens and drones do not have equal ranges of flight to their DCAs with the queens usually flying farther (Koeniger 1987). These mating areas are within a few kilometres of the mother colony to which the unsuccessful drones and successful queens return. The estimated distance from the colony for this activity is in a range of 4–5 km, requiring more than 10 min flight time (Ruttner and Ruttner 1966; Woyke 1956; Böttcher 1975; Loper et al. 1988). DCA locations are constant over years suggesting that geophysical and geographic properties are orientation cues. Drones from different colonies may find the same DCA and drones may switch DCA sites on the same flight (Loper et al. 1988).

5.1.2.5 Overdispersion of Colonies

Mature colonies of many social insects belonging to the same or closely related species tend to be overdispersed, or spaced in such a way that the distances between them are not random but approach uniform distribution, at least within species or between closely related species (Hubell and Johnson 1977; Hölldobler and Wilson 1990). This results in rather regular arrays of nests, and such local dispersion patterns have been found in a large number of ant species occupying a wide variety of habitats in temperate and tropical regions. In the case of cavity-dwelling honeybees, there seems to be little chance for overdispersion because of their use of already existing cavities.

5.1.3 Distribution of Honeybees

The current understanding of distributions of all the species and races of honeybees (*Apis*) is presented in excellent detail by Ruttner (1988). However, at the ecological level, honeybee populations in the temperate zones have not been well studied. There are few data from which to estimate carrying capacities of any ecosystem (see also Sect. 5.2.4). Taber (1980) estimated only two colonies/km^2 in natural populations of honeybees. New attempts are being made to estimate numbers of colonies of wild honeybees in some areas. Wenner et al. (1990) are studying feral colonies on an island off the California coast, by systematically removing all the nests. In the Amazon basin, the density of honeybees is estimated at 10 to as high as 108 colonies/km^2 (Kerr 1984; Roubik 1983, 1987; Taylor 1985). Yet, none of these estimates includes migrating swarms, nor has there been a careful analysis to find all the nests in a single square kilometre (Roubik 1989). Makela et al. (1990) developed a model of distribution via computer simulation incorporating object-orientated models with the programmed characteristics of energy use and behaviour of the honeybee colony dictating how they behave over time (Fig. 5.6). More field data are needed, however, to improve these models and better our understanding of colony distribution.

Selection pressures promoting the evolution of superorganisms worked toward their permanence as perennial highly eusocial colonies

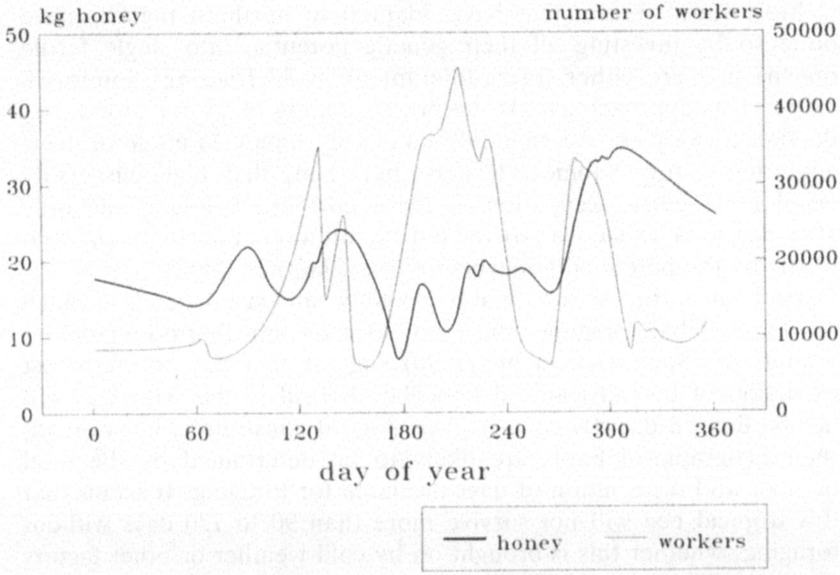

Fig. 5.6. Typical simulation of population dynamics in a single colony of European honeybees when nectar and pollen were available in varying amounts during the foraging season. The number of workers is seen to rise with emerging brood and precipitously fall when swarms leave. The amount of stored honey is shown to be affected by food consumption of the bees and incoming rate of nectar (primarily). (After Makela et al. 1990; Makela, pers. comm.)

(Michener 1974; Roubik 1989). Roubik (1989) conjectures that the transition from subsocial, semi-social, and quasi-social existence of colonial organisms to the high eusociality with its rigid social roles and permanence found in superorganisms is a tropical phenomenon. He argues that cold weather is likely to be a limitation to permanence in colonies. Only *A. mellifera*, *A. koschevnekovi*, and *A. cerana* which nest in protective cavities survive the temperate winters as intact units. There is a number of tropical species of stingless bees that can tolerate short-term exposure to low temperature. Some of these species exist in areas at altitude, such as in the Colombian Andes as high as 3000 m or in mountains of Taiwan and Borneo where they may experience subfreezing temperatures (Roubik 1989). Thus, it is the duration of continuous low temperature and the associated effects on availability of food, rather than the cold per se that is more likely to be limiting permanence.

Most social insects that have adapted to northern regimes have done so by investing all their genetic potential into single fertile queens that are either freeze-tolerant or avoid freezing (Southwick 1991). This approach avoids the precise timing of winter brood production (Seeley and Vischer 1985) and very high cost mode of overwintering entire colonies. It must be, then, that high eusociality developed before occupation of these northern habitats, and only after perennial existence evolved did exploitation of northern environments by the perennial honeybees ensue (Ruttner 1988).

How far north the perennial bee colony can exist depends as much on its available foraging time as it does on any thermo-regulatory limitations. Southwick et al. (1990) suggest that the northernmost excursion of the Africanized honeybee hybrid in the Americas will not be limited directly by any physiological constraints, but instead, their geographical limits are likely to be determined by the total number and distribution of days available for foraging. It seems that this tropical bee will not survive more than 90 to 120 days without foraging, whether this is brought on by cold weather or other factors (Woyke 1973; Villa 1987; Villa et al. 1991).

The spread of superoganismic colonies was probably favoured by improved competitive ability combined with increased effectiveness in food gathering, protection from enemies, reduced impact of climatic variables especially by improved modes of homesite construction (e.g. multiple parallel combs), initiation of new colonies provisioned from resources of the mother colony and, probably, division of labour. Great advantage is gained with a large number of individuals available, giving the ability to capitalize on sudden increases in standing crop of the nectar food resource, to provide better defence, and to tolerate exposure to low environmental temperatures. Odum (1971) suggests, however, that there can be colonies with too many individuals, over and above some illusive optimum size.

5.1.4 Intracolonial Structure

Within the colony, there is a characteristic morphological pattern within the confines of the dimensions of the nest cavity. During the foraging season, the colony organization has the worker brood cells more or less centrally located and surrounded in an arc on the upper portion comprised of cells filled with packed pollen. The large pollen band develops around the brood especially profusely during the time of intense pollen foraging (Camazine 1991; Camazine et al. 1991). Many of these cells are covered with a thin layer of honey. Another concentric arc of honey-filled comb is adjacent to the pollen. Drone brood cells are usually near the comb edges, and queen cells, when present, are usually vertically attached at the comb periphery. The placement of eggs by the laying queen, and nectar and pollen by the workers in the appropriate portion of the pattern is done in a way that maintains the pattern. It is likely that the pattern formation and preservation is effected by individual bees operating under local influences on the comb. Using simple local decisions by individual workers, Camazine (1991), and Camazine et al. (1991) were able to reproduce the pattern on a computer simulation. The net result is the maintenance of the global physical structure actually seen in real colonies in nature with a seemingly planned configuration of comb utilization.

5.2 Community Structure

5.2.1 Plant/Bee Symbioses

5.2.1.1 Foraging and Pollination

Intracolonial social structure is highly dependent on the availability of food within the foraging range of the superorganism. Maintenance of the intact superorganism, then, is highly food-dependent and fits the central-place foraging theory in which net energy gain is maximized (Orians and Pearson 1979). The model leads to a basic proposition that "*natural selection shapes behavior and, in the case of social insects, does so by acting at the level of colonies*" (Hölldobler and Wilson 1990, p. 387). Individuals move out into the field to search for and recover food not for themselves but to fill the stores of the colony. The model assumes that the greater the energy expense in obtaining food, the more choosy the forager should be, or the greater return the forager should seek. If a food patch is far away, it would be expected to be less attractive than a closer patch of similar quality because of the energetic costs in getting there and back.

Honeybees are especially flexible in their approach to foraging for food. As might be expected from the successful introduction of the species in many habitats, they are generalists without tight associations maintained for extended periods with particular species of flowers. Instead, the foragers may be constant on a floral resource for days, until the resource is depleted to a point below which they are recruited to another better resource (Ribbands 1949; von Frisch 1967; Seeley 1986; Sneyd and Camazine 1991). The constancy in collection usually applies to the type of resource, whether it be nectar or pollen. Few studies have delineated the foraging preference and its changes through the season. An early study by Parker (1926) and a later one by Free (1960) on fruit-tree pollination indicated that slightly more than half of the foragers returned with nectar loads and no pollen, about one quarter with pollen and no nectar, and less than a quarter with both commodities. Which flowers were visited during any foraging bout and on successive bouts is readily determined by examination of pollen loads on the returning bees. It is more difficult to determine the floral visitation in nectar collectors. Workers foraging for pollen specialize on pollen collection from the particular floral species on any single trip and on subsequent trips. Examination of pollen loads reveal that 87 to 100% of the pollen is from a single source, indicating that only one flower species is visited (Free 1963). The size of the foraging area is probably determined by nectar or pollen availability as affected by floral production, number and age of open flowers, competition from other floral feeders and local weather conditions (Beutler 1953; Percival 1961; von Frisch 1967; Shuel 1975; Southwick 1986; Southwick and Southwick 1983; see Sect. 5.1.2.3).

Many other bee species are specialists to the point of tight association with only one or two floral species. An example of the latter is the classical case of some male euglossine bees which interpret the morphology of certain orchid blossoms (e.g. *Ophrys speculum*) which they pollinate to be females of the species (Correvon and Pouyanne 1916; Faegri and van der Pijl 1979). The model is so good at mimicking the female bee that the male attempts to copulate with the flower and in so doing becomes covered with the pollen. It then travels to another flower and repeats the act transferring pollen and effecting pollination in the process.

5.2.1.2 Energy Flow and Material Turnover

The flow of energy through colonies of bees and the concomitant turnover of materials by the colony contribute substantially to the dynamics of community structure in ecosystems. It is an important element in understanding processes that build complex ecological communities. The amounts of protein and carbohydrate harvested, and the energy used by the bee colonies resident in a hectare of tropical rainforest were calculated by Roubik (1989). The biomass and rates of return of nectar, pollen and resins by foragers in 37 species of stingless bees (189 nests) were averaged over the number of colonies and used to calculated the approximate amount of material collected. Colony cycling was determined using measured sugar compositions of the honeys, and dervied energy and pollen protein values (averages of 70% sugar, and pollen containing 3.5% nitrogen; energy equivalents from Roubik 1983; Southwick and Pimentel 1981; Stanley and Linskens 1974). The bees occupying a hectare of tropical lowland harvest $6 \cdot 10^6$ kJ of carbohydrate annually. Each year 8.76 kg of pollen is collected containing about 5 liters of nitrogen. The energy equivalent in pollen for the hectare is about $22 \cdot 10^5$ kJ. As costs of wax production and foraging itself were not considered, the calculated values are likely to be conservative underestimates. The total energy for primary producers given for tropical forests in Odum's (1971) treatment of ecology is on the order of 54 million kJ for each hectare. Thus, the bees consume 11% of the net primary production in this habitat! This route of energy flow is almost never accounted for in treatments of energy balance in ecosystems.

In the process of natural colony activity, dead bees, frass (faeces), wax and other litter are subsequently scattered over the surrounding forest floor by the bees. Using the figure of 100 dead workers of *A. mellifera* being disposed of each day (Gary 1960; Bühlmann 1985) along with those that die in the field while foraging, a complete turnover of worker populations of tropical honeybees occur each 40–50 days (Roubik 1989). Therefore, during the year about seven times the adult population is scattered about the area of each nest site. Adult populations average 2000–3000 bees for stingless bee colonies, amounting to a few hundred kilograms of dead bees per square kilometre with an energy content of nearly 10^7 kJ (Roubik 1989). In addition, the colony litter ejected from nests consitutes about 1.8 t/yr/km^2. Honeybees distribute their waste in cleansing flights after accumulating excrements in their recta for extended periods of nest confinement during non-foraging periods of dearth. They may do this as groups of workers collectively voiding liquid frass in the air over the nearby ground area. If the 30 000 bees in a colony each voided 30 mg in this fashion, the total of nearly 10 kg would be released each year. Thus, the total contribution in bee cycling of materials over 1 km^2 in the tropical forest amounts to about 3 metric tons of material containing 5 liters of nitrogen and 15 million kJ of energy. Considering this value in the light of the energy trapped and made available by primary producers in tropical rainforests estimated by Odum above, we find that the bees alone cycle back into the ecosystem from their nests about 0.3% of the total primary production. This is a significant, usually unaccounted for contribution to the ecosystem.

5.2.1.3 Honeydew

When sap-feeding Homoptera including scale insects (Coccidae), spittle insects (Cercopidae) mealybugs (Pseudococcidae), leafhoppers (Cicadellidae) and especially aphids (Aphidadae) feed on the phloem sap of plants, they obtain food for themselves and eliminate excess material derived from digestive processes (honeydew). The expelled honeydew lands on nearby plant parts, and on the ground below. These droplets have high sugar contents (mostly fructose and glucose with some maltose, trehalose, and trisaccahrides) and some amino acids (White et al. 1962; Dixon 1985), and are attractive to many animals including ants and bees. Bees harvest honeydew and carry it back to the nest where it is stored in the honeycomb. The amount of honeydew collected by foragers is dependent on the availability of floral nectar which is normally collected in preference.

5.2.2 Selective Forces and Diversity

Honeybees originated in the tropics (Ruttner 1968, 1988) and were superorganisms before they colonized temperate zones. It would have been impossible for the bees to exploit the habitats now enjoyed if they had remained as semi-social entities with solitary individuals, such as fertile queens, diapausing or resistant eggs or pupae over-wintering (or over non-foraging season) as nearly all hymenopterans do. The honeybee over-wintering superorganism form is a peculiar and unique mode of confronting severe trials encountered in the environment. Extant populations of superorganisms are not as varied as one might expect when compared with the wide diversity in social and solitary species. Perhaps because of the common local decision approach to solving global problems, all superorganisms tend toward a similar structure. Tests on tropical feral honeybees in America that have never experienced cold conditions reveal that they use the same modes of thermoregulation as northern races (Southwick et al. 1990). On exposure to temperatures near freezing, they form clusters quite capable of withstanding extended periods of cold. They do differ, however, in their efficiency of solving the global problem of

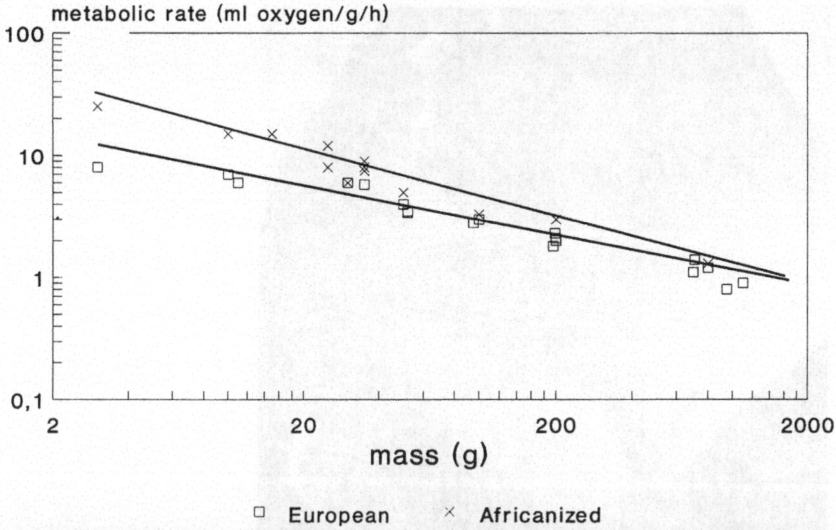

Fig. 5.7. The difference between metabolic rates of Africanized groups and European groups of honeybees as a function of environmental temperature. (From Southwick et al. 1990)

keeping warm and expend more energy at cool temperatures than temperate colonies (*A.m. carnica*) (Fig. 5.7). The difference between tropical and temperate colonies is not so much in the approach and mechanisms of facing the thermal challenge, but is in degree instead.

Superorganismic honeybees may reduce bee species richness, although in tropical forests stingless bee colonies live together with plenty of other species of bees. The ecological advantages of the superorganismic forms seemingly give them a superior edge over the solitary and semi-social species. This has been amply shown through the effects on native bee species when competing in natural populations in the tropics (Roubik 1978, 1989), and by introductions of *Apis* colonies into national parklands in Australia (Pyke and Blazer 1982).

5.2.2.1 Intraspecific Competition

In eusocial bee species, even with little evidence of intraspecific aggression, nest locations would be expected to be distributed over geographical areas allowing for home range and ample foraging areas. This may usually be the case in natural settings, but there are incidents in which forage is apparently not a serious limiting factor

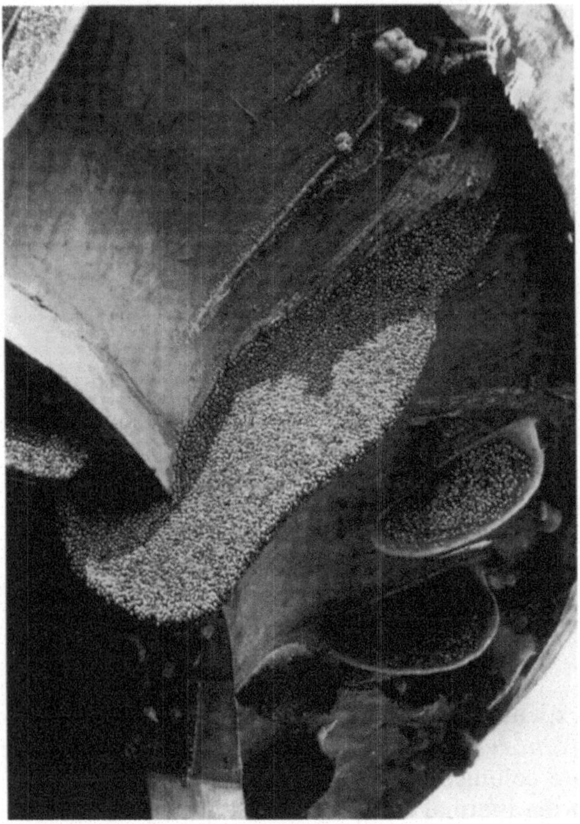

Fig. 5.8. Many *Apis dorsata* nests are often constructed in the same tree or, in this case, suspended from a water tower. (Courtesy of T.D. Seeley)

and numerous different nests are built in close proximity. In species requiring cavities as domiciles, homesites are most likely limiting. However, where cavities are not needed and the colonies are willing to build exposed nests, or even usually do so, no such limit is incurred and many nests coexist. Often several *A. dorsata* nests are constructed together in aggregations in the same tree or, as shown in Fig. 5.8, are suspended from a water tower. One study from India reported 156 colonies of *A. dorsata* in a single tree (Singh 1962). The overlap of foraging areas does not seem to hamper their success.

However, competition among honeybees is known, when nectar yield is low. Strong colonies will find and rob honey stores of weaker colonies (Gary 1966). Robbing is not carried out by masses of thieving foragers moving into the prey colony to steal the stores. Instead, it is done by low numbers of workers that enter the target colony one by one often without being noticed by the guards or fighting with them. Robbing workers may show a zig-zag flight pattern at the entrance that differs from normal returning foragers which fly straight into the entrance and proceed on inside. Cale et al. (1946) describes experienced robbers as being smooth, shiny and dark, probably from the wear and tear of conflicts with guard bees.

Intraspecific competition of *A. mellifera* foragers has been best studied for the feral honeybees in South and Central America. The overlapping populations of Africanized honeybees, and the relic populations of pure European races of *A. mellifera* show differences in active times of foraging during the day and under differing weather conditions. Members of the Africanized type forage earlier and in more inclement weather than do the European foragers. Yet this factor will only have effects at the colonial level if the food availability is a serious constraint (see Sect. 5.2.4.2).

Competition is not only restricted to foraging but has also been found in mating. Again, this has been best studied for the Africanized honeybees. The mating flight times and duration differ between the hybrids and Europeans, especially in the drones (Rinderer et al. 1985). Africanized drones leave the nest earlier in the day and at cooler temperatures than their European counterparts. They return earlier to the hive as well. However, their earlier mating flight results in more successful mating with virgin queens as they arrive in the drone congregation areas (Taylor and Rowell 1988; see Chap. 8).

5.2.2.2 Interspecific Competition Among Superorganisms

In order to co-exist, species must differ from one another sufficiently in their use of resources, primary nest sites and mating behaviour. Often, differences exist in body size and related size of food items preferred, such as in *A. dorsata* and *A. florea*. The large-bodied bees are not able to attain access to some of the small or deep corolla flowers that are exploited by the tiny bees; and the small bees are unable to force their way into the corolla of flowers requiring large heavy bees for mechanical movement of floral parts. Nevertheless, in Sri Lanka, Koeniger (1982) found that the smaller bees seemed to be dominant over the larger bees at artificial feeding dishes. In fact tiny stingless *Trigona* bees were most successful in excluding other bee species (*A. florea*, *A. cerana*, *A. dorsata*) from the food source.

Interspecific robbing among honeybees is extremely rare. Competition for nectar and pollen food resources mainly occurs at the feeding sites. Schaffer et al. (1983) and Pyke and Blazer (1982) found that introduced honeybee colonies successfully competed with native fauna to the detriment of the latter.

Although interspecific robbing is rare in a natural environment, *A. mellifera* colonies have been shown to efficiently rob *A. cerana* colonies (Koeniger 1982). Since both species are allopatric, this only can be observed in experimental situations. The robbing is particularly dramatic, because of interspecific "misunderstanding". When *A. cerana* guards defend their nest and attack an *A. mellifera* intruder, feeding behaviour instead of stinging is released in the *A. cerana* guard. The *A. mellifera* robber fills its honey crop, returns to the mother colony, and recruits further robbers that finally overwhelm the *A. cerana* guard force.

Interspecific mating competition of sympatric species is usually ruled out through strict separation of mating flight times and various anatomical constraints that prevent allospecific matings (see Chap. 6).

5.2.2.3 Competition with Individual Organisms

Nesting Sites. As nesting sites are one of the usual limiting factors for cavity-dwelling honeybee colonies, there is competition for these with other organisms. Not only are other bees, wasps and insects competitors, but also larger organisms, including birds and mammals, compete for the most favourable sites. Once occupied by a colony of bees, the home site is expected to be well protected by a guarding force of workers. However, when conditions, such as cool weather, require clustering of the colony, the entrance is left unguarded and organisms may enter the occupied cavity and set up housekeeping. A common intruder is the common white-footed mouse (*Peromyscus leucopus*). The mouse gains not only a protected nest site but also a food storehouse. Should the bee cluster move or the bees be disturbed by the intruder, they may kill it by stinging and then completely cover the body with propolis, preventing its putrification within the nest.

Food. Food availability, the second critical factor that limits colony growth and productivity, is also important to many other species of organisms, especially foraging insects. Many species of ants, solitary bees, wasps, butterflies, moths and other insects seek the highly nutritious flower nectar and pollen which are the only sources of food for the honeybee colony. Therefore, the competition for the floral resources is strong and may be indirect or direct. Indirect competition results from the feeding activities of nectarivores that seek food during times when the bee foragers are not flying. Thus, depletion of the resource can take place before the bees even get there (Roubik and Buchmann 1984). The resulting patchiness in resources requires that the forager force from a single colony be flexible in its selection of new floral patches. It also results in many floral patches being exploited at the same time by a single colony, with recruits willing to shift from one patch to another, and scout foragers searching for potential new foraging patches. Large colonies are adaptive in that, once they locate a good food patch, they can quickly mobilize a team of foragers to exploit it, thereby beating out other potential foragers, especially those of smaller colonies or of solitary habits.

Direct competition has been only rarely observed between honeybees foraging for nectar or pollen and other foraging species. Occasionally, when a honeybee encounters another insect feeder, it moves to a different blossom on the same plant with little interaction, and actual combat is extremely rare (Southwick 1983). Schaffer et al. (1983) found competition for nectar between introduced honeybees and native bees, and even ants at floral resources.

5.2.3 Diseases, Parasites, and Predators

5.2.3.1 Superorganism Pathology

The honeybee colony as a single entity is affected by a number of contagious diseases, predators, and parasites that would be unlikely to infest the individuals if they were living as solitary organisms. A normal colony consists of 30000 to 60000 individuals (mostly workers) during the foraging season, and perhaps half that number in the off-season in temperate climates. The continuous physical contact of the many individuals in the colony would be expected to facilitate the spread of disease within the group. Contact among colonies would be likely to spread contagious diseases from one group to another. Within the colony, food exchange involving all the bees sharing the available food (nectar and pollen) emphasizes the unity of the colonial entity. Its likely function is in the distribution of the chemical odours throughout the colony, which continually alert the members of the queen presence (Naumann et al. 1991) and specify the uniqueness of the group. This behaviour, of course, is ideal for the transmission of disease organisms from one part of the superorganism to another (Bailey 1981; Morse 1978).

The bees living in the crowded conditions of the superorganism have evolved mechanisms of protection for themselves and their stored food resources. They illustrate very hygienic patterns of behaviour (see Chap. 3). Even small particles of foreign material will be removed and discarded outside the confines of the colony. Individuals are also constantly cleaning their own bodies and will remove debris and ectoparasites. They also remove diseased larvae or pupae and discard them at some distance from the nest. Nevertheless, a large number of diseases, parasites and predators attack the superorganism. If the pest organism gains a foothold, the entire superorganism may be weakened or even destroyed. Worldwide, a number of mites infest colonies and exact great losses in commercial and natural populations. A number of the most common diseases and pests that attack superorganisms, but not isolated individuals, are listed in Table 5.1.

5.2.3.2 Specific Examples

We will elaborate on just a few of the problems encountered by the honeybee superorganism as listed in the table. More details of the specific diseases or pest organisms can be found in the cited references.

Disease (American Foulbrood). American Foulbrood, or AFB, is among the most serious diseases affecting honeybee colonies in the world (Morse 1978). Caused by *Bacillus larvae*, AFB is very contagious and spreads quickly within and among colonies. Because the spores are resistant to desiccation and can tolerate extremes in environmental conditions for extended periods, the spread of the disease is effected by spore movement, not just within the nest, but also through robbing and drifting bees from an infected nest to an uninfected one. A disease only of the developing larvae, AFB spores germinate after ingestion. They multiply in the midgut and penetrate the gut wall resulting in death of the larva. The colony level response includes removal of diseased and dying larvae and sanitation of their previously occupied cells.

Parasites (Varroa jacobsoni). The invasion into Africa, America, Australia, and Europe of the ectoparasite of honeybees, *Varroa jacobsoni* Oudemans in the last decade, has heightened the interest

Table 5.1. Principal diseases, parasites, and predators specific to the honeybee superorganism[a]

Agent	Name	Disease organism	Symptoms	Affected subject
Virus	Sacbrood	Sacbrood virus	Inability to moult	Brood and young adult
	Kashmir disease	Kashmir bee virus	Kills within 3 days	Adult
	Viral paralysis	Paralysis virus	Bloat, paralysis, kills	Adult
Bacteria	Septicemia	*Bacillus apisepticus*	Muscle degeneration	Adult
	American foulbrood	*B. larvae*	Larval death	Larvae
	European foulbrood	*B. pluton*	Larval death	Young larvae
	Powdery Scale	*B. pulvifaciens*	Larval death	Larvae
Protozoa	Nosema	*Nosema apis*	Dysentery	Adult
	Amoebae disease	*Malpighamoeba mellificae*	Gut malfunction	Adult
Fungi	Chalkbrood	*Ascosphaera apis*	Larval mummies	Larvae
Mites	Varroatosis	*Varroa jacobsoni*	Impairs development	Larvae/adult
	Tracheal disease	*Acarapis woodi*	Tracheal infestation	Adult
Insects	Greater wax moth	*Galleria mellonella*	Silk thread entrapment	Combs
	Lesser wax moth	*Achroia grisella*	Perforate cappings	Larvae
	Bee lice	*Braula* spp.	Ectoparasite	Adult
	Ants	Formicidae	Predation	Larvae, pupae, stores
	Wasps	Vespidae	Predation	Larvae, pupae, adults
	Bees	Meliponidae	Rob	Stores, take over nest site
		Apis spp	Rob	Stores
Mammals	Mice	Mus spp	Destroy nest	Nest construction
	Skunks	Mustelidae	Predator	Feed on bees, honey, pollen
	Bears,	Tayras, etc.	Rob	Stores, brood, colony
	Beekeepers	*Homo sapiens*	Rob	Honey, brood, colony
	Birds	Insectivore spp.	Predation	Drones mainly

[a] Bailey (1981), Morse (1978), Roubik (1989).

and study in this organism. It is probably the most dangerous parasite of the honeybee colony as, once infected, a whole colony usually dies within 3 years. *V. jacobsoni* is an Asian mite that infests *Apis cerana* nests naturally and is a well-adapted parasite, infesting but not killing the host colony (Peng et al. 1987; Southwick 1990). However, when it infests a colony of temperate region *A. mellifera*, this parasite kills the host colony by shortening the life spans of individuals and affecting the development of larvae (Kovac and Crailsheim 1988). Recent evidence indicates that the gravid female mite is attracted to bee (especially drone) larvae by an interspecific chemical kairomone (LeConte et al. 1989). The disease is spread by adult mites riding on the body surface of adult workers, drones and even queens, and by being transferred to other bees and colonies. A colony level response is particularly well expressed in *A. cerana* (but not in *A. mellifera*), which is not severely damaged by the mite (Peng et al. 1987). In *A. cerana* colonies, the population of mites remains small partly because workers show shared hygienic behaviour including removal of the adult parasites from their own bodies, from the bodies of other workers, and from larvae by grasping with the mandibles. The removed mites are taken away from the hive. The grasping worker may bite with sufficient force to sever and kill individual mites (Peng 1988). Workers may join forces to remove mites from a single sister in "group cleaning behaviour".

Functional Cleptoparasites (Honey Hunters and Beekeepers). We include here the description of another parasite called a functional cleptoparasite, the beekeeper, because functionally the effect is the same as of a natural parasite. A beekeeper may or may not manage a colony to maximize its production of honey stores. His parasitic qualities come from his stealing substantial portions of the stored food with some unintentional (unqualified beekeeper) or intentional kill of bees (honey hunter). Beekeepers in some cultures do not attempt to manage bee colonies but simply harvest the "crop" (e.g. *A. laboriosa*, Underwood 1990). Human-managed colonies can be profitably exploited when the beekeeper manipulates storage space by honey harvesting, thereby controlling swarming. If the beekeeper leaves adequate stores, the colony survives the non-foraging season and is little affected, and he would behave as a "well-adapted parasite". However, even without obvious effects, the beekeeper is forcing the bees to forage much more than they normally would, as evidenced by the increased quantities of food stored in managed hives (ca. 40–60 kg in temperate regions) than in natural nests (ca. 15–20 kg). Even natural nests are attractive to this parasite, especially in certain human cultures, such as found in Nepal and portions of Africa. The colonial response, at least in African *A. mellifera scutellata*, has been the adaptation to fierce defensive behaviour when the nest as a whole is parasitized.

Predators (Vespids). There are numerous predators that feed directly on individual bees or consume bees and/or their products at the nest. At least 40 species of mammals attack and devour bees and their stored food (Morse 1978; Roubik 1989). A typical predator that consumes workers both in the field and at the colony entrance are the wasps of the genus *Vespa*. Some vespid species enter the honeybee nest and remove larvae and pupae, may devour workers and destroy the nest (Matsuura and Sakagami 1973; Matsuura 1985). The wasps bite off the head and the abdomen and fly back to their own nests to feed the bee thorax to their wasp larvae. The two wasp species, *V. orientalis* and *V. mandarina* are formidable predators of bee colonies in Japan and will kill guard bees and then enter the nest and feed on brood and adults, which they take back to feed their own larvae. When *A. cerana japonica* is attacked by hornets (*Vespa simillima xanthoptera*), a coordinated group defence is used (Ono et al. 1987). Individual hornets are balled by many of the small bees, and the hornets become sufficiently overheated to be killed (see Chap. 4).

5.3 Superorganism Size

For normal social order to exist in a honeybee colony, there must be a minimum of about 200 workers along with a fertile queen (Michener 1974; Morse 1978). With lower numbers of workers, the social entity is broken down and the colony cannot survive. Even with this low number of individuals as a minimum, survival of the superorganism is only possible under the most favourable environmental conditions.

5.3.1 Temporal Variation in Group Size

When many workers are foraging, the group size resident in the nest is reduced by the number in the forager force. In honeybee colonies occupying regions with a temperate climate, and in stingless bees in the tropics, the forager force may comprise 10 to 30% of the colony worker population (Inoue et al. 1985; Roubik 1989). This foraging activity only reduces the number dwelling within the nest cavity temporally within the time period of a single day. In fact, the members of the group out foraging are still part of the whole (see Chap. 4). Group size influences the rate at which food resource patches are discovered. Time to discover a food patch decreases as group size increases (Pulliam and Caraco 1984). As resources become more transitory, the advantage of social cooperation becomes greater and the portion of a patch harvested by a single individual forager becomes less sensitive to the reduction in per capita food availability brought about by increases in foraging group size (Pulliam and Millikan 1982). When food patches last long enough to be consumed completely, the presence of additional group members decreases per capita food consumption. If, on the other hand, the patches disappear in less time than normally is required for the food in the patch to be completely consumed, the presence of additional group members has little effect on per capita consumption. The rapidly disappearing patches increase the importance of reduced search times.

5.3.1.1 Environmental Determinants of Group Size

Spatial and temporal distributions of resources influence greatly the group sizes of superorganisms. Most colonies of honeybees in the higher latitudes have an annual cycle in their populations from a low of about 10000–12000 bees near the end of the non-foraging season to a high of about 20000–40000 bees or more in the peak foraging season when swarming takes place (Nolan 1925; Jeffree 1955; reviewed by Seeley 1985; Fig. 6.6). Colonies with fewer than about 16000 workers entering winter conditions are unlikely to survive extended cold winters of the northern climate. Colonies in the warm season in these areas show rapid spring growth that is tightly coupled with the availability of protein (pollen) resources. The colony population size follows an annual cycle but seems to be keyed to photoperiod. Day-length influences population size, probably through a set of interrelated events including time available for foraging and the resulting influx of food for storage, and effects on the queen laying rates. Kefuss (1978) showed that the increasing photoperiod in late winter and very early spring influences brood-rearing activities.

In tropical environments, honeybee colonies are limited in their foraging time not by cold weather but by the onset of the wet and dry seasons. The population of colonies in an area fluctuates according to the availability of floral resources. In Panama there is a distinct dry season from January through April, followed by a wet season during which nearly all the precipitation occurs. Populations of colonies increased and peaked regularly each year (Boreham and Roubik 1987; Roubik and Boreham 1990). The annual peaks were likely to be reproductive swarms because they occurred during times of floral abundance and peak colony pollen harvest. On the Pacific side of the isthmus, cumulative monthly totals show an annual peak in numbers of colonies during the dry season (January to April), and on the Atlantic coast, two annual peaks, one in the dry season and another early in the rainy season (June through July). The peak numbers of swarms and colonies correspond directly with availability of floral food resources, as recorded on Barro Colorado Island by Croat (1978). The floral phenology is tightly linked with the precipitation with great plant growth commencing with the onset of the rainy season. Later in the rainy season and early in the dry season flowering occurs. It is only after the onset of flowering with its concomitant supply of food resources that proliferation of honeybee colonies can take place. Borham and Roubik (1987) suggested that the second rainy season peak in swarming on the Atlantic side was caused by colony absconding due to regular flooding of low nest sites, and reproductive increases from exploitation of renewed early floral resources. The onset of heavy rains in May triggers the flowering of several principal honeybee pollen sources on the Atlantic coast. The lack of the second colony peak on the Pacific coast may be due to less flooding and to the more extensive deforestation in that region.

5.3.1.2 Intrinsic Determinants of Group Size

In the Panama study by Boreham and Roubik (1987), a number of swarms consisting of more than 50 000 Africanized bees were found early in the rainy season on the Atlantic side of the isthmus of Panama. The average reproductive swarms are generally 10 000 to 15 000 bees. In Africa, a large number of small colonies sometimes combine forming very large swarms containing up to 1 million bees and with as many as 500 mated queens (Silberrad 1976; Hepburn and Whiffler 1988). The megaswarms in Panama may well have been the result of similar coalescence of a number of smaller swarms. Although no examinations were made for multiple queens in these megaswarms, there were many small swarms normally seen in the early rainy season.

5.3.2 *Fitness and Group Size*

5.3.2.1 r and K Selection

MacArthur and Wilson (1967) found that the fitness of a population of organisms is dependent on the stage of colonization of a geographical area and population growth forms. In newly available areas, those species with high rates of reproduction and growth are more likely to survive. In late successional stages, those species with lower growth potential and with better facility for competitive survival will have the advantage. For bee colonies inhabiting a new area, their rapid growth and swarming provide material for "r selection", yet their long-term continuance under better equilibrium conditions is insured by their increased competitiveness with endemic species that are solitary. In the former case, r-selection is promoted because of the rapid increase within the colony which then throws off reproductive swarms to occupy new home sites. In the latter case, their superorganismic mode of living gives them an advantage under "K selection" conditions. The population of colonies in any area is determined by resource availability. With adequate resources during the foraging season, the doubling time may be considerably less than a year, even down to a few weeks (with up to eight reproductive swarms leaving a parent colony in one season). This compares to the doubling times seen in some non-colonial insects such as the rice weevil (*Calandra*), which doubles its population in 6 weeks, or small mammals like voles (*Microtus*) which have a doubling time of 8 weeks (Odum 1971). If the colonial population is expanding rapidly, there will be a large number of young colonies in the area. This has been the scenario at the forefront of the expanding population of Africanized honeybees as it moved through northern South America and Central America into northern Mexico (Taylor 1977, 1985; Roubik 1989). Once well established, the age structure of the colony population may be markedly different, with most niches occupied by sessile "old" colonies.

5.3.2.2 Queen Fecundity

The size of the colonial group is ultimately determined by the fecundity of the laying queen. Even with adequate supplies of food stores, the colony cannot grow faster than that permitted by the rate of queen ova production. When a queen's rate of egg laying is limited for some reason, the workers will raise a new queen to replace her (supersedure), thereby attempting to insure the maintenance of an optimum size colony with adequate foragers and house bees. Thus, colony fitness is determined in part by group size and by the queen fecundity.

5.3.2.3 Periods of Dearth

When foraging is not possible and the honeybee colony must rely on food stores, its size is important in at least two ways. In the long term, the size must not be too great as to too rapidly consume the stores collected prior to the season of dearth. Yet, it must be large enough to have provided a large enough forager force to obtain the necessary quantity of food during foraging to supply adequate stores for the period of dearth. In the shorter-term scenario, the colony size must be adequate to provide the adequate thermal properties to prevent excessive heat loss and concomitant heat production and utilization of honey stores, at least when exposed to the northern temperate winters. Southwick (1988, 1985a,c) showed that there is a strong dependency of heat production, heat retention, and ability to maintain high core temperature on winter cluster size (numbers of bees). Mass-specific heat production decreases dramatically with increasing colony size. Furthermore, he found that when clusters over a wide range of sizes are tested at a moderate temperature (15 °C) and a more demanding low temperature (2 °C), the metabolic cost curve rotates clockwise about a cluster size of about 16 000 bees (1.67 kg). That means that although it costs more metabolically for small clusters to thermoregulate at colder temperatures, it actually cost less (per unit mass or per bee) for larger colonies to do so. Therefore, it would be advantageous for bees overwintering at low environmental temperatures (ca. <5 °C) to establish large colonies ($>16\,000$ bees) entering the period of hive confinement. In central Germany, where the average daily temperature in the coldest month is about 2 °C, managed colonies overwinter at about 1.6 kg (Koeniger, pers. comm.). This must be more than coincidence.

5.3.2.4 Defence and Group Size

Defensive behaviour is related to group size, with larger groups spending more effort on defence after disturbance than small groups. In a study of defensive behaviour of honeybees, Southwick and Moritz (1985) found a group effect such that as the number of bees increased, the defensive response (as measured by a metabolic test) increased PER BEE. The rate of increased response was greatest in the smallest groups (1 to 100 bees) and became uniform for larger numbers of bees in the group (up to 10000 bees). The defensive response was elicited in these tests by a timed exposure to a precisely measured aliquot of the alarm pheromone component, isopentyl acetate. They concluded that these results may show why individual bees, when separated from the group, as when they are foraging in the field, are unlikely to sting. The same individual bees, however, when they are a part of the large group are quite likely to sting. This evidence gives support to the idea that the bees are quite effective in protecting the colony unit against predation, but are poor at preventing predation on isolated individuals. Very large groups can be effective at repelling a would-be predator at little cost to the colony. Thus sociality is, at least in part, a response to predator pressure.

5.3.3 Optimization Model of Group Size

The dynamics of group size in organisms have been modelled using descriptive population biology (e.g. the Birth-Immigration-Death-Emigration process, BIDE; Pulliam and Caraco 1984). Such models have limited applicability to superorganisms such as colonies of honeybees. However, these authors go further into fitness consequences of group size. One aspect of group size noted in a previous section is the effect of larger colonies having greater rates of discovery and exploitation of patches of food than smaller colonies. Also noted was the dependence of colony group size on overwintering potential. In a sense, it seems that bigger is better, but as Odum (1971, p. 209) noted, "*bee or termite colonies, can get too big for their own good*". The greater the number of individuals in the group, the greater are the benefits, up to a point. There comes a point at which even greater numbers become detrimental. In the active season, Naumann et al. (1991) note that swarming may be the result in honeybee colonies that have grown too large for the adequate dissemination of queen substance to all members, resulting in the production of queen cells. This implies some optimum size, as illustrated in Fig. 5.9. This is basically a reiteration of Allee's principle (Allee et al. 1949) that the number of individuals in a group that results in optimum population growth and survival varies with conditions such that there can be undercrowding or overcrowding densities that may be limiting. In the active foraging season, a very large colony within the limits of queen chemical dispersion would be adaptive; while in

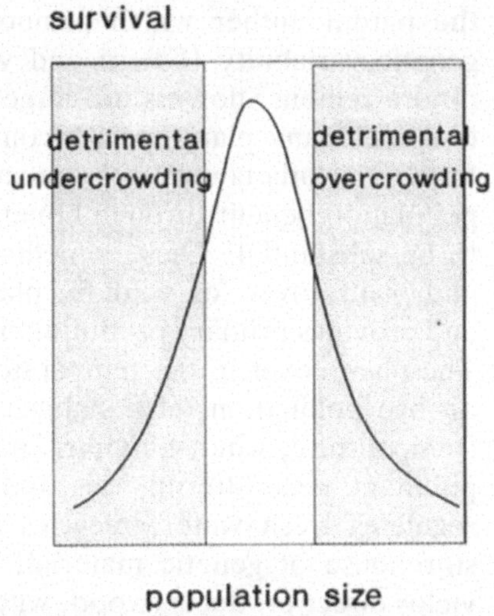

Fig. 5.9. Optimum population size. This demonstrates Allee's principle (Allee et al. 1949) that the number of individuals in a group that results in optimum population growth and survival varies with conditions such that there can be undercrowding or overcrowding densities that may be limiting

non-foraging periods, large colonies would also be adaptive, at least in northern regions, limited only by food stores. We know of no attempts to determine optimal colony sizes for any race or species of eusocial insect, but this area promises rewarding research.

5.4 Ecological Impact of Superorganismic Honeybees

5.4.1 Natural Areas

Of all the land area in the world, about 29% is forested (6% is tropical rain-forests where most species of animals and plants live), 25% is meadows and pastures, 10% is cropland, 34% is deserts, rock/tundra and marshes, and 2% is protected wilderness, wildlife refuges, parks, highways and urban districts (Wilson 1990; Davis 1991; Phillips 1991). This land supports thousands of species of flowering plants which require cross-pollination, mostly by social bees. The main advantage for entomophilous flowers over those that are self-fertile is the possible pollen dispersal over vast distances from the parent anther which promotes outcrossing, thereby increasing genetic variability (Faegri and van der Piji 1979). In the northern tundra regions, flowers are large and showy yet pollinator density is low and many plant species compensate by being facultatively self-fertile, yet others must rely on social bees (Barbour et al. 1976). The pollination benefit through honeybees for natural ecosystems is likely to be substantial. These benefits include fruit and seed set for food and plant cover for wildlife, plant cover that also reduces erosion, and provides primary prodution injecting oxygen into the atmosphere. The plant cover in the temperate regions, which is highly dependent on bee pollination, also stabilizes the biosphere, moderates climate (temperature, wind, humidity, impinging solar radiation), serves as a pollutant sink filtering out and detoxifying poisonous substances, regulates fresh water, recycles nutrients, decomposes wastes, is a storehouse of genetic material, provides for sport hunting/wildlife, yields direct products (wood, wild berries and nuts), provides secondary pollination resources in the vicinity of agricultural crops (with the ever-increasing loss of habitat, this is most important for survival of wild colonies of bees), and provides aesthetic value.

Pesticides have taken a toll and reduced the natural pollination fauna. Honeybee colonies have partly counteracted this loss. In some regions of the world, where so much natural habitat of wild bee species has been destroyed, honeybees are imported for effective pollination of agricultural fields. In fact, their competitive edge is so great that they may crowd out other bee species (Pleasants 1980, 1981; Roubik 1978, Schaffer et al. 1983). Foraging honey bees have even been observed taking pollen from the bodies of bumblebees (Plowright and Laverty 1984). Competition among nectar feeders, resource partitioning, and community structure are all affected by introduced honeybees.

5.4.2 Agricultural Land

Hundreds of species of agricultural plants in over 40 plant families worldwide are pollinated, at least in part, by bees and other insects (Crane and Walker 1984; Free 1970; McGregor 1976; Southwick and Southwick 1989a,b). Many agricultural crop plants require outcrossing to produce viable fruit, and others show hybrid vigour with the production of a better fruit crop. The yields of many fruit, vegetable, seed, and nut crops would drop substantially without the pollination by honeybees (Morse 1988; Olmstead and Wooten 1987).

It has been estimated that honeybees alone account for as much as 80% or more of all crop insect pollination (Camazine and Morse 1988). The demand for pollination services is increasing with the move toward the large monoculture form of agriculture and the rising demand for hybrid seed production (Free 1970; Robinson et al. 1989). Without the contribution of honeybees, many popular crop products would be missing from the consumer market. Honeybees are now distributed worldwide because of their cosmopolitan foraging behaviour, adaptability to wide ranges of environments (primarily because of their colonial life style), and migratory activities of amateur and professional beekeepers (Ruttner 1988).

Chapter 6 Reproduction

If we admire the truly wonderful power of scent by which the males of many insects find their females, can we admire the production for this single purpose of thousands of drones, which are utterly useless to the community for any other end, and which are ultimately slaughtered by their industrious sterile sisters? (Darwin 1859)

6.1 Individual Reproduction

6.1.1 Sexual Reproduction

Reproduction of honeybee colonies can be partitioned into two distinct categories: individual and colonial reproduction. Both are equally important and very specific in honeybees and will be discussed separately in this section.

6.1.1.1 Sex Determination and Sex Ratios

The determination of sex forms the basis for any sexual reproduction. Sex determination in honeybees is based on a haplo-diplo system, typical for all Hymenoptera (Crozier 1975). The male sex results from unfertilized eggs. The queen has the option whether to release some sperms during oviposition and produce a fertilized female egg or to lay an unfertilized male egg (see Chap. 7). At first glance it may appear as if the queen has complete control over sex determination. If we study the mechanism of queen egg laying more carefully this becomes less obvious. Koeniger (1970) showed that, before oviposition, the queen determines the size of the cell with her front legs. Whenever she encounters a wide diameter drone cell she deposits an unfertilized drone egg. If it is a smaller worker or a queen cell, then she will lay a fertilized egg. Obviously, the cell size is an important cue to sex determination. The queen, however, has no means to change or interfere in cell construction. It is solely the workers that construct the combs and they limit drone cell construction to an average of about 10% and a maximum of 35% of the total comb area (Weiss 1962; Free 1967; Taber and Owens 1970; Seeley and Morse 1976; Lee and Winston 1985). Therefore, the workers have considerable influence on the maximum number of drones possibly produced in the colony. Furthermore, the workers must prepare the cells before they can be utilized by the queen for oviposition. They intensively clean and polish each cell before an egg is laid in the cell. The workers are in actual control and the queen has only the choice of a veto. She can refuse to lay in drone or worker cells produced by the worker caste but she cannot actively increase the number of drones above the constraints set by the colony.

Besides the construction of drone, worker, and queen cells, the workers have an even more powerful tool to control the sex ratio in the colony. After oviposition, it is a long way of development for the egg to become a fertile sexual reproductive, and workers performing brood care have very precise control over the allocation of resources on the various sexes. In honeybees, it seems that the worker force is physically in control over the actual production of sexual reproductives. This becomes obvious because size and the nutritional status of the colony affects the regulation of drone production (Free and Williams 1975). Small colonies with poor pollen and honey supplies will severely cut back on their drone production (Velthuis 1990).

The problem of who is in control of regulating the allocation of colonial and individual resources in each sex, has been discussed already in Chapter 2. The queen should favour a 1:1 ratio, as classically predicted by Fisher (1930), but the workers should prefer a 3:1 (female:male) ratio (Trivers and Hare 1976) because they are much more closely related to sisters ($G = 0.75$) than to brothers ($G = 0.25$, see Chap. 7 for a detailed definition of G). Although the theory may be strong for several monogynous monandrous ant species, there seem to be some important constraints in the case of honeybees. First polyandry weakens the arguments of Trivers and Hare (1976) and the relationship asymmetry will drastically decrease from 3:1 approaching 1:1 depending on the number of drones that mated a queen. The more males that mate with a queen, the smaller the relationship among the female offspring. If there is no kin recognition, the optimal worker investment ratio, sr, in a polyandrous mating system can be calculated in an outbred population in a uniform environment according to Charnow (1978) with

$$sr = \frac{(z + 2)/(2z + 2)}{z/(2z + 2)} \tag{6.1}$$

where z = number of drones inseminating a queen.

This simplifies in an investment ratio of (z + 2)/z females : males. This investment bias is derived by simply taking the ratio of the average pedigree coefficient of relationship, G (see Chapter 7 for a definition), of the workers to their sisters, and their "brother" drones (Moritz 1985b).

Obviously, the intracolonial relationship asymmetry and the potential conflict between queen and worker over the investment decreases with increasing number of matings (Moritz 1985b) but is certainly not zero. If there is kin recognition among patrilines, there is actually an additional source of conflict not only between queen and workers but also among patrilines. Craig (1980) suggested that the actual colony strategy might reflect a "compromise" between queen and workers by invoking selection at the colony level. Indeed, it seems we may leave out important aspects if we try to explain the investment ratio in honeybee colonies exclusively on the basis of individual selection.

A first approach to include colony-level selection into sex ratio theory has recently been proposed by Boomsma and Grafen (1990, 1991). They focus on the obvious bimodal distribution of sex investment ratios among colonies in various ant species (Pamilo and Rosengren 1983; Nonacs 1986). They found colonies that predominantly reared males, whereas others specialized in females. Taylor (1981) concluded that queenright honeybee colonies should be selected to compensate for the overall male bias in the population as a result of frequency-dependent selection in the population. Grafen (1986) found that selection can occur at the level of separate colonies if there is a difference in the cost of producing an individual of either sex between colonies. A nagging constraint in this model is that the workers needed some sort of environmental cue to determine the actual cost for producing a sexual reproductive. The theory was extended by Boomsma and Grafen (1990) when they argued that variation in relationship asymmetries among colonies could select for bimodal sex ratios. Again, the workers need to be able to evaluate and capitalize on the intracolonial relationship. Their prediction is that colonies with low relationship asymmetries should specialize in males, whereas colonies with high asymmetries should focus on female production. The model of Boomsma and Grafen (1991) can be easily extended to any higher degree of polyandry and also to the honeybee system. Becuase of the extreme polyandry in honeybees both the average relationship asymmetries in colonies and the between colony variance are low. The prediction would therefore be that all queenright colonies invest similarly in each sex, close to a 1:1 ratio. Yet, some constraints have to be considered when applying the model to honeybee populations. The assessment of the relationship asymmetry by the individual workers may be difficult, because even with cumulative information, errors will occur since the information of each additional nest member is not independent (Ratnieks 1990b). Therefore, a reasonably large diversity of discrimination cues is necessary, which has not been found in honeybees so far (Moritz 1991). Furthermore, kin recognition in combination with nepotism tends to destroy the model (Boomsma and Grafen 1990), and though it seems unlikely that nepotism at the worker level is significant in honeybees, this can definitely not be excluded for worker–queen interactions (see Sect. 8.2.2). A final complication arises from the swarming behaviour of honeybees which imposes additional constraints on the model.

Another approach to explain sex ratio selection on the colony level has been suggested by Frank (1987). He studied the effect of Local Mate Competition (males disperse less than females, Hamilton 1967) and Local Resource Competition (females disperse less than males, Clark 1978) on optimal sex ratio strategies. Frank's model predicts that even a small imbalance between local mate and resource competition causes optimal investment ratios selected at the colony level. The rationale is that large colonies lose more inclusive fitness due to competition between related sexual reproductive offspring than do small colonies. As a consequence, the larger colony should invest more in the competing sex, whereas the smaller colonies should do the opposite. This prediction nicely agrees with the situation in honeybees, where indeed small colonies tend to neglect drone rearing and invest exclusively in females. Large colonies produce large amounts of drones (Free and Williams 1975) but the investment in females is not necessarily proportionally larger. Only a few queens are reared each season in both strong and weak colonies. As we will see later in this chapter (Sect. 6.1.1.3), drones disperse further than queens and indeed Frank's model predicts that under the local resource competition (males disperse better than females) scenario the average colony sex ratio should become more male-biased at increasing colony productivity.

All sex ratio theories for honeybees are hampered by a cumbersome lack of empirical data. Numerical sex ratio data are readily available (Weiss 1962) yet precise investment ratios in both sexes have never been determined in accurate quantitative analyses in honeybees. In fact, it seems almost impossible to dissect colonial investment into drones and queens. The main problem is how to allocate the resources of the parent nest (bees, combs, stores, etc.) after swarming. If this is treated exclusively as an investment in a new queen, Macevicz (1979) determined investment ratios of 1:15.8 (male:female). But is it realistic to count the parent nest solely as investment in the new queen? This approach seems a little oversimplified, because how can we discriminate between parent nest resources allocated in drones and those allocated to the new queen? If only 1% of all individuals in a swarm is drones (Avitabile and Kasinkas 1977) then the majority of the drones is still in the parent nest after swarming. Drones are unable to forage for themselves and only stay fertile if they are well attended by the workers. How can we determine how much of the worker force is allocated to drone rearing and feeding? It does not seem realistic to count all resources of the parent nest just as investment into the new queen. Colony demography is far too complex to work with such oversimplifications even as a rough base line. If the combs of the parent colony are not considered as investment in both queen and drones, then the figures of Macevicz (1979) reduce to 1:2.6. This is still a strongly female-biased figure because Macevicz counts the worker force that remains in the parent colony as plain investment in the new queen. What about the workers that are maintaining the drones in the colony? What sort of investment are the honey stores that the drones feed on? Certainly not all of this can be attributed as investment into the new queen. In light of the large numbers of drones produced (between 5000 to 20000; Weiss 1962; Allen 1965a,b; Page 1981) and the few number of queens (up to 10; Allen 1956; Winston 1980) the concept of a female-biased sex ratio in honeybees looks very counterintuitive under a superorganismic approach. Macevicz's (1979) estimates on Weiss's (1962) data are not helpful in increasing our understanding of investment sex ratios in honeybee colonies. The ideas actually hindered the development of alternative theories on the problem of investment ratios in honeybees since they masked the role of the colony in natural selection processes and led Seeley (1985) to the conclusion that Macevicz's (1979) estimates "... *suggest a female bias in the investment ratio, unlike the sex ratio itself which is vastly male biased*" (Seeley 1985, p. 53).

The problem obviously arose because the view was too restricted on individual reproduction. Fisher's (1930) sex ratio theory is designed for large outbreeding populations of gonochores (individuals with different sex). But how does Fisher's theory fit if we consider the honeybee colony as the major unit of interest and not the individual? What if we just look upon drones and queens as gametes of a hermaphrodite or co-sexual colony? This important question we will readdress later in the chapter. Under the co-sexual colony model we have a problem similar to that of most higher plants. Lloyd (1988) points out that it is extremely difficult to partition investment in co-sexual organisms into male and female contributions. Are so-called accessory investment structures of a flower, like nectar, to be considered as male or female investment? A similar problem is given when the investment of sex in a honeybee colony is to be determined. Lloyd (1988) shows that the resource allocation of hermaphrodite organisms in each sex is determined only by its fitness or reproductive value. This is obvious because the relatedness of either sex to the gametes is equal. When we look at the honeybee colony we find this to be different. Let us consider a randomly chosen allele of a randomly chosen worker of a colony. The probability that this allele descended from the queen is 1/2. The probability that it is from a specific male is 1/2z (z = number of males which inseminated the queen). Offspring drones only carry the queen's genes. Therefore, the average probability that a randomly picked allele in a random individual of the worker force of a colony is also found in an offspring sexual reproductive is

$$r = (1/2) \times p_m + (1/2) \times p_f, \tag{6.2}$$

where p_m = probability of identical mother queen gene in worker,
p_f = probability of identical father drone gene in worker;

for drones we obtain $p_m = 1/2$ and $p_f = 0$, thus,

$$r_d = (1/2) \times (1/2) + (1/2) \times 0 = 0.25.$$

The probability that the same allele is found in a virgin queen is

$$r_q = (1/2) \times (1/2) + (1/z) \times (1/2) = 0.5(0.5 + 1/z)$$

because $p_m = 1/2$ and $p_f = 1/z$.

Obviously, with increasing z, r_q equals r_d and the investment should become equal in both sexes. Given a 17-drone mating, as estimated by Adams et al. (1977), the ratio under this model should be 1:1.23 (male:female). Because the relationship asymmetry is small, the reproductive value of a drone should be of much more significance for the actual numeric biased sex ratios rather than their relationship to the parent colony. If we look at the reproductive value of drones, a plausible adaptive reason for a strongly male-biased sex ratio becomes very obvious. Queens can mate with more than 20 drones, which requires a 20-fold male number to begin with. In the drone congregation area, it is the drones that have to detect, locate, and pursue the queen. There is a great surplus of drones that saturates the drone congregation area before the queens arrive. The abundance of drones minimizes mating times and the probability of queen losses during their nuptial flight (Gary and Martson 1971; Koeniger 1988a; Fig. 6.1). The majority of the drones will not succeed in mating and will be expelled from the colony at the end of the mating season. Obviously, the reproductive value of a drone is dramatically smaller than that of a queen and in this light it seems the colony may be close to a genetic 1:1 investment strategy with its strong phenotypic 1000:1 male bias in the actual numeric sex ratio.

Fig. 6.1. Queen and comet tail of drones on a drone congregation area. (Courtesy of G. Koeniger)

6.1.1.2 The Queen

The queen is the only female sexual reproductive of the colony. Typically, the freshly emerged virgin, if she does not join an after-swarm, screens the colony for other rival queens to kill, and is assisted by the workers in this task (Caron and Greve 1979). The honeybee colony presents us with an extreme case of monogyny (Winston 1987). Only under rare circumstances is more than one queen tolerated in the colony. If after-swarms issue with several virgin queens, monogyny is restored *before* the first nuptial flight. Indeed, this is a very odd and risky strategy that is particularly well expressed in European honeybees. Some African subspecies (e.g. *Apis mellifera lamarckii*) are known to produce many virgin queens that coexist peacefully in the colony (Ruttner 1988). Not until the first virgin returns successfully inseminated from a mating flight are the other queens killed by the workers. The strategy of strict monogyny in the critical phase of swarming, when no operating queen is in the colony, is even more surprising when we look at the risk of a queen in getting lost on a mating flight. Tiesler (1972) showed that only 70%–80% of all queens kept in small mating colonies successfully return from their nuptial flight under fair weather conditions in western Europe. If we transfer this data to full-size colonies (which may be erroneous, but there is no better estimate), this means that about 25% of all colonies become hopelessly queen-less because there is no brood left in the colony to rear emergency queens. The colonies will produce laying workers only and finally die later in the season. Even with a thorough analysis, it remains unclear why the maintenance of monogyny in the colony during the swarming season should be adaptive. Natural selection at the superorganism level should clearly favour a polygynous phase until a newly fertilized queen can take over the colony. This obviously has been realized in *A.m. lamarckii*. However, also under strict individual level selection, it is not very evident why there should be competition among the virgin queens before any of them are even close to a reproductive stage. If a virgin queen dies on her nuptial flight she still gains inclusive fitness through a sister queen in the colony, if she has not killed her beforehand. The lethal queen fighting would only pay if the risk of getting killed after mating is significantly larger than the gain through inclusive fitness. Thus, killing of an inferior queen, with the loss of inclusive fitness may be adaptive if the risk of being killed by this queen after mating is high.

Once the queen has killed all her competitors, she prepares for her mating flight. During the first 5 days after her emergence the glandular activity develops and the production of royal pheromones in various glands is increased (Butler 1960; Butler and Fairey 1964; Butler et al. 1964; Pain 1961). The composition of the fatty acid secretions of the mandibular glands changes in the first days of life from a more worker-like signal to a typical queen like one (Crewe 1982, 1988). Freshly emerged queens have a high proportion of 10-hydroxy-2-decenoic acid in the mandibular gland secretions, which is a typical worker compound. Also, the tergal gland activity increases during these first days of life. They secrete important compounds that are used as mating attractant for drones (Renner and Baumann 1964; Renner and Vierling 1977; Vierling and Renner 1977) and as recognition labels for the workers (Velthuis 1970, 1985; Moritz and Crewe 1988a). Hammann (1957) reported that virgin queens preparing for mating smell strongly aromatic and clearly different from egg-laying queens.

After a few orientation flights (up to 10 min) the queen starts for her nuptial flight 6 to 7 days after emergence. Workers take active part in this event and push and pull the queen around as if they want to chase her out of the hive for mating (Huber 1792). Indeed, queens not attacked by workers do not leave for nuptial flights (Hammann 1957). Another behavioural trait is typical for the queen in this premating phase: she spreads her sting chamber and vibrates her abdomen apparently in order to stop the molesting by the workers (Alber et al. 1955; Hammann 1957). She releases a pheromone with her faeces, which is used during fighting among queens (Page and Erickson 1986a) and stops the agonistic behaviour of workers. Instead, workers begin autogrooming (Post et al. 1986; Page et al. 1987). Older queens apparently do not produce this pheromone, since workers do not show the behavioural changes from faeces of queens that are 2–3 weeks old.

After mating is completed the queen only flies again during swarming and does not perform additional mating flights even if her semen supply is diminishing. She stores the semen in a spermatheca throughout her lifetime and has no possibility to refill it with new semen. Once the semen is in the spermatheca it is thoroughly mixed (Laidlaw and Page 1984; Moritz 1986a) and clustering of semen of individual drones as stressed by Taber (1955) seems to be the exception. Queens with decreasing semen stores produce an increasing amount of drone brood in worker cells and are usually superseded by a new queen that is reared by the colony. In these cases one can sometimes find more than one queen in the colony for that short period.

6.1.1.3 The Drones

The arrhenotokous parthenogenetically produced, haploid male sexual reproductives (drones) are often misnamed as a caste. For example, even in Winston's recent textbook on honeybee biology, the surprised reader can find that: "*the primary differentiation in bee castes is male versus female*" (Winston 1987, p. 64). Clearly we would not maintain such a classification in human populations and place men and women into different castes. Not only because of the long way of women's liberation there are good reasons not to confuse caste and sex! Drones obviously represent the male sex but not a specific caste as defined by Wilson (1971): "*Polymorphism is defined . . . as the coexistence of two or more functional castes within the* **same sex**" (emphasis ours).

Because the only obvious task of drones is plain reproduction, they are sometimes referred to as the male gametes of the colony. This definition sounds actually quite nice under a superorganismic concept and looks suitable at first glance. Male gametes of individual organisms are small, appear in large number and move towards the female gamete which is less mobile. As we will see later, the analogy of drone and gamete only partly holds true and certain modifications have to be made to genetically adopt the gamete definition.

Freshly emerged drones are usually fed by workers and they stay in the warmth of the centre of the brood nest until they are 5 days old (Mindt 1962). They have not been found to participate in any particular social tasks in the colony and their major activities besides feeding and grooming are a general inactivity. When they are older they leave the brood nest and are found on the honeycombs where they feed themselves on the stores but still need the additional care by workers. During this in-hive period of life the semen of the drones migrates from the testes into the seminal vesicles where they undergo their final maturation (Mindt 1962). After a few orientation flights and not before the age of 12 days the drones have mature spermatozoa and take off for their first mating flights (usually more than 25 min; Drescher 1968).

Sexually mature drones fly off in large numbers on warm, sunny days and aggregate in drone congregation areas to mate with a queen (Ruttner and Ruttner 1972). Their flight range is usually between 5 and 6 km (Ruttner and Ruttner 1966; Böttcher 1975). It is still unknown how drones locate such aggregation areas. Because these mating areas are constant over the years, geographic or geophysical properties are believed to be the orientation cues for drones and queens (Ruttner and Ruttner 1966; Praagh et al. 1976; Loper et al. 1988). Only on these mating areas can drones be attracted to queens. As soon as the queen leaves the area she looses attractivity, as could be demonstrated in experiments with tethered queens (Ruttner and Ruttner 1963, 1965; Gerig 1971). The drones are caught within seemingly magic boundaries on the area and until now it is still completely unclear which environmental cues cause the drones to circle in the air at a certain location and wait for the queen.

To take the magic component out of the mating behaviour of drones an attractive "neutral mating model" has been suggested by the group of Taylor (Taylor and Rowell 1988; Rowell and Taylor 1988). They argue that queens fly further for mating than drones, and by this avoid inbreeding under a random distribution of drones with a shorter mean flight distance and a standard deviation. An appealing theory for all those cases in which drones do not aggregate on specific drone congregation areas. Yet, in most cases the drones seem to assemble in a non-random lek-like fashion on congregation areas. Moreover, there is strong evidence that drones mostly fly further than queens (Ruttner and Ruttner 1966, 1972; Böttcher 1975). Drones from different apiaries are found on the same drone congregation area, which also conflicts with the random mating model. The existence of drone congregation areas was also nicely documented with aid of radar analysis. Loper et al. (1988) found that even on a single mating flight drones can switch between congregation areas. The presently available empirical data indicate that assembly of drones on lek-like sites, and not random distribution is the rule for mating in honeybees, but other mechanisms certainly can not be excluded.

Due to the large flight range the honeybee can increase its effective population size substantially. The mating range of drones may easily exceed 110 km^2 and that of queens 30 km^2 (Ruttner and Ruttner 1972). Under experimental conditions Peer (1957) observed mating distances of more than 16 km. In spite of a low density of two colonies per square kilometre in natural populations (Taber 1980; Visscher and Seeley 1982) the number of colonies within the mating range can be close to 300. If we use a conservative estimate of 200 colonies (N_f) within the mating range and an eight-drone-mating with each queen on the average ($N_m = 8N_f$), the effective population size is $N_e = 736.36$ using Wright's (1933) formula for the effective size of male haploid populations (see also Chap. 7).

$$N_e = \frac{9N_fN_m}{2N_f + 4N_m}. \tag{6.3}$$

This agrees well with the theoretically based expectations for honeybee populations of Yokoyama and Nei (1979), ranging from $N_e = 428$ to $N_e = 1093$. Thus, despite the low actual population density, the genetically effective population size is considerably large.

6.1.1.4 Mating Behaviour

Mating of queens and drones is a spectacular event in *Apis mellifera* and has been behaviourally and physiologically analyzed in detail by G. Koeniger (1986, 1988a,b; Fig. 6.1). Prior to copulation the queens are pursued in flight by a comet of hundreds of drones. The drones fly below the queen level focussing her with the dorsal front part of the large compound eye at an angle between 0° and 40° (Gary 1963; van Praagh et al. 1976). Mating takes place during flight 20 m up in the air where the drone grasps a queen with his legs and inserts his endophallus into the vaginal duct. The drone presses his complete haemolymph volume into his endophallus during eversion and dies of shock in fractions of a second. Up to 17 drones (Adams et al. 1977) can mate the queen and they die immediately after mating. The successful, but yet dead drone leaves his mating sign, a part of the endophallus, in the vaginal duct of the queen. The mating sign, however, does not function as a mating plug as has been incorrectly argued by Thornhill and Alcock (1983). A subsequent drone can easily remove the previous mating sign with special structures of his endophallus (Koeniger 1984, 1986). The mating sign actually seems to function opposite to a mating plug that is supposed to prevent further matings. Queens with a mating sign are more attractive to drones than non-mated queens and therefore it substantially promotes multiple mating (Koeniger 1990). Koeniger (1986) suggests that optical cues might be important for the increased attractiveness of queens with mating signs. Queen dummies marked with aluminium foil at the rear end proved to be more attractive than unmarked dummies. This theory is particularly appealing because the mating sign is fluorescent with emission peaks around 320 nm and 460 nm (Koeniger 1987), a wavelength at which drones are extremely sensitive (Autrum and von Zwehl 1962). Furthermore, the mating sign shows bright reflections in ultraviolet wavelengths between 320–360 nm (Koeniger 1990).

Such a function of a mating sign is of course surprising. Under individual selection theory there seem to be no obvious reasons for a drone to make a queen more attractive to other competitors after successful mating. The classical understanding of the mating sign

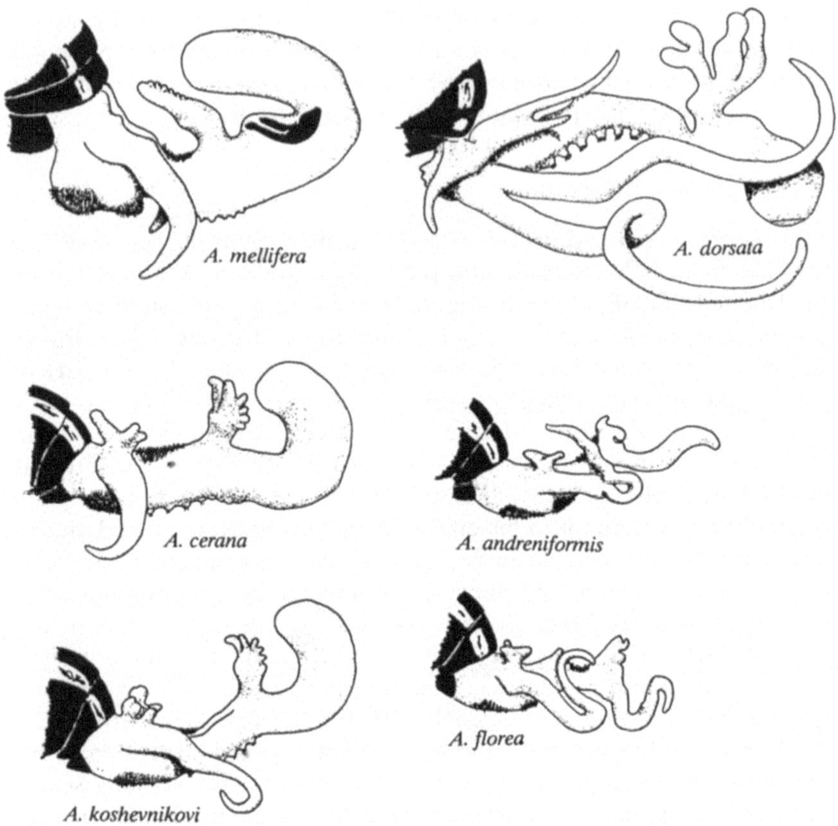

Fig. 6.2. Endophalli of the cavity breeding honeybees *Apis mellifera*, *A. cerana*, *A. koshevnikovi* (*left from top to bottom*) and the free-breeding species *A. dorsata*, *A. andreniformis*, *A. florea*. The fine structured endophallus of the dwarf honeybees (*A. andreniformis*, *A. florea*) has been suggested to deposit the semen in the queen so that it directly enters the spermatheca. (Redrawn and modified after Koeniger 1991)

is that it functions as a mating plug to prevent further matings (Thronhill and Alcock 1983). Clearly, it is in the interest of the male to prevent further matings under a plain individualistic selection concept. The individual drone would be genetically more effective if all of the queen's offspring would also be his. The adaptive value of the increased queen attractiveness must lie somewhere else. The behaviour and mating physiology of the drone has been selected either via the queen or more likely via the colony (Koeniger 1990). In principle, all theories that plausibly explain the evolution of poly-

Table 6.1. Number of Sperms in drones and freshly mated queens, and the number of matings for the currently known honeybee species

Species	Number of sperms × 10^6			Number of matings
	per drone	in oviduct	in spermatheca	
A. mellifera[a]	10–11	<170	4.7	<17
A. cerana[b]	1–2	10–20	1.4	<30
A. koshevnikovi	?	?	?	?
A. florea[c]	0.4	0	0.9–1.4	<4
A. andreniformis[d]	0.1	0	>1	>10
A. dorsata[d]	2.5	?	3.6	>1

[a] According to Woyke (1966).
[b] According to Woyke (1975).
[c] According to Koeniger et al. (1989).
[d] According to Koeniger et al. (1990).

andry (Crozier and Page 1985) also fit the evolution of changing a mating plug into an attractiveness label. This leads us to a concept that indeed in some ancient bee ancestor the mating sign was a plug, but through evolution the function of this plug completely changed, leaving out the individual drone's interests. The selection mechanism may be very simple: if the queen that mates with a single drone produces a colony that is less fit than that of a queen that mates with many drones, then natural selection should favour drones that grant the colony the larger fitness. We will return to this problem in some detail in chapter 8.

Multiple mating may have evolved a different drone adaptation in *Apis florea* and *Apis andreniformis*. In these species the semen is directly injected into the spermatheca of the queen via a specially designed endophallus (Koeniger et al. 1989; Koeniger and Koeniger 1990a,b; Fig. 6.2). There is no "waste" of semen like in *A. mellifera* or *A. cerana* where the semen is deposited in the lateral oviducts and only about 10% migrates into the spermatheca.

6.1.2 Parthenogenesis

6.1.2.1 Production of Drones by the Queen

In addition to sexual reproduction, parthenogenetic reproduction takes place in the honeybee colony. Obviously, the sex determination

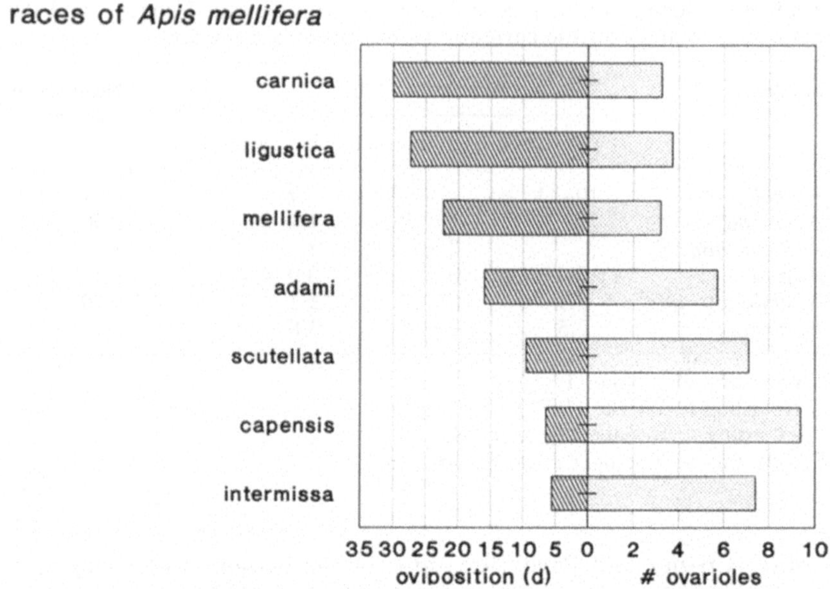

Fig. 6.3. Ovary development and number of ovarioles of laying workers in various races of *Apis mellifera*. (Data from Ruttner and Hesse 1981)

system is based on the combined reproductive strategy of sexual and parthenogenetic reproduction. The unfertilized eggs laid by the queen develop into male drones (see Chap. 7 on genetics), thus reflecting arrhenotokous parthenogenesis. Drone eggs are laid whenever the queen encounters a large drone cell. She measures the width of the cell with her forelegs (Koeniger 1970) and deposits an unfertilized egg. Mistakes may happen if the queen runs out of semen in her spermatheca. She then deposits unfertilized eggs also in worker cells. The drones emerging from these cells are smaller than those developing in regular drone cells. If the gamete of the queen and the drone's sperm have the same sex allele, then fertilized eggs are laid that develop into diploid drones (Woyke 1962, 1963a, 1965). These diploid drones are recognized by the workers, cannibalized in early larval stages and later appear as unsealed brood cells in the otherwise capped brood comb (see Chap. 8).

6.1.2.2 Production of Drones by Workers

Besides drone production of the queen, workers too are able to produce drone offspring parthenogenetically. Typically, laying workers appear whenever a colony becomes queenless and there is no possibility to rear a queen from remaining brood. Page and Erickson (1988) argue that the laying worker phenomenon is even more common. Under queenright conditions they found worker produced drones in 38% of all colonies tested in their study. The number of offspring produced by laying workers can be considerable. Up to 6000 drones can be produced by a single queenless colony in contrast to 20000 by a queenright colony throughout a season (Page and Metcalf 1984; Page and Erickson 1988). This figure is significant at the population level and has an impact on optimal investment sex ratios as pointed out by Page and Metcalf (1984).

There is substantial racial variation for the time of worker ovary development in queenless colonies. European honeybees take 15 to 31 days until the first egg-laying workers appear (Ruttner and Hesse 1981). African workers show a substantially faster ovary development ranging from 5 to 11 days (Fig. 6.3). Also, the number of eggs laid varies substantially among subspecies. *Apis mellifera adami* from Crete, for example, produces almost 30 times as many eggs in 48 h as *A.m. mellifera* from southern France (Ruttner and Hesse 1981). Because worker-laid eggs are haploid in all honeybee subspecies except the Cape honeybee, drones are produced by laying workers. Although laying workers under natural conditions prefer drone comb for their eggs (Page and Erickson 1988), they also oviposit in worker cells. Typically, many eggs are deposited per cell, and a few larvae may simultaneously hatch, but only one will develop to the imaginal stage.

Although the colony will finally die, the adaptive significance of the reproductive strategy of laying workers can be easily understood (in contrast to Robinson et al. 1990). The worker-produced drones are able to transfer genes to the next generation: an obvious fitness advantage compared to colonies that do not produce laying workers and have no sexual offspring at all after the irreplaceable loss of the queen (see Chap. 8). The produced drones of course only will form a selective advantage if they significantly contribute to the gene pool and find virgin queens with which to mate. When queen-produced drones (developed in drone cells and in small worker cells) are in direct competition, there seems to be a reproductive advantage for those developing in drone cells. Berg (1988) showed that large drones

Fig. 6.4. Distribution of thelytoky in the South African honeybee population. (From Hepburn and Crewe 1991)

from drone cells have a longer flight duration than the small drones that developed in worker cells. Sometimes beekeepers argue that worker-produced drones are mainly reared in worker cells and the drones would be insignificant for natural selection. Yet this seems to be an artefact from a lack of drone combs in commercially managed colonies. There are repeated reports that not only the queen but also the workers preferably lay eggs in drone cells rather than worker cells (Free and Williams 1974; Page and Metcalf 1984; Page and Erickson 1988). The significance of the worker-produced drones relates not only to the number of drones and to their size, but also to the timing of drone production. Queens only produce drones in large numbers in the main reproductive season. Therefore, worker-produced drones outside the prime mating season may still have a reasonable chance for mating a queen. Also in off-season, old queens can be superseded by new ones or emergency queens will be reared by the colony because the old one died. In all these cases the worker-produced drone force could be of great significance.

6.1.2.3 Production of Females by Workers

Female offspring are diploid and only develops from fertilized eggs which are usually only laid by the queen. However, in one subspecies

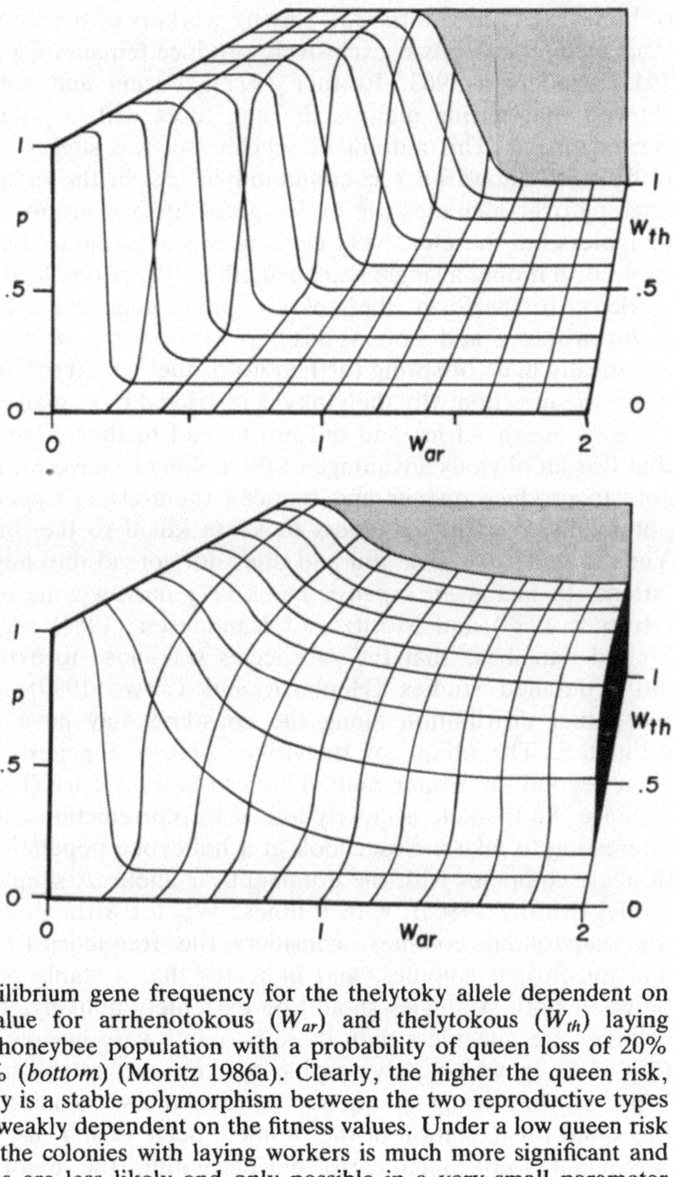

Fig. 6.5. Equilibrium gene frequency for the thelytoky allele dependent on the fitness value for arrhenotokous (W_{ar}) and thelytokous (W_{th}) laying workers in a honeybee population with a probability of queen loss of 20% (*top*) and 80% (*bottom*) (Moritz 1986a). Clearly, the higher the queen risk, the more likely is a stable polymorphism between the two reproductive types which is only weakly dependent on the fitness values. Under a low queen risk the fitness of the colonies with laying workers is much more significant and polymorphisms are less likely and only possible in a very small parameter area

of honeybees this is different. Laying workers of the Cape honeybee, *Apis mellifera capensis*, exclusively produce female offspring (Onions 1912; Anderson 1963; Ruttner 1977). Verma and Ruttner (1983) showed that during meiosis the egg fuses with a polar body in a central fusion. This automictic mechanism was suggested by Tucker (1958) and allows no recombination of loci in the offspring, unless crossing-over mediates the exchange of linkage groups. Mother and daughters are therefore very close to a genetic clone. Ruttner (1988) showed that only a single recessive allele, th, at one locus determines workers to perform thelytokous parthenogenesis. Heterozygous Ar/th workers and homzygous Ar/Ar workers produce parthenogenetically male offspring (arrhenotoky allele = Ar). This, of course, poses the question why thelytoky is restricted to a small habitat in the Cape of South Africa and did not spread further. One might argue that it is an obvious advantage of the colony to have workers that are able to produce queens and requeen themselves. *Capensis* colonies potentially live forever unless they are killed to the "last" worker. Yet the thelytoky gene did and does not spread into adjacent populations. In fact *A.m. capensis* has a very narrow zone of abundance (Ruttner 1977) and Moritz and Kauhausen (1984) suggested on a limited data base that the subspecies was close to extinction. Yet, more detailed studies (Hepburn and Crewe 1990) indicate that the actual distribution along the coastline may have been underestimated. The ability of thelytokous parthenogenesis is abundant even beyond the winter rainfall area in South Africa (Fig. 6.4).

Since the th allele is closely linked to reproduction and fitness it is interesting to take a closer look at a honeybee population where the th allele competes with the dominant Ar allele. A simple population model (Moritz 1986c), with a fitness W_{ar} for arrhentokous and W_{th} for thelytokous colonies, considers the frequencies of queenless and queenright colonies, and indicates that a stable polymorphism between both strategies should be extremely unlikely. If 20% of all colonies are queenless than there is only a very narrow fitness array (Fig. 6.5) in which polymorphic systems can establish themselves. With an increasing frequency of queenless colonies, the fitness threshold for a fixation of the th allele decreases. Thus, obviously, if the queen is not much at risk during mating (the major incident for colonies to become hopelessly queenless), it is advantageous to produce drones. Under a high queen risk thelytoky is the superiour strategy. The unstable climate at the Cape of Good Hope supports this hypothesis: sudden weather changes in the mating season (Schulze

1974) and strong winds up to 13 m/s (Tribe 1983) may cause a high frequency of queen losses during mating.

6.1.3 *Negative Reproduction*

6.1.3.1 Brood Cannibalism

Colony growth is subject to extra and intracolonial environmental conditions. Under certain restrictive circumstances, eggs and brood at all stages can be cannibalized by the workers. Cannibalism of drone brood is particularly common whenever the nutritional conditions of the colony require protein input but pollen cannot be foraged because of poor weather (Weiss 1984). Worker brood is only cannibalized on rare emergency occasions. If, for example, the temperature in the brood nest has dropped substantially below 34 °C for an extended period of time, and the worker brood is damaged, workers clean the cells by removing and eating the larvae: a phenomenon called hygienic behavior that gained particular importance through the work of Rothenbuhler (1964a) on resistance to American foulbrood (*Bacillus larvae*). Oophagy or cannibalism of young larval stages also occurs frequently in colonies with laying workers where multiple larvae are found per cell (Page and Erickson 1988). The workers tolerate multiple larvae in brood cells only until 3 days after eclosion. Older larvae are always found singly in cells. Page et al. (1989a) argue that selective egg/larval cannibalism might occur in connection with kin recognition. However, any empirical evidence in support of this hypothesis (which is only meaningful for the female brood) is lacking. In light of the mode of drone and diseased brood elimination, it seems the bees have other priorities rather than kinship in brood removal.

Cannibalism is obligatory for diploid drone larvae (Woyke 1962, 1967). They are eaten immediately after eclosion from the egg. Woyke (1967) postulated a so-called "eat me" pheromone that is mainly produced on the first day of larval life, and to a lesser extent on the second day. Older larvae are not recognized and are reared to the imaginal stage in the colony. Also in *A. cerana* diploid drone larvae are cannibalized, however at a slightly further stage of development (day 2–3). The nature and composition of this "cannibalism" pheromone unfortunately has not been further analyzed. Woyke

(1967) added diploid larvae to brood cells containing worker larvae which caused the workers to cannibalize both worker and diploid drone larvae. The use of solvent-extracted diploid drone larvae, however, did not cause cannibalism of worker larvae. From this expreiment Woyke concluded that the pheromone should be some compound in the lipid fraction. Indeed lipids and fatty acids have been shown to be important for brood recognition in honeybees (Koeniger and Veith 1983; Le Conte et al. 1989; Moritz et al. 1991). Nevertheless, it seems more plausible that the diploid drones lack a certain substance in their first days of life rather than produce a special compound to be recognized by the workers. Distortion of the "normal" genetic make-up seems more likely to delay normal gene action rather than to induce the production of new physiologically active compounds. Diploid drones could smell just odd, because an important compound is missing, causing exactly the phenomenon observed by Woyke (1967; see also Chap. 7).

6.1.3.2 Worker Policing

Oophagy of worker-laid eggs could also be a phenomenon caused by "worker policing" and the polyandrous mating system of the queen. Workers discriminate between worker- and queen-produced eggs and remove the eggs laid by the workers (Ratnieks and Visscher 1989). Ratnieks (1988) presented an appealing theory why workers should prevent each other from reproducing and form a "drone police", as Seger (1989) puts it. The basis for his ideas are set by Alexander and Borgia (1978) who state that a numerically insignificant "outlaw" worker is unlikely to win against the collective composed colony when there is conflict between the optimal evolutionary strategy of the individual and the colony. Ratnieks (1988) modelled a situation where one putative "police" allele invades a population of non-policers. The police allele, which is only expressed in the workers, can invade either via the queen or via the queen's mates. In the case of the queen route, the probabilities that the allele is passed on to the next generation are 0.5 (for the queen's sons) and 0.25 (for worker-produced drones). If the allele is introduced by a drone, then it is only transmitted to the next generation by worker-produced drones with a probability of 0.5/z (z = number of mates). The queen-produced drones do not carry the allele. If p^2 is considered to be close to zero, then the frequency of the police allele in the males after one generation, p_m^*, can be calculated by the following approximation:

$$p_m^* = 2p_0(0.5)e_1 + 2p_0(0.25)(1 - e_1) + zp_0(0.5/z)(1 - e_2), \quad (6.4)$$

where e_1 = proportion of drones produced by heterozygous queen,
e_2 = proportion of drones produced by workers that received the allele via a drone,
p_0 = initial frequency of police allele.

This condenses to

$$p_m^* = p_0 + 0.5p(e_1 - e_2). \quad (6.5)$$

The police allele only affects the male population and $p_f^* = p_f$. Because the allele frequency can be calculated in male-haploid populations as $(2p_f + p_m)/3$ (Wright 1969), the police allele will increase whenever $e_1 > e_2$. Since the queen produces many more drones than workers (Page and Erickson 1988), this condition is clearly given in honeybees. Ratnieks (1988) found that the allele will go to fixation as soon as the queen mates with more than two drones, and re-invasion of a "non-police" allele is only possible if the queen mates with less than two males on the average. Interestingly, nepotism relaxes the invasion conditions for the "police" allele. This surprising result becomes plausible if we think of the following scenario. When a police allele is present in one patriline, these "police" workers will preferentially prevent the non-police workers from reproduction, thus increasing the probability for police offspring. Ratnieks (1988) points out very clearly that worker policing is unlikely to be a cause for polyandry but certainly would be a plausible result of natural selection in a population of social organisms with polyandrous matings. Policing alone needs some other selective mechanism to cause polyandry.

6.2 Colonial Reproduction

6.2.1 The Seasonal Development of a Colony

6.2.1.1 The Yearly Life Cycle in a Temperate Climate

In addition to the individual reproductive strategies discussed above, the colony as a whole undergoes a reproductive cycle of growth and stasis (Fig. 6.6). The winter phase is certainly characterized by a conservative strategy of the colony with only limited production of brood. At the end of winter, the first brood is produced but still in

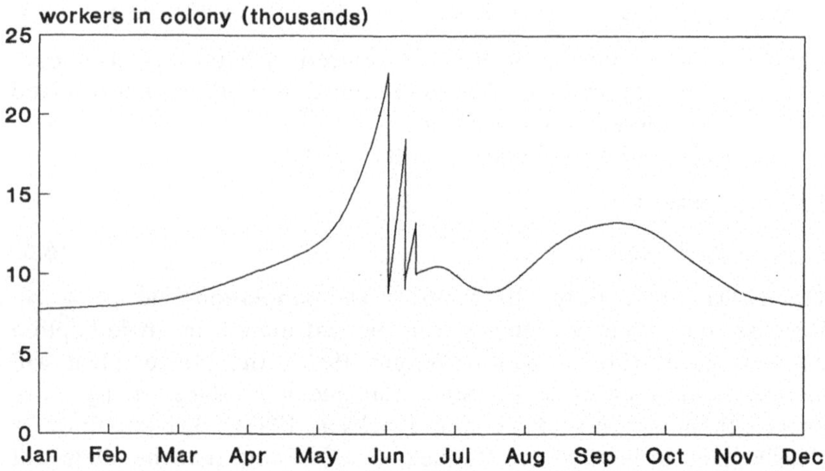

Fig. 6.6. Typical population dynamics of a honeybee colony through a year in temperate climate. The swarming season is characterized by the oscillating worker number in the colony (May–June). (Data of Winston 1989)

small numbers. Day-length seems to be an important parameter in this initial phase of colony growth (Kefuss 1978). Explosive development starts as soon as protein in the form of pollen is abundant and can be foraged. The brood nest then quickly expands and males are produced. With increasing size the colony will prepare for a fission and all its investment is in the production of a swarm. The peak brood production is reached shortly before swarming (Seeley and Visscher 1985) with 30 000 or more sealed brood cells. A few days before swarming the number of brood declines because the old queen stops oviposition some days before swarming. After the swarm has issued, the production of new brood drops sharply. The new queen has to emerge, kill her rivals and get mated before brood production can proceed. Brood production is then high again until the photoperiod shortens and foraging conditions worsen. After swarming the colony switches its strategy and prepares for developing a unit which can survive the harsh winter conditions. The nectar and pollen stores are filled to maximum capacity utilizing even those areas that were previously used for brood production. The wax production and construction of new comb decline. The long-lived winter bees are reared and the drones are expelled from the colony as soon as the favourable conditions of the peak season are over. The colony re-

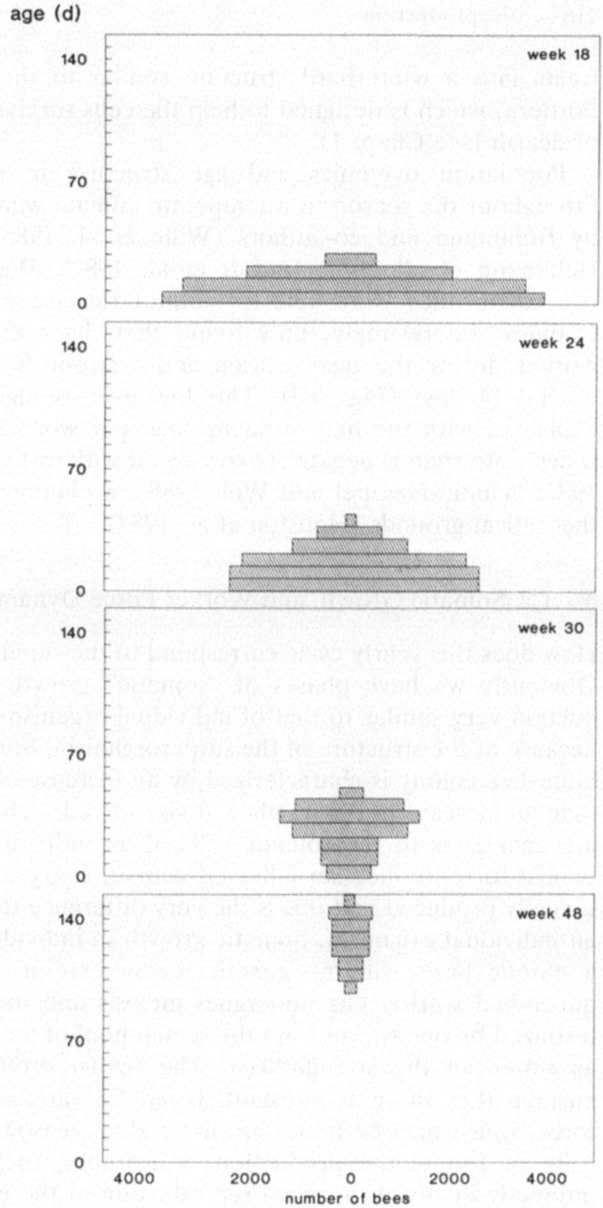

Fig. 6.7. Colony age profiles in several time phases of the year. The width of the *shaded bars* indicates the number of bees in a given age class (intervals in weeks). At the beginning of the season in early spring (*top*: *week 18*) mostly young bees crowd the colony. Notice the low average age of less than 2 weeks. At the end of the year (*bottom*: *week 48*) the colony is characterized by a majority of bees older than 100 days. (Data of Bühlmann 1985)

treats into a winterhard structure similar to the Gemmula of the Porifera, which is designed to help the cells survive extensive periods of dearth (see Chap. 1).

Population dynamics and age structure in honeybee colonies throughout the season in a temperate climate were intensely studied by Bühlmann and co-authors (Wille et al. 1985; Bühlmann 1985; Bühlmann et al. 1987; Imdorf et al. 1987). Based on brood and bee counts they were able to monitor the age profile of honeybee colonies. Interestingly, they found that the average life-span of a worker during the peak season and a strong honey flow does not exceed 14 days (Fig. 6.7). This low average figure may be partly explained with the high foraging load per worker during peak flow times. Life-span is negatively correlated with foraging time (Neukirch 1982; Schmid-Hempel and Wolf 1988): a phenomenon predicted on theoretical grounds (Houston et al. 1988).

6.2.1.2 Somatic Growth and Worker Force Dynamics

How does this yearly cycle correspond to the superorganism concept? Obviously we have phases of "somatic" growth and sexual reproduction very similar to that of individual organisms, but yet different because of the structure of the surperorganism. Somatic growth in the honeybee colony is characterized by an increase of individuals rather than an increase in the number or size of cells. The worker bees thus are analogous to the somatic cells of an individual, which are also unable to reproduce sexually. Of course, every individual worker is sexually produced and this is the very difference to somatic growth in an individual organism. Somatic growth in individuals only works on a mitotic basis with no genetic recombination. In contrast, every queen-laid worker egg undergoes meiosis and recombination and is fertilized by one sperm from the semen pool of many different drones as stored in the spermatheca. The sexual production of workers ensures that there is substantial genetic variability in the worker force, which may be important for various reasons (see Chap. 8). In spite of the sexual production of workers, they clearly are only indirectly involved in sexual reproduction of the colony. If we try to search for specific "gametes" of the colony, the workers are certainly not these because they cannot be inseminated by a drone. Their only way of individual reproduction is asexual via parthenogenesis. Since the workers form the vehicle to ensure the survival of the sexual reproductives, the analogies between colony growth and somatic growth of individuals is apparent. The soma of an individual organism is from an evolutionary point of view just the envelope for the gametes or genes (Dawkins 1976), the colony serves as an envelope of genotypes.

6.2.2 *Sexual Reproduction of Colonies*

6.2.2.1 Production of Gametes and the Zygote

If we can identify somatic sub-units in a superorganism, we also would expect to find gametic subunits and a zygotic structure. Intuitively, the sexual reproductives, both queens and drones, seem to fit this definition. However, does this really hold true after looking closely at the corresponding functions of gametes in individuals?

The gametes form the basis for sexuality. For the production of gametes, at some stage in the life history of an organism, diploid mother cells undergo a meiotic division in which genetic material is reduced in each haploid gamete compared to the diploid somatic cell set. The reduction and recombination of genes in the somatic cells thus seems to be the crucial part. Since we have identified the workers as forming the "soma" of the colony, it is easy to test whether drones and queens fulfil these "gamete" requirements. As we will see in the genetics chapter, individual drone and worker siblings have one-quarter of their genes in common, workers and sister queens one-quarter (half sister) or three-quarters (super sister) respectively. The workers' brother-drones only carry genes from the common mother queen, and the father drone contribution is reduced. However, in contrast to the organismic gamete, the drones genes are eliminated completely and there is no recombination. In workers' sister queens, the father drone contribution is on the average reduced by 1/z (z = number of drones that inseminated the queen) and the mother queen contribution by 1/2. Thus, obviously, the "reduction" takes place when the sexual reproductives are produced, but to a different extent than in true organismic gamete formation. To a limited degree, the recombination prerequisite seems to hold also for both drones and queens. The brother drones result from meiotic processes in the mother queen that allow recombination of grand parental genes. The sister queens undergo both sexual and meiotic recombination. Thus, in brief: the drones and queens carry a limited subsample of genetic material of the colony in order to sexually spread the colony's genes in the population. By this definition, even zygote formation finds its analogy under a superorganismic view. The mated queen holds all genes that will later be found in the "somatic" worker force of the new colony; a clear parallel to the classical zygote definition for individual organisms. Finally, evolutionary theory predicts that there should be only two types of gametes – one as small as possible and one as large as possible. A third, intermediate size of gamete apparently cannot coexist with the other two in an evolutionary equilibrium (reviewed by Stearns 1988). It is hard to judge whether drones are as small as possible and queens are as large as possible. One would not expect a large difference in actual size between them, because biophysical requirements have to be met to result in functional flying organisms. These constraints certainly reduce the plasticity of "gamete" polymorphism. Nevertheless, the production of a single drone seems to be not very "expensive", because they are produced in large numbers. In contrast, only very few queens are reared by the colony. The sex ratio nicely fits the micro – macro gamete requirement (in terms of production costs and not actual size) in a superorganism analogy. We do not want to state that sexuality has evolved independently for superorganisms, but we want to point out that the rules and constraints that are valid for sexuality and gamete production in individual organisms also hold true for superorganisms.

6.2.2.2 The Sex of a Colony

Since gametes and zygotes have been analogized to drones and virgin queens respectively, one might think of the type of sexuality that honeybee population maintain at the superorganismic level. Clearly, we can distinguish between male (micro) and female (macro) gametes and therefore have *anisogamy*. The microgametes are cheap and can be produced in abundance, which perfectly fits the drones. Once in the drone congregation area, they are the ones to locate and approach the female "gamete". Typically, the macrogametes produce zygotes that rapidly grow and survive well. A definition that, with some limitations, fits the newly mated queen building up a new colony (see Fig. 6.6). The queen, physically of about the same size as the drone, is only "macro" if the workers are counted as investment in the queen. The analogy is more obvious if we look at the mating of *Scaptotrigona postica*. Here the drones, in the classical definition of male gametes, approach the larger virgin queen, just in front of the nest entrance (Engels and Imperatriz-Fonseca 1990). The drones are obviously attracted by pheromonal cues (Engels 1988; Engels and

Engels 1988) and may remain over many days in a tight assembly in front of the colony that contains the virgin queen. The larger virgin female will only very briefly conduct a nuptial flight, because she seems to be only singly mated. After insemination she further gains in size and becomes physiogastric.

Since the colony can produce both types of "gametes" at the same time, they are clearly co-sexual. Like in many plants, they functionally have a self-incompatibility. Queens mated with their brother drones are partly infertile because of the genetic load at the sex locus and the production of diploid drones (Woyke 1962). In nature "selfing" is avoided, because drones and queens of the same colony mostly fly to different drone congregation areas (Ruttner and Ruttner 1972). This may be an elegant adaptation to avoid matings of nest members and inbreeding.

A definite sex of a colony can easily be attributed in those cases when the queen is lost and cannot be replaced. If laying workers take over, the colony may become "male", or, in the case of the Cape honeybee, "female". In the latter case it resumes co-sexuality as soon as a new queen is raised, mated and starts drone production. Velthuis (1990) argued that even queenright colonies have distinct phases of sexuality. Indeed, the timing of the main drone production is well before the rearing of queens and not necessarily synchronized with the production of new queens. In fact, it seems that the honeybee colony in spite of its vastly numeric bias for female nest members, is male (in terms of genetic effect on the population) most of the time during the foraging season. Drones can be found throughout the season, whereas virgin queens are restricted to the brief swarming season, with the exception of new queens superseding the old one.

As we have seen, there are many analogies between individual sexuality and sexual processes on the superorganismic level. Within limits, colonies may even have a sex. But can we conclude that colonies also mate with each other? From acknowledging drones and queens as gametes of the colony this might look acceptable. However, on a closer view, it seems this statement does not hold. The problem arises from the queen mating with a large number of drones from different colonies. The "female gamete" unites with a large number of male gametes from very different superorganisms. The situation is similar to the sexuality of a cherry tree with self-incompatibility. Here too, pollen of very many individuals can be transferred to the female gametes produced by the tree. The female organism somehow "mates" with a large subset of the male population instead of with a specific individual. The same holds true for the Porifera example discussed in Chapter 1. The male gametes of a wide collection of conspecific individuals are filtered into the ascon cavity where the female gamete cells are awaiting fertilization. Thus, although colonies do not mate with each other, their mode of mating is not an exclusive trait of superorganisms, and can readily be found in individual organisms.

Although we find strong similarities between the honeybee colony and sponges and plant systems here, other reasons than the mating system have caused the German beekeepers to talk about a bee colony as a "Bienenstock", a name derived from the plant breeders' language who use the definition "stock" for perennial plants. Indeed the asexual rather than the sexual processes of honeybee colonies guided them in their definition. As we will see in the next section, this was not necessarily a good idea.

6.2.3 Asexual Reproduction of Colonies

At first glance, the asexual reproduction of honeybee colonies seems to be even more conspicuously related to organismic biology than the sexual mode of reproduction. The analogy between offshoot of a plant and the honeybee swarm seems to be so obvious that it caused the German, Dutch and Nordic beekeepers to use the nomenclature typical for plant breeders rather than animal breeders when talking about bee breeding (Morgenthaler 1951). For example, when they split a colony to make two new ones, they speak of an "Ableger" (German), "aflegger" (Dutch), aflaegger (Danish), avlaegger (Swedish) (= offshoot). But is it really justified to draw this analogy? A closer look at the mode of swarming and its genetic consequences shows that the analogies are far less clear than those for sexual reproductive mechanisms.

6.2.3.1 Swarming Behaviour

Clearly, swarming represents the main reproductive phase (sexual and asexual) in the life cycle of the honeybee colony. The close connection between sexual and asexual processes is obvious because the result of swarming is a virgin queen that needs to be mated. Only the primary swarm represents a sort of asexual reproduction, because the old queen leaves the old nesting site with up to 60% of the colony's adult worker force (Martin 1963; Getz et al. 1982; Winston

1980). It maintains its genetic composition. The remaining new colony (at the old site) undergoes sexual recombination as soon as the new queen is mated.

From the evolutionary and ecological point of view, swarming bears a lot of problems because resources have to be split. How many swarms should the colony produce? How many bees should join the swarm or stay behind? Which bees should join and which remain? Since the worker decisions are very closely linked to the colony fitness, it seems obvious that alleles which caused the workers to resolve the problem incorrectly became quickly extinct. Although the colonial superorganisms may tolerate suboptimal divisions for specific environmental situations, there definitely are thresholds that are critical for colony survival. A good example is given by tropical bees which cannot surivive in a temperate climate because they form too many swarms with an extremely limited number of individuals. These small units are unable to develop to a sufficient winterhard colony within the necessary time scale required in the temperate climate (Southwick et al. 1990).

It is certainly one of the most intriguing problems of honeybee biology to investigate on which information basis individual worker bees becide whether to leave with the queen or to stay with the old nest. The detailed knowledge of swarming mechanisms would not only increase our understanding of honeybee ecology but also greatly help practical bee management. The Africanized bee expansion (see Chaps. 5 and 8) shows how the adaptive swarming behaviour of African honeybees is superior to the maladapted swarming behaviour of the European honeybees in tropical South America.

6.2.3.2 How Many Swarms?

The number of swarms produced is of immediate consequence for the fitness of the honeybee. It basically boils down to the problem of whether to produce few large colonies or many small ones. Theoretically, this is a well-known and often addressed problem in evolutionary ecology (Stearns 1976). Brood size should be such that a maximum number of sexually mature survivors are produced. Thus, the bee colony should produce as many colonies as possible. However, the smaller the swarm size, the smaller the chance of survival under poor conditions (see Chap. 3), and obviously there is somewhere a point at which no more swarms should be produced because they become too small to ever reach maturity. Maturity in the case of bee colonies means to successfully survive a period of

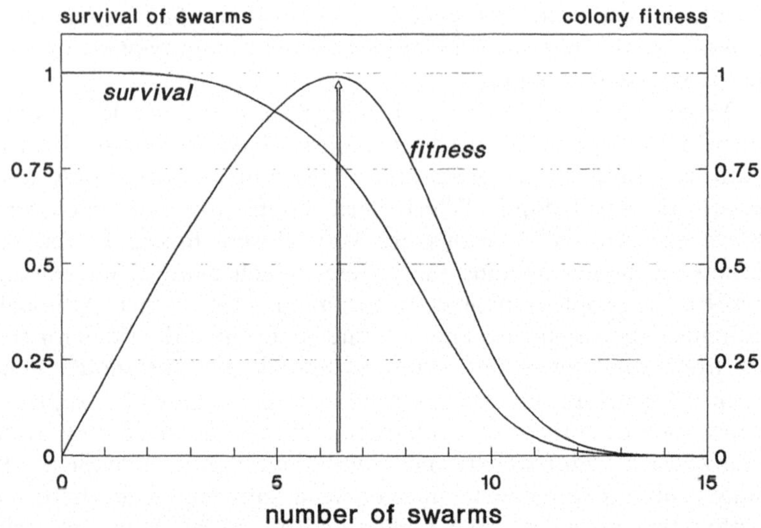

Fig. 6.8. Relationship between the number of offspring swarms produced by a colony and the survival of such an offspring swarm through the first season of life. An arbitrary (but certainly possible) survival curve (*survival*) of swarms produced by a colony of honeybees is shown as a hypothetical example to reveal the relationship between swarm survival and total colony fitness. Clearly, as more swarms are produced they decrease in size and the probability for individual swarm survival declines. The net fitness curve (*fitness*) is calculated from the product of the number of swarms times the individual swarm survival and plotted in relative values of the maximum reproductive output (4.7 surviving swarms = 1). It is proportional to the area under the survival curve. Natural selection should favour a swarm number that maximizes the net colony fitness (*arrow*)

dearth and swarm again in the next season. Colonies in the temperate climate produce a limited number of swarms (usually one or two; Seeley 1978). This ensures that both parent colony and swarm remain large enough to regain a sufficient worker force for survival through the long winter phase. Tropical honeybees have very different swarming tactics. They typically produce many small swarms, up to 17 per season in the well-studied case of the Africanized honeybee in South America (Otis 1982). The importance of a well-adapted swarming behaviour is clear when we look at the dynamics of the Africanized bee process in South America (see Chaps. 5 and 8). European honeybees in South America became virtually extinct within a few years in the feral population, apparently mainly due to their unadaptive swarming behaviour. Figure 6.8 shows how production of swarms and colony survival affect the swarming strategy: the lower the survival rate the smaller is the optimal number of swarms produced to achieve maximum reproductive output.

6.2.3.3 The Primary Swarm: Join the Old or the New Queen?

Using an individual selection model, Bulmer (1983) studied how a honeybee colony should divide its worker force, n, between the old and the new queen. If x_m is the proportion of workers that joins the mother and $x_s = 1 - x_m$ remains in the nest to join their sister, and $W_s(x_s)$ and $W_m(x_m)$ are the probabilities that these colonies survive to maturity, then the workers should segregate such that the following is valid:

$$G_{w,m}W'_m(\hat{x}_m) = G_{w,s}W'_s(\hat{x}_s) = W_n/2, \tag{6.6}$$

where $G_{w,m}$ = relationship worker – mother-queen,
$G_{w,s}$ = relationship worker – sister-queen,
$W'_m(\hat{x}_m)$ = derivative of fitness value when x_m is at equilibrium,
$W'_s(\hat{x}_s)$ = derivative of fitness value when x_s is at equilibrium,
W_n = fitness of undivided superorganism.

Many values of x_m will satisfy Eq. (6.6), yet the strategy that maximizes W_n via an optimal x_m should be favoured by natural selection. The empirical data, however, are still lacking, although it is certainly possible to estimate the fitness parameters in natural populations and use these to compare expected and natural values for x_m.

Another curiosity of primary swarming behaviour is the fate of the old mother queen. In stingless bees, the old fertile and inseminated queen, which has proved its ability to bring a colony to its swarming status, remains in the old nest to be successful for another season (Kerr 1951; Michener 1974; Engels and Imperatriz-Fonseca 1990). The young virgin queens leave the nest to a new nesting cavity, which was previously furbished with nesting material and provisions by workers of the parent nest. In such a swarming procedure, the old queen is never, and the young queen is only very briefly at risk. In honeybees this is quite different. The old queen has to stop oviposition in a very favourable time of the year to reduce her ovaries and to regain her flight ability. The swarm does not immediately move to a new nest but instead initially settles in the open, exposed to wind, rain and other unfavourable weather conditions and mishaps (see Chap. 4). As soon as the swarm has settled there is a complete separation from the old nest. The bees seem to forget the old nesting site as soon as the swarm has established itself around the queen. Only then the majority of the scout-workers starts to search for new nesting sites, although there may be some scouting before the swarm issues (Lindauer 1955). It can take many days until a final decision on the new home is made. In many cases there may be no suitable nesting site available, which in cold climate means the death of the swarm as soon as winter starts. The risk of losing the fertile queen is obvious and from an anthropomorphic point of view it seems a completely absurd strategy. If we move our household from one place to another, we look for a new home first and then load the removal van. The honeybees for as yet unknown reasons take another way: first order the van; move out and forget to load most of the furniture; park the van in the next no-parking zone; hope for the best to get no ticket, and look for an adequate home.

6.2.3.4 The Afterswarms: Join the Super or Half Sister?

To explain worker segregation in afterswarms, another approach was put forward by Getz et al. (1982). They tested if kin recognition interfered with decision making of the workers whether to join the afterswarm or remain in the nest. Queens homozygous for cordovan (cd/cd) were inseminated each with a wild-type (+) and a cordovan drone (cd). In this way, colonies were established in which two patrilines could be phenotypically distinguished (cd/+ or cd/cd). Getz et al. (1982) examined the afterswarms that were produced by two of these colonies and found that the mutant workers preferred to associate with their swarmed supersister queen. They interpreted that kin recognition occurred during swarming and that this was a cue for the segregation of workers between afterswarm and mother colony. Two problems arose: The first is a systematic one that the authors acknowledged themselves. It could have been that cordovan bees originated from stock that had a higher swarming tendency. A reciprocal result with the cordovan bees in the hive and the wild types in the tree would have helped. The second equally serious point is that the experiment was not repeatable (Getz and Winston, unpublished results, cited in Winston 1987).

It remains an open question how the workers behave in the afterswarm situation where their average relationship is equal to both the leaving and the remaining queen. Using Bulmer's (1983) concept, the workers should favour a different ratio than for primary swarms because they are, on the average, more closely related to the mother queen than to sister queens. The scarce data on swarming dynamics do not necessarily indicate such a differentiation in primary and after-swarm strategy. Although afterswarms in Winston's (1980) study had 75% of the number of bees of the primary swarm, the absolute number of bees in the colony during take-off of the after-swarm is less, so that the actual investment split is very similar to that during primary swarming. Winston (1987) suggests that the worker force in the colony at the time of the first after-swarm regained up to 75% of that of the size when the primary swarm issued, because in the meantime further brood emerged (Fig. 6.9).

6.2.3.5 Is Swarming = Colony Budding?

Although beekeepers tend to handle swarming like a budding process, the analogy is only partly true. Budding is strict asexual reproduction in which a new organism is produced by the parent organism. As we have seen above, only primary swarms fulfil the requirement of strict asexuality. They represent a genetic fission in that all genes present in the parent colony are represented in both the swarm and the remaining colony. This, however, is the only "budding" analogy. Budding usually means the production of a small bud from a larger

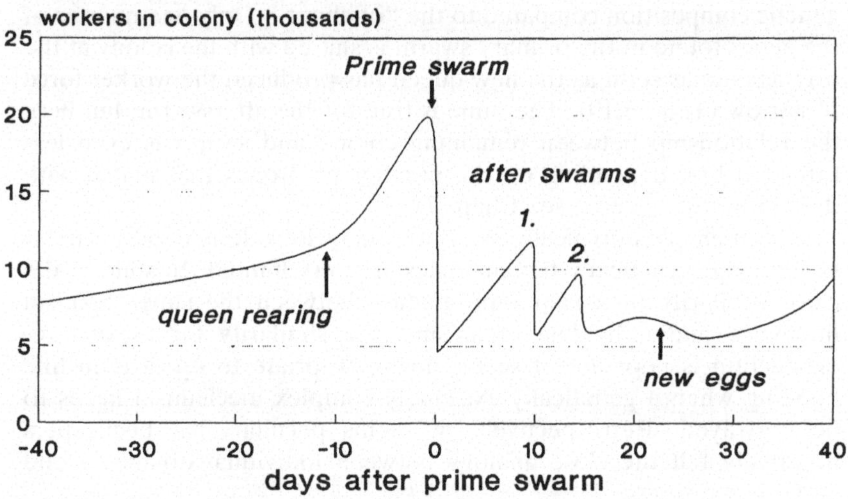

Fig. 6.9. Number of bees in the colony after repeated swarming

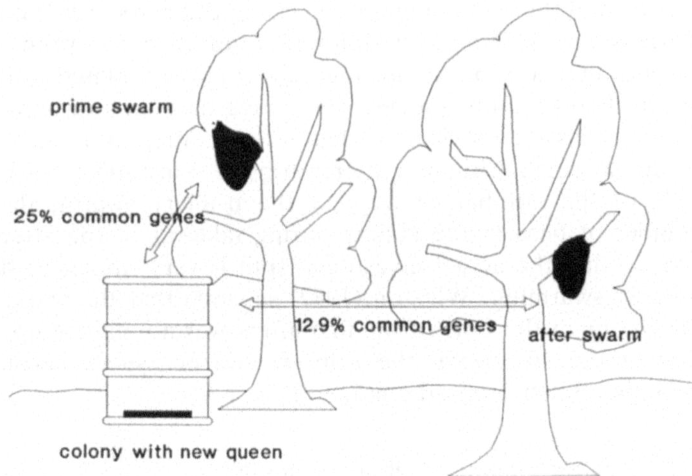

Fig. 6.10. Relationship between the remaining colony with a new queen, prime, and after swarm

parent. Since investments in the leaving swarm and the parent colony are rather similar, this weakens the analogy. Furthermore, in plants and also in our sponge example, the parent organism retains its genetic integrity during budding. This is very different in honeybee swarming. The "parent" organism (in terms of the beekeeper, the remaining colony with the virgin queen) substantially changes its genetic composition compared to the "offshoot". Only one-quarter of the genes found in the primary swarm is shared with the colony at the original site as soon as the new queen has produced the worker force of her own (Fig. 6.10). The same is true for the afterswarm, but here the relationship between remaining colony and swarm is even less (about 0.129, depending on the number of drones that mated with the old mother queen, see Chap. 7).

Although phenotypically swarming may look functionally similar to budding, genetically the analogies are very limited. In spite of the close similarity of sexual reproduction between the superorganism honeybee and individual organisms, the similarity for asexual reproduction is poor and it seems not appropriate to use a term like budding when a genetically extremely complex mechanism needs to be portrayed. Retrospectively, it seems peculiar that beekeepers intuitively felt the close analogy between individual organisms and the superorganism, but actually used one of the few exceptions where the analogy is weak in their nomenclature. The swarm is not an offshoot of a colony, it represents a mix of asexual and sexual reproduction that is only found in superorganisms and is a very special way of reproduction typical for many eusocial insect societies.

Chapter 7 Genetics

7.1 Gene Expression vs Genotype Expression

7.1.1 The Limbs and the Stomach Model

Tempore, quo in homine non, ut nunc, omnia in unum consentientia, sed singulis membris suum cuique consilium, suus sermo fuerit, indignatas reliquas partes sua cura, suo labore ac ministerio ventri omnia quaeri, ventrem in medio quietum nihil aliud quam datis voluptatibus frui; conspirasse inde, ne manus ad os cibum ferrent, nec os acciperet datum nec dentes conficerent. hac ira dum ventrem fame domare vellent, ipsa una membra totumque corpus ad extremam tabem venisse. inde apparuisse ventris quoque haud segne ministerium esse, nec magis ali quam alere eum, reddentem in omnes corporis partes hunc, qou vivimus vigemusque, divisum pariter in venas, maturum confecto cibo sanguinem.

[In the days when man's members were not all in harmony, as is now the case, but had each its own will and language, the limbs thought it unfair to care for the belly, to work for him and provide everything, while the belly was resting in the midst and did nothing else but enjoy the good things bestowed upon it. They therefore conspired that the hands should pass no food to the mouth, nor the mouth accept anything offered, nor the teeth grind anything what they received. While they sought in this angry spirit to starve the stomach to submission, the whole body got weak and miserable. Hence it had become clear that laziness was not the task of the stomach. It was fed, as it nourished the rest, by giving out to all parts of the body that by which we live and thrive, when it has been devided equally amongst the veins and is enriched with digested food – that is the blood.] (Agrippa Menenius in order to convince the plebeians of the need of unity and to achieve their return from mount Aventin to the city of Rome. Cited in Livy, *Ab urbe condita* II 32,9–32,12).

Higher organisms consist of various organs which, through miraculous coordination, allow the organism to survive and reproduce. Coordination is achieved via hormonal and neuronal regulation systems. Although their cells carry the complete genetic code, the organs themselves cannot survive outside the organism, which is the central unit of natural selection. All cells comprising the organism originate from a single zygotic cell and they usually contain the identical sets of genetic information. However, there certainly is variation among cells as far as gene expression is concerned. Without a doubt, genes expressed in an interneuron of the central nervous system are very different from those inducing secretion of digestive enzymes in the midgut. Moreover, gene expressions in one organ can have various interactions with gene expressions in other organs. A most dramatic example of this is complete epistatic effects, in which gene action in one organ suppresses the expression of another gene at a very different location in the organism. Such non-linear interactions between the gene products are often crucial in allowing the organs to function properly and to form an operational unit, the organism. Without the products of genes in limbs, the stomach is quite useless, and, yet, stomachs (unless they grow too fat) inevitably are helpful for limb action.

7.1.2 Intraorganismic Genetic Variation – the Principal Difference

The genetic uniformity of cells of an individual organism is often highlighted as the major difference between individual organisms and superorganisms (Wilson 1971; Starr 1979; Getz 1991). Of course, a superorganism is composed of genetically different distinct organisms, and, consequently, not every gene of the superorganism is expressed in every suborganism, not only because the gene is not active, but because it is not present. Although this is a principal difference, this alone does not seem to be the fundamental divergence between classical individuals and superorganisms. In an individual organism not every gene is expressed in each somatic cell. Phenotypically there is no difference for the cell between a present but not active gene and a non-existent gene. We can expect interactions among genotypes of members of a colony of social insects, similar to the interactions between genes in an intact individual. Also, it is reasonable to assume that the expression of certain genotypes is epistatic over others. A good example may be the regulation of reproduction in a honeybee colony. The queen suppresses the ovary development of workers in the colony, even though the workers have the potential to produce offspring of their own. The crucial point lies in the probabilities for somatic genes to be represented in the reproductive cells. It is here, where we find the major and most important difference between individual organisms and superorganisms.

7.1.3 Gametes – the Functional Difference

Irrespective of whether a cell is in the limb or in the stomach, half of its genes will be represented in any given gamete of the diploid organism. This chance of being represented in a gamete is equal for all genes in each cell. This is the very basis for the understanding of the operation and evolution of organisms. In superorganisms, the picture is completely different. The dominant reproductive and not the sterile caste produces the offspring. This is an obvious conceptual difference between an individual organism and a superorganism. In individual organisms every gene has the same chance to be transmitted from one generation to the next. In superorganisms, some genes (those in reproductives) have a higher probability to be transferred to the next generation than others (those in sterile workers). It is this that often causes biologists to completely reject superorganism models. To us, this difference is one of the very reasons why we call a colony of eusocial insects a superorganism and not an individual.

7.2 Genetic Basis of Honeybees

7.2.1 Male Haploidy

Before we can tackle problems directly related to the genetic peculiarities resulting from superorganisms, we have to define the genetic system of honeybees in detail. Since Dzierzon (1845) we know that male honeybees result from unfertilized eggs. Male haploidy is accepted as a general characteristic for all Hymenoptera (Sanderson 1932). The chromosome number of $n = 16$ ($2n = 32$) determined by Petrunkewitsch (1901) was confirmed by many other studies (Armbruster 1913; Doncaster 1906, 1907; Jegen 1920; Mark and Copeland 1906; Meves 1903, 1907) before and after 1940 (Hachinohe and Oshini 1952; Hoshiba and Kusanagi 1978; Hoshiba 1979; Hoshiba et al. 1981; Hoshiba and Okada 1986; Sanderson and Hall 1948, 1951; Verma et al. 1982). This chromosome number seems to be valid for the *Apis* species *A. mellifera*, *A. cerana*, *A. florea* and *A. dorsata* (Fahrenhorst 1977a,b) in contrast to earlier reports (Deodikar and Thakar 1966; Thakar and Deodikar 1966) which claimed a set of only eight chromosomes for *A. florea* and *A. dorsata*.

The somatic cells of drones do not remain haploid, but undergo various endomitotic duplications during development, and become highly polyploid depending on the tissue. This was first observed in pharyngeal glands (Painter 1945) and found repeated support in subsequent studies (Lerer and Dixon 1975; Merriam and Ris 1954; Sanderson and Hall 1951; Stekolshchikov 1971). Risler (1954) reported that drones have a similar DNA cell content as cells of corresponding female tissue (however, see critique of O'Brien 1956). Mello and Takahashi (1971) found that DNA content in the larval silk gland and Malphigian tubules was even larger in drones than in workers. The increased DNA content could of course also reflect polyteny, the extension of DNA per chromosome, as shown for honeybee epidermis cells by Risler and Romer (1968). The germ line of the drone remains haploid and does not undergo functional meiotic division (Meves 1903, 1907) and the terminology *male haploid* makes perfect sense genetically.

Although there is no functional meiosis, the processes during spermatogenesis are clearly derived from meiotic mechanisms (Sharma et al. 1961). After pro- and metaphase the chromosomes form a close cluster. There is some disagreement about the further processes involved in the rudimentary first division. Some authors report the separation of a small bud of cytoplasm from the cell (Meves 1907; Hachinoe and Oshini 1952; Woyke and Skowronek 1967) whereas others do not (Sanderson and Hall 1948; Wolf 1960). Hoage and Kessel (1968) found in their electron microscope study that the spermatocytes remain connected by intercellular bridges. Smith and Peacock (1957) argued that buds in their squashed sections arose from such bridges, but Sharma et al. (1961) could clearly discriminate between buds and bridges in *Apis cerana*. Therefore, the true mechanism of the first division still remains uncertain, particularly since the modes of spermatogenesis are highly variable within the Hymenoptera (Crozier 1975).

The second division is unanimously described as unequal in spermatogenesis in honeybees (Hachinoe and Oshini 1952; Hoage and Kessel 1968; Kerr and Laidlaw 1956; Kumbkarni 1964; Sharma et al. 1961; Wolf 1960). While most authors assume that only one sperm originates per spermatocyte, others (Mark and Copeland 1906, 1907; Hoage and Kessel 1968) suggest that two sperms are produced per spermatocyte. Anyway, the drone's gametes are genetically exactly identical and this is a feature which will have strong implications for our further considerations about the genetics of the honeybee colony as a superorganism.

7.2.2 *Sex Determination*

7.2.2.1 The Sex Locus

Sex in honeybees is not determined by a chromosomal switch mechanism as in most other insects. Intuitively, one would actually assume that no genetic model could possibly explain the determination of sex in a male haploid population (Bridges 1925). Nevertheless, several theories have been developed which try to explain sex determination on a genetic basis in male haploids (see Crozier 1975 or Woyke and Hillesheim 1989 for a review). Sex in honeybees is apparently determined by a single sex locus with a wide variety of alleles rather than by sex chromosomes. Homozygosity or hemizygosity at this locus results in male organisms. Heterozygous larvae develop into females (queens or workers, depending on the larval diet). This multiple allele single locus model was first discovered in *Bracon hebetor* (Whiting 1939, 1940, 1943) but it also fits for honeybees. It was Mackensen (1951) who found a locus which was viable in hemizygous drones but lethal if homozygous (Fig. 7.1). In inbred colonies, the queen will produce fertilized eggs in which the sex locus becomes homozygous and consequently develop into drones.

7.2.2.2 Diploid Males

The diploid drones which result from fertilized eggs are detected by the workers in the early larval stage and are cannibalized (see Chap. 6). The scattered brood nest, due to the removed diploid drone larvae after hatching from the egg, is typical for inbred colonies (Woyke 1962). In natural mating, it is difficult to identify brood losses due to diploid drones. Because of the polyandrous mating system, the average inbreeding in the population has to be very high before effects become obvious at the level of the brood pattern. This also has implications for the evolution of polyandry in honeybees (Page 1980; see Chap. 8).

The first evidence of diploid tissue in drones was found by Rothenbuhler (1957) and Drescher and Rothenbuhler (1964) in

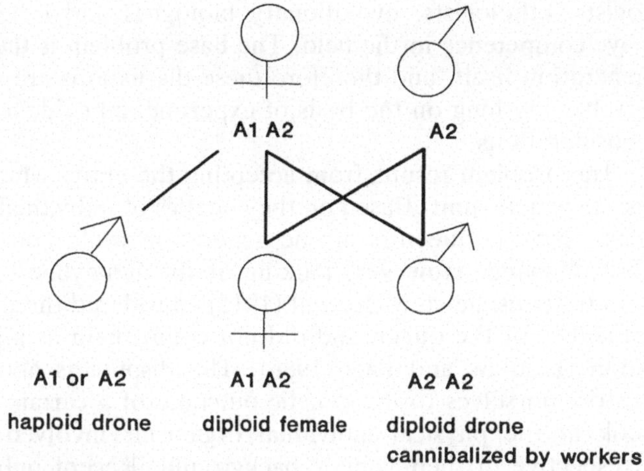

Fig. 7.1. Sex determination in the honeybee

mosaic individuals. The strongest support for the one locus model in sex determination of honeybees was contributed by Woyke (1962, 1963a). He produced diploid drones by rearing larvae in an incubator on royal jelly and transferring them into the colony 3 days after hatching from the cell (Woyke 1969). At that developmental stage the diploid larvae were beyond the critical time of being cannibalized (6–48 h after hatching) and developed into intact male organisms. Using genetic markers, Woyke (1965) could show that indeed diploid biparental drones appeared as expected under the one locus model.

Interestingly, diploid drones have a spermatogenesis similar to that of haploid drones (Woyke and Skowronek 1967). Consequently, diploid drones have diploid sperms. Insemination of queens with diploid drones have not been successful so far and there are no reports of any triploid female offspring, as reported for *Bombus atratus* (Garofalo and Kerr 1975).

7.2.3 The Terminology Problem

7.2.3.1 What Is a Generation?

In diploid populations with distinct individual organisms the definition of a generation seems to be a pointless question. In the genetics of superorganisms, however, the term "generation" can cause fierce debates and discussions among geneticists, sociobiologists, ethologists, evolutionary biologists and others who feel to have competence in the field. The base-problem is the definition of a generation itself, and therefore these discussions are very unlikely to resolve anything on the basis of experimental evidence or theoretical considerations.

The problem results from accepting the entity which is considered as the genetic unit. Based on the genetics of individuals, it seems very clear that the queen is in one generation while her offspring are in the following. However, looking at the honeybee colony from the superorganismic view, Newell (1915) considered the drones as "flying gametes" of the queen and did not count them as a separate generation (Laidlaw and Page 1986). The dispute is about whether we restrict ourselves to the genetic outcome of a certain mating, or just look at the physical individual organisms involved in the mating irrespective of their genetic background. Recent publications in the field of sociogenetics seem to reject the latter (Breed and Page 1989; Laidlaw and Page 1986). From a genetic point of view, it is indeed more meaningful to use definitions which explain the genetics rather than aspects in mating behaviour and reproductive physiology.

To avoid semantic confusion about definitions, we will stick to a system in which only females (queens or workers) can represent a generation, and we will not consider the drones as a generation of its own. This results in a rather complex matrifilial system with overlapping generations in the colony, which is one of the prerequisites for eusociality (Wilson 1975).

7.2.3.2 Who Is Who in a Honeybee Family?

The difficulty in assigning individual organisms to meaningful generations also affects the interpretations of the relationships between the members of a colony (Polhemus et al. 1950). One is tempted to use the terminology from diploid populations to address family relationships in a honeybee colony. So, with no doubt, we could call the queen the mother of all other organisms in the colony and the workers certainly are her daughters. Again the drones pose a problem. Physically they are sons of the queen and brothers to the workers. Yet, from a genetics point of view it seems more correct to call them gametes of the queen to emphasize the different relationship between drone and queen on the one hand, and worker and queen on the other (Fig. 7.2). Also, the relationship between a worker and a drone of the same colony obviously is different than that between two sister workers (Fig. 7.3) particularly in the light of

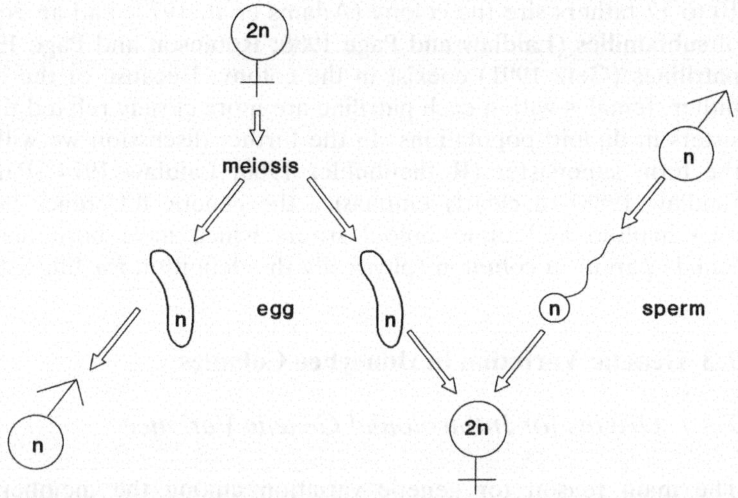

Fig. 7.2. The drones and also the semen they produce are genetically identical to the queens gametes. The only meiotic recombination occurs in the queens gametes

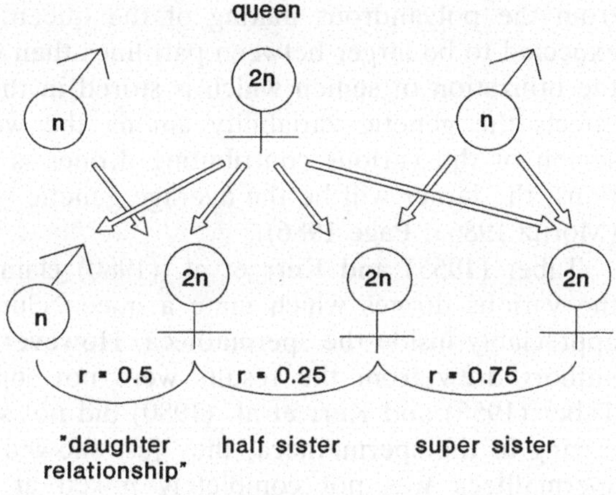

Fig. 7.3. Various relationships in a colony headed by a non-inbred queen with unrelated drones

the polyandrous mating system (see Chap. 8). Laidlaw and Eckert (1950) argued on this basis that genetic brother-sister matings are impossible in honeybees.

Another complication in the genetics of a honeybee colony is the multiple mating of the queen. As a consequence, not one but a set of

10 to 17 fathers sire the colony (Adams et al. 1977) and an equal set of subfamilies (Laidlaw and Page 1986; Robinson and Page 1988) or patrilines (Getz 1991) coexist in the colony. Because of the haploid father, females within each patriline are more closely related than full sisters in diploid populations. In the further discussion we will adopt the term supersister (Rothenbuhler 1960; Laidlaw 1974; Page and Laidlaw 1988) to clearly emphasize the genetic difference between male-haploid and diplo-diploid sisters which have both male and female parent in common (physically the definition for full sisters).

7.3 Genetic Variation in Honeybee Colonies

7.3.1 Origins for Intracolonial Genetic Variance

The main reason for genetic variation among the members of a colony results from the recombination of paternal and maternal chromosomes during meiosis in the queen's gametes. Both drones and workers are equally affected by meiotic variation. The second reason for variation only affects the female offspring since it results from the polyandrous mating of the queen. Genetic variability is expected to be larger between patrilines than within a patriline. Also the utilization of semen which is stored in the queen's spermatheca affects the genetic variability among the workers. The more the semen of the various contributing drones is used at equal proportions, the larger will be the average genetic variability in the colony (Moritz 1986a; Page 1986).

Taber (1955) and Kerr et al. (1980) claimed that the semen of the various drones which mate a queen clump and are not mixed appreciably inside the spermatheca. However, the conclusions these authors drew from the results were not supported by their data. Taber (1955) and Kerr et al. (1980) did not show that there was no mixing in the spermatheca, they just showed that the semen in the spermatheca was not completely mixed at random (Crozier and Brückner 1981; Page and Metcalf 1982; Page 1986). In fact, in most of the queens tested, semen of all types of drones used for insemination was utilized close to the expectation for random fertilization. This is a clear indication of substantial mixing in the spermatheca and not for clumping of semen. Also, in more recent studies the mixing of semen in the spermatheca has found further experimental support (Page et al. 1984; Laidlaw and Page 1984; Moritz 1983) revealing twofold: there is a mixing of semen in the spermatheca, and there is a lack of intraspermathecal sperm competition. There is a weak but significant effect of the insemination sequence on the representation of semen in the spermatheca (Moritz 1986a). Late injected semen seemed to be more efficient then early injected semen. The last male was about twice as successful as the first one. Combining these results with the data of Woyke (1983), one can conclude furthermore that the semen is not mixed in the oviducts.

7.3.2 Measures of Genetic Relationship

The genetic relationships among the members of the colony are of crucial importance for the understanding of natural selection and honeybee breeding. Different measures of genetic relatedness are appropriate for the genetic analysis which corresponds to the problem to be solved. Before we deal with specific problems in detail, it is necessary to briefly review the various possible descriptions of genetic relationship. This seems to be a trivial matter, but it is not, and very often it actually causes a lot of confusion. In principle, there are two different types of coefficients to describe relationship:

1. symmetrical coefficients in which the relationship between individual A and individual B is identical to the relationship between B and A;
2. asymmetrical coefficients in which the relation from A to B may differ from the relationship from B to A.

   Mathematically, the first coefficients represent correlations between A and B, whereas the latter can be interpreted as regression of A on B or B on A, respectively.

7.3.2.1 Coefficients of Relationship Based on Pedigree Information

Malecot's Coefficient of Coancestry. The simplest coefficient to evaluate genetic relationship has been derived by the French biomathematician Malecot (1948). He defined a coefficient of kinship (Malecot 1969), Phi, as the average probability that a randomly chosen allele at a given locus in individual A is identical by descent to a randomly chosen allele in individual B. Phi has also been termed as coefficient of co-ancestry (Falconer 1981), consanguinity (Crow and Kimura 1970), parentage (Kempthorne 1969), or transmission (Spiess 1977). Let us first consider the classical diploid case (Fig. 7.4):

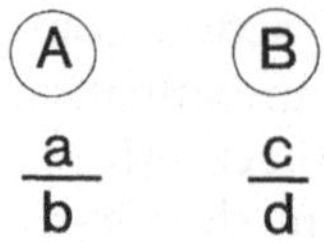

Fig. 7.4. Two individuals *A* and *B* with alleles *a*, *b*, *c* and *d* respectively

Let A have the alleles a and b at a given locus and B the alleles c and d at the same locus. ϕ is then simply the average probability p for a = c, a = d and b = c, b = d, which computes as follows

$$\phi = [p(a = c) + p(a = d) + p(b = c) + p(b = d)]/4 \quad (7.1)$$

or in a more general form

$$\phi = \Sigma\, p_{ij}/2n. \quad (7.2)$$

where p_{ij} is the probability for the ith allele of individual A to be identical by descent with the jth allele in B. The number of alleles at the locus under consideration is given by n. This definition can also be used in male haploid populations.

The specific feature of ϕ_{AB} is that it exactly reflects the inbreeding coefficient, F, of any diploid offspring produced by A and B. The coefficient F is the probability of a randomly chosen gene to be identical by descent to the other gene at the same locus in the same individual. Although F is not explicitly defined for haploid organisms, it is desirable to have an F value for drones. In the light of the polyploidization of somatic tissue in drones and the genetic uniformity of the gametes, it seems plausible to define the inbreeding coefficient of drones as F = 1.

Wright's Coefficient of Relationship. A little more complex than Phi, though more widely used, is Wright's (1922) coefficient of relationship r. Although r has often been interpreted as the *proportion of common alleles shared by descent in two individuals* (Crow and Kimura 1970; Wilson 1971; Spiess 1977), Pamilo and Crozier (1982) showed that this is not exactly true. The coefficient, r, actually represents the geometric mean between the two possible asymmetrical pedigree coefficients of relationship G between two individuals (see below). The coefficient of relationship between two individuals A and B can be easily derived from ϕ_{AB}

$$r_{AB} = \frac{2\phi_{AB}}{[(1 + F_A)(1 + F_B)]^{1/2}}, \quad (7.3)$$

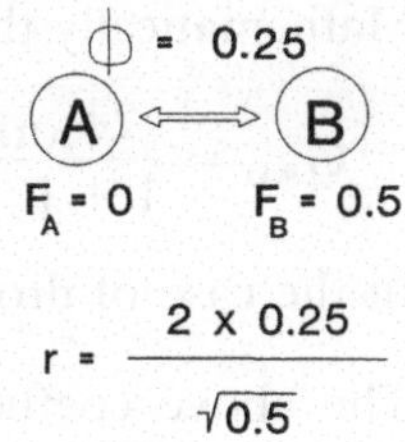

Fig. 7.5 Relationship between the coefficient of coancestry ϕ, the coefficient of relationship r, and the inbreeding coefficient F. B is an inbred individual ($F_B = 0.5$), whereas A is non-inbred

where F_A and F_B are the inbreeding coefficients of A and B respectively.

This definition is not equivalent to the fraction of alleles shared by two individuals through common descent. Such a proportion would be expressed through $2\phi_{AB}/\{[(1 + F_A) + (1 + F_B)]/2\}$ with the denominator consisting of the arithmetic rather than the geometric mean. The crucial difference to the coefficient of kinship, Phi, is that the inbreeding coefficient or, more precisely, the relation of the individual to its own gametes is of importance. Figure 7.5 gives an example for calculating r.

Asymmetrical Coefficients of Relationship. In sociobiology it is often important to express directionality in relationship. Common questions concerning the evolution of altruistic behaviour are: How certain can individual A be that it has its genes represented in individual B? Should A invest in its own offspring or in related offspring produced by B?

It is obvious that we need a definition for relationship which expresses directionality if we are going to deal with such problems. It could be that all of B's genes are from A, but A has in addition to those genes transmitted to B other genes which are not represented in B. This is the typical case for the relationship between queens and their offspring drones. All of the drone's (B) genes are inherited from queen (A), but only half of the genes of queen (A) are present in drone (B).

The pedigree coefficient of relationship, G_{AB}, expresses such asymmetry in relationship (Crozier 1970; Crozier and Pamilo 1980). G_{AB} is defined as the proportion of A's genes identical by descent with genes present in B, or in other format

$$G_{AB} = \frac{\text{mean identity by descent of B's genes with A's gametes}}{\text{mean identity by descent of B's genes with its own gametes}}.$$

Mathematically this can be written as follows:

$$G_{AB} = \frac{2\phi_{AB}}{1 + F_A}. \tag{7.4}$$

In the case of drones F_A is equal 1 and $G_{AB} = Phi_{AB}$.

The above coefficients are extremely helpful for theoretical considerations of intra- and intercolonial relationships. Furthermore, they are useful whenever exact breeding pedigree information is available. However, they are less useful for the estimation of relationships in natural populations, because pedigree information is usually not available.

7.3.2.2 Relationships in Natural Populations

In natural colonies pedigree information is not easily accessible and relationships generally are estimated via marker alleles which are supposed to be selectively neutral. Especially with the recent development in genetic finger-printing, the method becomes more powerful than it was using isozyme markers. Nevertheless, isozyme studies gain in accuracy when multiple loci are studied (Pamilo and Crozier 1982). In honeybees, however, only very few polymorphic allozyme systems have been described (Table 7.12), and multiple polymorphic loci are usually not available (Sylvester 1986). DNA sequence information or the use of probes which reveal high interindividual genetic variability (Hall 1986, 1988; Severson et al. 1988) seem to be more promising for the estimation of relationships in honeybee populations.

Regression Coefficient. Hamilton (1972) defined the regression coefficient of relatedness b_{AB}. Like G, it contains information about the direction of relatedness and reflects the regression of group A on group B:

$$b_{AB} = \frac{Cov_{AB}}{s_B^2}, \tag{7.5}$$

where Cov_{AB} is the genotypic covariance between A and B and s_B^2 the variance of B. The regression coefficient, b_{AB}, is estimated by using the technique demonstrated in a simple example in Table 7.1. The allele frequency of each worker of the nest is compared to the allele frequency in the rest of the nest. Thus, we get pairs of values for each worker: allele frequency in worker – allele frequency in rest

Table 7.1. Estimating genetic relationship from field data using the regression coefficient (Pamilo and Crozier 1982). The regression of the allele frequency of A in the group on the individual is determined. For comparison, the average of the individual coefficients of relationship [see Eq. (7.3)] for each group member is also listed. This represents the accurate relationship information, given that all common alleles are identical by descent

	Genotype	Frequency of A		Average allele identity to rest group
		Individual	Rest of group	
	AA	1	3/4	3/4
Group 1	AB	1/2	1	1/2
	AA	1	3/4	3/4
	AB	1/2	0	1/2
Group 2	BB	0	1/6	3/4
	BB	0	1/6	3/4
		b = 0.58		$\bar{r}$ = 0.67

of nest. Using information from many colonies we are able to determine the regression coefficient of relatedness.

The information obtained in this procedure is very different than that from the pedigree estimates. The regression b is an estimate for a genetic relationship based on a finite sample size and therefore subject to statistical confidence limits. Pamilo and Crozier (1982) point out that a value of b = 0.75 does not necessarily reflect actual supersisters, but can result from inbreeding in previous generations.

Identity by Descent. Another very elegant technique to estimate relationship in natural populations of social insects is given by Queller and Goodnight (1989). They modified a method originally proposed by Grafen (1985) as an estimator for relationship. The procedure is similar to the regression method of Pamilo (1984) since it also compares the frequency of a given allele in one individual with the frequencies in the nest and the overall population. The estimator is simple but powerful, since it is more accurate than those discussed above (Queller and Goodnight 1989). It has an amazing generality and operates under many constraints (for example small and unequal group sizes) where other estimators suffer from biases or imprecision. However, Pamilo (1990) concluded from computer simulations that his (1984) regression estimator and Queller and Goodnight's (1989) estimate give very similar results. Using the identity by descent method we can describe relationship, r, as follows

Table 7.2. Example for the use of Queller's and Goodnight's (1989) estimator for relationship. Let us take the first (AB) individual of group 2 for a stepwise explanation of the table. The individual carries allele A at the first allelic position and B at the second. The intraindividual allele frequency for the gene at the first allelic position (column 1) is 1/2 (since the other allele is B). The same is true for the allele frequency of the gene at the second allelic position in this individual. If we look at the other group members in group 2 (thus excluding individual 1) we see that the allele frequency for A (1st position) is zero and the frequency for B (2nd position) is 1. Furthermore, the frequency for A is 5/6 in the other group (group 1) and 1/6 for B

	Genotype	P_{ij}		$P_{i(-j)}$		$P_{(-i)}$	
	AA	1	1	3/4	3/4	1/6	1/6
Group 1	AB	1/2	1/2	1	0	1/6	5/6
	AA	1	1	3/4	3/4	1/6	1/6
	AB	1/2	1/2	0	1	5/6	1/6
Group 2	BB	1	1	3/4	3/4	1/6	1/6
	BB	1	1	3/4	3/4	1/6	1/6

The estimator for the relationship is thus:

$$\frac{(1 - 1/6) + (0 - 5/6) + 4(3/4 - 1/6) + (1 - 1/6) + (0 - 5/6) + 4(3/4 - 1/6)}{(1/2 - 1/6) + (1/2 - 5/6) + 4(1 - 1/6) + (1/2 - 1/6) + (1/2 - 5/6) + 4(1 - 1/6)} = \frac{7}{10}.$$

$$r = \frac{\sum_i \sum_j \sum_k \sum_a p_{i(-j)} - p_{(-i)}}{\sum_i \sum_j \sum_k \sum_a p_{ij} - p_{(-i)}} \tag{7.6}$$

where $p_{i(-j)}$ = allele frequency of group i excluding individual j,
$p_{(-i)}$ = allele frequency of population excluding group i,
p_{ij} = allele frequency of individual j in group i;

the indices: i for groups, j for individuals, k for loci, a for allelic positions. (1 in males, 2 in females, this is not the number of allelomorphs in the population!)

The same example as above reveals the efficiency of the estimator. Let us take two groups, each with three diploid individuals of genotypes as given in Table 7.2. We then determine the allele frequencies P_{ij} in each individual $_{ij}$. Next, the gene frequencies, $p_{i(-j)}$, for an allele in individual j in group i excluding individual j are determined. Finally, the gene frequency, $p_{(-i)}$, is calculated by determining the average gene frequency of the allele under consideration in the total sample excluding group i. In this example we estimate a relationship of 0.70 which is very close to the average intracolonial relationship under the assumptions in Table 7.2 ($\bar{r} = 0.67$).

7.3.2.3 Weighted Coefficients of Relatedness

In sociobiology it is sometimes not only necessary to determine the proportion of identical genes between individuals but it also may become important to estimate the actual number of identical genes. Pamilo and Crozier (1982) derived weighted coefficients of relatedness which accounted for the difference in the reproductive value of males and females (Hamilton 1972). The haploid males contribute only half as much to further generations as females. The number of identical genes carried by A and B can be expressed by

$$G'_{AB} = G_{AB}\frac{v_A}{v_B} \qquad (7.7)$$

and

$$b'_{AB} = b_{AB}\frac{v_A}{v_B} \qquad (7.8)$$

where v = reproductive value.

In the male haploid honeybees, the reproductive value is v = 1 for males and v = for females. Then, b′ can be calculated directly from molecular allele information by using the absolute numbers instead of the relative frequency data.

7.3.3 Genetic Relationship Between Members of a Colony

Since we now know how to determine genetic relationship in theory and from empirical studies, we are able to focus on the actual relationship in the colony. We will need both a quantitative genetic and a sociobiological approach in the further discussion of the genetics of the superorganism. Therefore it is necessary to discuss all of the above pedigree coefficients in the following paragraphs. For the sake of simplicity of the following models, we assume that the drones which mate with the queen are unrelated to each other and to the queen. Neither queen nor workers are inbred.

7.3.3.1 Relationship Between Queen and Her Offspring

The relationship between queen and offspring can be analyzed using the pedigree in Fig. 7.6. Since all coefficients are based on Malecot's coefficient of kinship, we first determine $\phi_{\text{queen-offspring}}$, then the inbreeding coefficient, and finally derive the other parameters from these values, using Eqs. (7.2) and (7.3).

$$\begin{aligned}\phi_{\text{queen-worker}} &= [p(a = c) + p(a = d) + p(b = c) \\ &\quad + p(b = d)]/4 \qquad (7.9)\\ &= [1/2 + 0 + 1/2 + 0]/4 \\ &= 1/4.\end{aligned}$$

The probability for a queen's gene to be transferred to the worker is 1/2 $[p(a = c) = p(b = c) = 1/2]$. Because the drones are unrelated to the queen, there is no similarity in the paternal part of the worker's and queen's genes $[p(b = d) = 0]$ and between maternal and paternal genes $[p(a = d) = 0]$. Since the queen and workers are assumed not to be inbred, both $r_{\text{queen-worker}}$ and $G_{\text{queen-worker}}$ are therefore equal to $2 \times \phi = 1/2$.

The relationship between the queen and her offspring drones generates larger differences between the three coefficients. $\phi_{\text{queen-drone}} = 1/2$ is easily obtained and equal to the relationship between an individual with its own gametes:

$$\begin{aligned}\phi_{\text{queen-drone}} &= [p(a = c) + p(b = c)]/2 \qquad (7.10)\\ &= [1/2 + 1/2]/2 \\ &= 1/2.\end{aligned}$$

The calculation of $r_{\text{queen-drone}}$ is a little less clear because it includes the inbreeding coefficient of the drones and $2 * \phi$ is divided by the square root of 2:

$$r_{\text{queen-drone}} = \frac{2 \times 0.5}{\sqrt{(1 + 0)(1 + 1)}} = 1/\sqrt{2} = 0.707.$$

$G_{\text{queen-drone}} = 1/2$ is different from $G_{\text{drone-queen}} = 1$, which indicates the asymmetry in the relationship between a queen and her offspring drones. It reflects that only half of the queen's genes are represented in the drone but all of the drone's genes are present in the queen.

$$G_{\text{queen-drone}} = \frac{2 * 0.5}{1 + 1} = 1/2 \qquad (7.11)$$

$$G_{\text{drone-queen}} = \frac{2 * 0.5}{1 + 0} = 1. \qquad (7.12)$$

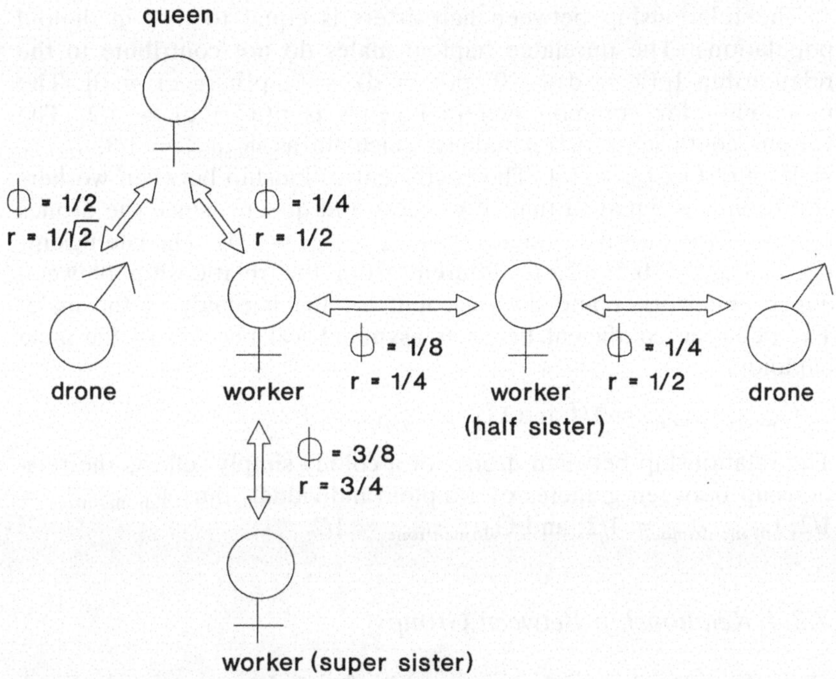

Fig. 7.6. Symmetric measures of relatedness in a colony of honeybees. Super-sister workers have a common father drone

7.3.3.2 Relationship Between Workers and Drones

The polyandrous mating system results in two different categories of relationships between workers, those who have the same father drone in common (supersisters) and those who are sired by different drones (half-sisters). Figure 7.6 elucidates the pedigree for the determination of worker relationship. $\phi_{supersister}$ equals 3/8. The probability of having a paternal allele in common is $p(b = d) = 1$. The probability for common maternal alleles is $p(a = c) = 1/2$. Since paternal and maternal genes are assumed not to be identical; $p(a = d) = 0$ and $p(b = c) = 0$. Because the workers are not supposed to be inbred, we obtain the famous but frequently misinterpreted coefficient of relationship $r_{supersister} = 3/4$ for supersisters in Hymenoptera. $G_{supersister}$, which is the coefficient mostly meant when discussing relationships on sociobiological terms, also equals 3/4 if there is no inbreeding.

The relationship between half-sisters is equal to that in diploid populations. The unrelated haploid males do not contribute to the relationship [p(a = d) = 0, p(b = d) = 0, p(b = c) = 0]. The probability for common maternal genes is p(a = c) = 1/2. The various coefficients of relatedness calculate as $\phi_{halfsib} = 1/8$, $r_{halfsib} = 1/4$ and $G_{halfsib} = 1/4$. The coefficient of kinship between workers and drones is equal to that of workers and queens, since the drones represent the queen's gametes ($Phi_{worker\text{-}drone} = 1/4$). The coefficient, $r_{worker\text{-}drone} = 0.5/\sqrt{2}$, is different from the relationship between queen and worker and does account for the haploidy of the male. The pedigree coefficient becomes asymmetrical because of the male haploidy:

$$G_{worker \rightarrow drone} = 1/4 \text{ and } G_{drone \rightarrow worker} = 1/2.$$

The relationship between drones of a colony simply reflects the relationship between gametes of a diploid individual, thus $\phi_{drone\text{-}drone} = 1/2$, $r_{drone\text{-}drone} = 1/2$, and $G_{drone\text{-}drone} = 1/2$.

7.3.4 Relationship Between Groups

Honeybees live in social groups and it often is important to deal with the group as a genetic unit rather than the individual worker in a colony. In principle, there are two ways to handle this problem. One is to look at the group as a composition of many individuals, the other is to look at the group as one superorganism with intra-organismic genetic variation. Though at first both approaches appear very similar, genetic relationships in terms of expressing the probability for common alleles which are identical by descent differ substantially.

7.3.4.1 Average Individual Relationship

The average individual relationship can be calculated on the basis of the information of the previous paragraphs. If we have two groups A and B with n and m group members respectively, we can calculate the average individual pedigree coefficient of relationship as follows:

$$\bar{G}_{AB} = \sum_{i}^{n} \sum_{j}^{m} G_{i,j}/nm, \tag{7.13}$$

where $G_{i,j}$ is the individual relationship between group member i in group A and member j in group B.

Thus $\bar{G}_{AB}$ is nothing else but the arithmetic mean of the individual relationships of all possible combinations between pairs of members of two different groups. The same "average individual" notation can be used for the other coefficients of relationship, Phi and r.

The average individual relationship is an important parameter for problems related to kin selection and inclusive fitness theory. It also has implications for the worker-queen conflict about the investment in each sex (see Chap. 8). Generally, the average coefficients of relatedness are important if properties of the individual workers are to be analyzed. As soon as group characters are studied, it seems to be more meaningful to deal with the group as the genetic unit of interest rather than with the individual components of the group.

7.3.4.2 Group Relationship

The quantitative genetic analysis of social behaviour (i.e. characters which are only expressed by a group and not by individuals) has to rely on relationships between groups. Since we usually are mainly interested in selectable traits, it is the additive gene effects and, in this case, the sum of all additive gene effects in the group which become important. The non-additive effects due to interactions between genes and also between genotypes (!) are not selectable (see Sect. 7.4). This does not mean that non-additive gene effects are unimportant for processes in natural selection. In fact, Moritz and Southwick (1987) and Moritz and Hillesheim (1989) showed that these non-linear interactions can be crucial for the expression of social behaviour of groups.

If we have to address something like a group relationship, it seems most meaningful to stick to the definitions we already made for individual organisms. If in the individual case we asked for the probability of genes identical by descent in two organisms, we now ask for the same probability in two groups of organisms. The one group is thus treated as a bag full of genes, and we look at the other group as yet another bag of genes and we search for identical genes in both bags. If we use the superscript * to indicate the group parameters (in contrast to the individual parameters), we obtain, in analogy to Malecot's coefficient of kinship, the parameters for groups defined by Moritz (1986b) and reviewed by Owen (1989b). Let us first look at the female's contribution to similarities between two groups.

$$\mathrm{Phi}^{*}_{\mathrm{female}} = \sum_{i}^{2^n} \sum_{j}^{2^m} p(i, n)\, p(j, mn)\, s_f(i, j), \qquad (7.14)$$

where n = group size of group A, m = group size of group B,
p(i, n) = probability of allelic combination i in group A,
p(j, m) = probability of allelic combination j in group B,
s_f(i, j) = frequency of alleles identical by descent which are transmitted via the queen in allelic combination i and j.

The terms p(i, n) and p(j, m) are dependent on group size and follow, in the case of the queen, a binomial distribution (because she is diploid).

Since this may be confusing for the non-geneticist, let us calculate a simple example to understand the system more easily. Let us have two very elementary groups each consisting of two workers, the smallest possible group size. Both groups are taken from the same colony and have the same mother queen. Let the mother queen have allele *x* and *y* at a given locus and let her produce two sets of two gametes which either carry allele *x* or *y*. The queen will distribute both alleles at random to the eggs, which will be fertilized and develop into the four workers which form our two groups. Table 7.3 shows the possible combinations for alleles of a given locus in both groups and the corresponding probabilities for descent identity. There will be a certain probability that by chance both gametes, which end up in one group, carry the same *x* allele and none of the *y* allele. This probability is smaller for identical alleles in both gametes ($p_{(1,2)} = p_{(3,2)} = 0.25$) than for an equal distribution of one *x* and one *y* allele ($p_{(2,2)} = 0.5$). The probabilities of certain combinations of the two alleles in the two gametes are calculated by a binomial distribution (shown in Table 7.3). These probabilities alone do not tell us

Table 7.3. Probabilities for identity by descent of maternal genes in two, 2-worker-groups, A and B, from the same colony. x and y are maternal alleles, and n represents the unknown paternal alleles. Under monandry we obtain Phi* = (1 + 0.625)/2 = 0.8125; the frequency of the corresponding genotypic makeup of the group follows a binomial distribution

Frequency	Group B	Group A			
		Frequency	xn/xn 0.25	xn/yn 0.5	yn/yn 0.25
0.25	xn/xn		1	0.5	0
0.50	xn/yn		0.5	1	0.5
0.25	yn/yn		0	0.5	1
ϕ^*_{fcmale} (maternal genes only) = 0.625					

anything about the genetic relationship. We need a measure which expresses the frequency of descent-identical alleles in both groups. The probability that a female allele in group A is also found in B is s_f. Thus, if we have the case of x,x in group A and y,y in B, then s_f will equal 0. If all the four gametes carried the x allele, then $s_f = 1$ and so on.

With this in mind, it is obvious that the larger the groups, the larger the probability that they contain the same set of alleles. With an infinite group size of A and B, the probability to find the same female alleles in both groups is 1.

The same approach used for female genes can be used to describe the probability for identical male genes in two groups.

$$\phi^*_{male} = \sum_i^{d^n} \sum_j^{d^m} p(i, n)\, p(j, n)\, s(i, j), \tag{7.15}$$

where n = group size of group A,
m = group size of group B,
d = number of drones per mating,
p(i, n) = probability of allelic combination i in group A,
p(j, n) = probability of allelic combination j in group B,
s(i, j) = probability of alleles identical by descent which are transmitted via the drone in allelic combination i and j.

Again the probabilities p(i, n) and p(j, m) are dependent on group size but they represent a multinomial frequency distribution if the queen has mated with more than two drones. Figure 7.7a,b illustrate the relationships between two subgroups of a colony, depending on group size and the number of drones which mate with a queen. Two mating types, which are commonly used in honeybee breeding are shown: (a) all drones originate from the same queen mother, and (b) all drones are unrelated. In case (a) the probability for identical male genes asymptotes to 0.5 for an infinite number of drones whereas in case (b) this probability tends to zero. The average of both male and female values represents the overall ϕ^* group relationship.

Since the definition for ϕ^* has been made, it is possible to redefine the other measures of individual relationship for groups by using Eqs. (7.3) and (7.4) (Moritz 1988a). The meaning of these new coefficients, however, are different than those described by the individual model. Therefore, one might feel more comfortable looking at the definition of G for individuals first and then transferring its genetic meaning to the group situation, instead of simply solving the

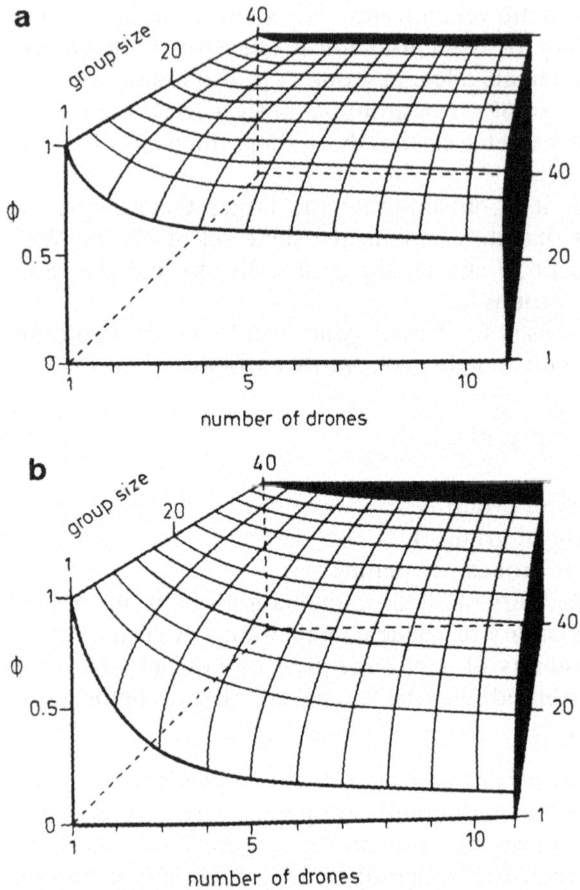

Fig. 7.7a,b. Estimates for ϕ (probability for a male gene to be identical by descent in two groups of workers) dependent on the number of workers in the group (group size) and the number of males which mated the queen. **a** All drones are offspring of one mother queen; **b** all drones are unrelated. The case of two drones in **b** represents the course of Phi' (probability of a queen gene to be identical by descent in two groups of workers)

problem algebraically, as done previously (Moritz 1988a). G was defined as the proportion of A's genes identical by descent with genes present in X. In order to transfer this definition to groups, we have to treat A and B as groups instead of individuals. Thus the ratio between the mean identity of group A's genes with potential group B's gametes (ϕ^*_{AB}) and the mean identity of B's genes with its own potential gametes (ϕ^*_B) has to be determined.

Table 7.4. Probabilities for identity by descent between genes in group B and its own potential gametes. a and b = maternal alleles, c = paternal allele, frq. = frequency, p = probability for identity. The total is obtained by adding the probabilities multiplied by the corresponding frequencies

Frequency	Composition	Possible gametes					
		a		b		c	
		Frequency	p	Frequency	p	Frequency	p
0.25	ac/ac	0.5	0.5	0	0	0.5	0.5
0.50	ac/bc	0.25	0.25	0.25	0.25	0.5	0.5
0.25	bc/bc	0	0	0.5	0.5	0.5	0.5
Total		0.4375					

Table 7.5. Probabilities for identity by descent between genes in group B and the potential gametes in group A. Notation as in Table 7.4

Frequency	Group B composition	Group A composition Frequency	ac/ac 0.25		ac/bc 0.5			bc/bc 0.25	
		Possible gametes in group A Frequency	a 0.5	c 0.5	a 0.25	b 0.25	c 0.5	b 0.5	c 0.5
0.25	a c a c		0.5	0.5	0.25	0.25	0.5	0	0.5
0.50	a c b c		0.25	0.5	0.25	0.25	0.5	0.25	0.5
0.25	b c b c		0	0.5	0	0.5	0.5	0.5	0.5
Total		0.375							

Let us take our basic two-worker-group model again. The probabilities for allelic combinations and the corresponding descent identities are given in Tables 7.4 and 7.5. Interestingly, the information in Table 7.4 has a very close connection to the definition of the inbreeding coefficient of individuals. The relationship of an individual Y to its own gametes is $\phi_{YY} = (1 + F_Y)/2$ (F = inbreeding coefficient) and F_Y is proportional to ϕ_{YY}. Also in groups, ϕ^*_{BB} is a measure of genetic uniformity (we will discuss the "inbreeding coefficient" of groups again in Sect. 7.6.3). We see that $G^*_{AB} = 0.857$ is higher than the ϕ^*_{AB} coefficient (0.8125) similar to the individual

Table 7.6. r*, defined as proportion of identical alleles shared between two groups. Notation as in Table 7.5

Frequency	Group B composition	Group A			
		Frequency	0.25 ac ac	0.5 ac bc	0.25 bc bc
			proportion of identical alleles		
0.25	a c a c		1	0.75	0.5
0.50	a c b c		0.75	1	0.75
0.25	b c b c		0.5	0.75	1
Total		0.8125			

case. In general terms, we can formulate the pedigree coefficient of relationship for groups as

$$G^*_{AB} = \frac{\phi^*_{AB}}{\phi^*_B} \tag{7.16}$$

with the definitions as given in Eq. (7.1)

The definition of r* for groups seems to be a delicate problem, since there are several possible approaches which lead to different results. Again, it would be most helpful to stick closely to the definition of r for individuals. This, however, is not unambiguously clear because r has been addressed in various ways. We could define r as the geometric mean between the two possible G values (Pamilo and Crozier 1982); we could simply adopt the algebraic Eq. (7.2) (Moritz 1988a); or we could apply the frequently used verbal definition as the proportion of alleles shared by X and Y. Since the first two approaches can be calculated from the above derivations, let us concentrate on the third definition. Numerically it means, we first put all the genes of both groups together and then determine the proportion of identical genes of the total genes under consideration. For the sake of simplicity, let us ponder again the simple two worker groups (Table 7.6). If in the one group we have the alleles a and c in one worker and a c also in the other, and an a c, b c combination in the second group, then the proportion of shared alleles would be 6/8 (0.75), because we have eight alleles in total but only six of them are identically shared by both groups. If we calculate this for all possible combinations of alleles we obtain a value of 2Phi*, which ignores the intragroup relationship (i.e. the denominator in the other two definitions). The general formulation, sticking to the definition of Pamilo and Crozier (1982), is

$$r^*_{AB} = \sqrt{G^*_{AB} G^*_{BA}}, \tag{7.17}$$

which seems to be closer to Wright's original definition for individuals.

7.4 Quantitative Genetics

7.4.1 The Problems in Analyzing Social Characters

Quantitative genetics as the theoretical basis for animal breeding deals with the inheritance of traits which ideally are determined by a large number of genes and can be characterized with a mean and a variance (for an introduction into this field, read Falconer 1989 or Hill 1984). The expression of a trait (P) in a certain individual may deviate from the mean phenotypic expression (μ) in the population due to environmental effects (E) and gene effects (G):

$$P = \mu + E + G. \tag{7.18}$$

The gene effects combine from additive, dominant, and epistatic interactions between the loci which determine the expression of the phenotype. Thus the deviation from the population mean can be written as follows:

$$P - \mu = E + G_A + G_D + G_E, \tag{7.19}$$

where G_A = additive effects,
G_D = dominance effects,
G_E = epistatic effects.

The major aim of the quantitative geneticist is to estimate the impact of additive gene effects on the expression of a certain phenotype, because they represent the selectable part of the genetic variation in the population. For traits of individuals the above model has proven to be extremely useful. For social traits, however, the model has to be substantially extended in order to reflect not only interaction among genes but also among genotypes. A simple example shows that the individual model in Eq. (7. 12) does not suffice to explain gene expressions in social groups. Let us assume two groups of 10 000 bees, one consisting of 9999 workers and one queen, the other consisting of 10 000 workers only. This may look like a trivial example, but, in terms of quantitative genetics, it is not, and it is actually extremely complex and poorly understood. Even though in both groups we have exactly the same set of genes, we are to expect dramatic differences between the two groups. If we look at something like brood production it becomes obvious that there is more to a social phenotype than just gene expression and environment. Genotype–environment interactions (e.g. which determine the queen) and genotype–genotype interactions have crucial impact in this example. The complexity of a consequent extension of the individual model to a group model becomes apparent in a series of theoretical papers by Griffing (1967, 1968, 1969, 1982a,b). Sophisticated pedigree information is needed to estimate all interaction components invoked in these models, which severely restricts their feasibility in honeybee genetics. Subsequently, several attempts have been made to use simplified models (Chevalet and Cornuet 1982a; Oldroyd and Moran 1983; Moritz 1986b; Bienefeld 1988) which are helpful in estimating some of the most important quantitative genetic parameters. These models do not reveal the complete picture of quantitative genetics of a honeybee colony but they do allow some basic predictions about selection in honeybee populations, which have proven helpful in honeybee breeding in the past (Cornuet 1983; Moritz 1985a; Bienefeld 1988; Bienefeld and Pirchner 1990). In spite of the theoretical problems, selection on the colony level is clearly routine in bee-breeding practice. Wislon and Sober (1989, p. 350) obviously overlooked the breeding work in honeybees when they stated that "*community-level selection has never to our knowledge been attempted . . ., even though the procedure is straightforward and the economic benefits potentially great*".

7.4.2 The Classical Parameters

7.4.2.1 Repeatability

In animal husbandry, repeatability, R, is one of the crucial parameters for performance testing. It reflects whether a once-shown trait is a true character of the tested individual or just a "by chance" result. R is the ratio between the variance due to the true performance of the individual (V_P) and the total phenotypic variance, which is composed of V_P and the variance due to temporal differences in the environment, V_t, which also encloses imprecision in the data acquisition due to experimental variation.

$$R = \frac{V_P}{V_P + V_T} \tag{7.20}$$

where V_P = variance due to performance value,
V_T = temporal environmental variance.

R is usually estimated via an intraclass correlation (Pirchner 1979). Classical livestock is long-lived and allows for repeated tests under controlled conditions. In honeybee breeding this is quite different. On the one hand, V_t is extremely large because foraging conditions and nectar availability are inconstant and only poorly predictable (as we know from weather forecasts). Furthermore, the worker bees tested in spring are different individuals than those tested in summer. Thus, if we test the same colony over a year, we do not obtain a true replicate of information of the same individual, which is assumed in the above model. The workers of a colony will be super- and half-sisters to the workers tested 8 weeks earlier. The queen, on the other hand, is still present in the colony and might fit the model well. Does this affect the estimation of the repeatability?

The genetic differences between the colony at time, t_1, and at a later time, t_2, are critical for the estimation of the repeatability. If this variation is large, the estimation of R is rather meaningless, since both genetic and environmental variance affect the phenotype expression. If large colonies of honeybees are tested, this seems to be a minor problem. We have shown in Eq. (7. 13) that the relationship between large groups of the same colony is close to one. Thus, if we test a colony in spring and a second time in summer we have a situation very similar to individual organisms. The genotypic composition will be very similar, particularly in the light of the intense mixing of semen in the spermatheca of the queen (Laidlaw and Page 1984; Moritz 1986a). Problems arise only if there is not equal utilization of semen, which might occur under special conditions (Page 1986). For practical purposes, a healthy full-size colony can be considered genetically uniform for estimations of the repeatability as long as the same queen resides in the colony.

For small experimental groups, however, this does not hold true. The genetic variation between a subgroup sampled at two different times can be substantial, which obscures the estimate of the repeatability. In this case the same group should be tested repeatedly within the life-span of the group members to get the desired information. However, care has to be taken not to test behaviour which is affected by the age polythism system. Behaviour that depends on age can only be repeatedly tested if short-term measurements are possible.

7.4.2.2 Selectability

The major issue in quantitative genetics is whether a trait can be modified by selection or not. One wants to make predictions about the response to both artificial selection by breeders and natural selection in the wild. From quantitative genetic theory we know that selection can only operate on additive gene effects (Falconer 1981). All other non-linear interactions between genes (e.g. dominance and epistasis) are not selectable. The ratio between additive genetic variance and total phenotypic variance has often been referred to as the term heritability or h^2 (in the narrow sense):

$$h^2 = V_A/V_P, \tag{7.21}$$

where V_A = additive genetic variance,
V_P = total phenotypic variance.

This term, though very popular and widely used, is actually utterly misleading and often causes substantial confusion for non-geneticists. A character can be completely inheritable, exclusively determined by additive gene effects, independent of any environmental impact, and yet have a heritability of zero. The heritability of fitness, for example, should be close to zero in populations at equilibrium (Fisher 1930; Falconer 1989). The problem is that additive genetic *variation* is necessary for selection. If everything is the same, nothing can be selected. Thus, as soon as a character is at fixation in a population its "heritability" is zero even though it is completely inherited. To avoid this confusion, it seems helpful to talk about "selectability" instead (Crozier pers. comm.). This would clearly emphasize the actual meaning of h^2 rather than implying a false genetic interpretation.

7.4.2.3 Non-Linear Interactions

Non-linear interactions between genes in an individual are usually not selectable, because they result from unique gene combinations which are unlikely to be repeated in subsequent generations. In practical breeding schemes, only hybrid programmes rely on such non-linear interactions. The actual selection in terms of improving the genetic value is performed in the pure lines. Dominance effects in diploids, where one gene suppresses the expression of the other gene at the same locus are typical non-linear interactions. Epistatic effects can also account for non-linear interactions. Interestingly, in drones we cannot find any dominance effects because of the hemizygosity. Epistasis, however, will occur in both workers and drones (e.g. the *tan* allele of the *snow* locus is epistatic to *chartreuse*; Laidlaw et al.

1964; for a recent review of honeybee mutants see Woyke and Hillesheim 1989).

In addition to these non-linear gene interactions we get genotype and phenotype interactions in a honeybee colony. Particularly phenotype interactions are often affected by strong gene-environment effects and are of striking importance for the function of the colony. The queen, a typical example for gene – environment interaction, is critical for the operation of the colony. The presence of a queen and the production of queen substance is, for example, "epistatic" over the expression of genes for ovary development in workers (Butler and Fairey 1964). Such interactions have also been analyzed at the worker level. Moritz and Hillesheim (1985) reported a strong genetic component in the determination of reproductive ("dominant") workers in the Cape honeybee, *Apis mellifera capensis*. Such dominant workers produce 9-oxo-2-decanoic acid (9-ODA) (Hemmling et al. 1979; Crewe and Velthuis 1980; Velthuis et al. 1990) and suppress the ovary development in non-reproductive "subordinate" workers. Hillesheim et al. (1989) found that colonies with a large proportion of "subordinate" workers were superior to those with a large proportion of "dominant" workers in brood rearing, and hoarding. Another case of genotype interaction was found by Moritz and Southwick (1987). They found additive but also non-additive interactions and suggested a "group heterosis" for the reaction of experimental groups to alarm pheromones (Fig. 7.8). Genotypically mixed groups responded more strongly than non-mixed worker groups from a single colony. Recently, genotypic variance in the colony has been proven to be crucial for division of labour in the colony. Guarding, undertaking, and queen rearing have been shown to be subject to genetic variance between patrilines and the appearance of genetical specialists in the colony (Robinson and Page 1988; Frumhoff and Baker 1988; Page et al. 1989a,b; Calderone and Page 1988; Rothenbuhler and Page 1989). However, genotypic intragroup variance also may have negative effects on group behaviour. Moritz and Hillesheim (1989) found that the higher the number of patrilines in a group, the poorer was hoarding behaviour in small experimental groups.

7.4.3 Similarities Between Colony Members

Genetic variance components of a character are usually estimated via similarities, expressed as covariances, between related individuals (Falconer 1989). If we deal with superorganisms we clearly have to

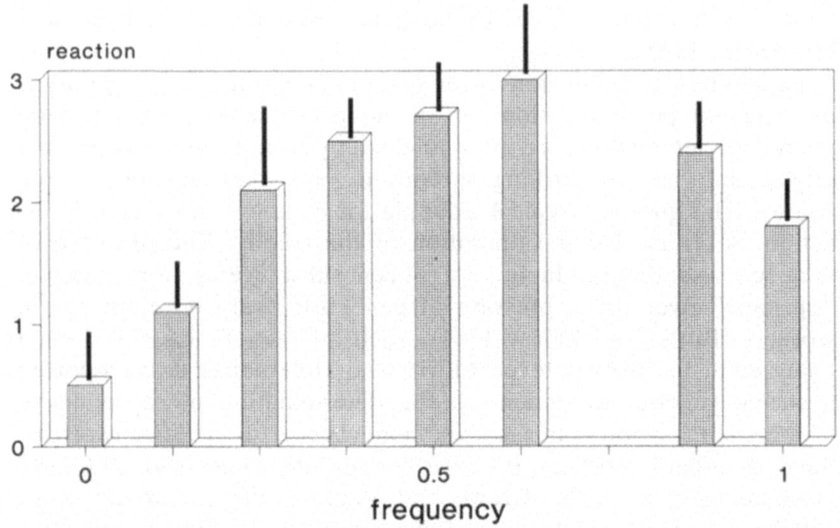

Fig. 7.8. Alarm reaction of mixed groups of honeybee workers to alarm pheromones. Each group consisted of two subgroups of workers from two colonies, A and B, with similar low defensive behaviour. The *frequency* of colony A workers in the group is given on the *horizontal axis*. The reactivity (*reaction*) of the mixed groups is higher than that of either pure group

follow a similar approach. However, we have to regard two cases: one is the covariance between the colony members, and the other is the covariance between groups or colonies of honeybees. In both cases the reasons for similarities may be very different and need special analyses.

7.4.3.1 Genetic Covariance Among Workers

Two individuals can "covary" (i.e. vary in the same direction because of similar phenotypes) either due to common genes or to common environmental conditions. The genetic co-variance is the parameter of major interest since it determines the estimate of the selectability. Genetic covariances can be estimated via an analysis of variance (intraclass correlation) or a regression analysis (regression coefficient). This estimate for similarity then has to be divided by the relationship between the two individuals in order to obtain h^2. Thus in general:

$$h^2 = \frac{\text{genetic similarity between two individuals}}{\text{relatedness between two individuals.}}$$

Similarity can be obtained by both additive and non-additive gene effects. Again, one of Malecot's (1948, 1969) definitions can be used:

$$\mathrm{Cov}_{AB} = [(\phi + \phi')/2] \times V_A + \phi \times \phi' \times V_D, \qquad (7.22)$$

where ϕ = probability for female descent identical genes in A and B,
ϕ' = probability for male descent identical genes in A and B,
V_A = additive genetic variance,
V_D = dominance variance.

Workers in a colony are either super- or half-sibs. Correspondingly, their expected genetic covariances are different. For half-sisters we obtain

$$\mathrm{Cov}_{\mathrm{halfsib}} = [(1/2 + 0)/2] \times V_A + 1/2 \times 0 \times V_D = 0.25\, V_A. \qquad (7.23)$$

Thus, the similarity among half-sisters is due to one-quarter of additive gene effects. There can be no correlation due to gene interactions if the drones are unrelated and have no genes in common.

The supersisters have the male-transferred genes in common which results in the following expected covariance:

$$\begin{aligned}\mathrm{Cov}_{\mathrm{supersib}} &= [(1/2 + 1)/2] \times V_A + 1/2 \times 1 \times V_D \\ &= 0.75 V_A + 0.5 V_D. \qquad (7.24)\end{aligned}$$

The similarity between supersisters is affected by both additive and non-additive interactions. With classical statistical analysis it is easy to empirically determine the covariance between super- and half-sister workers. The problem is to discriminate half- and full-sib workers. In experimental research this can be done with the use of genetic markers (for example Moritz and Hillesheim 1989). In studies on natural populations and for the analysis of practical breeding schemes, this is not possible and one usually takes the average individual intracolony relationship (Sect. 3.4.1) in Malecot's equation (Rinderer 1977; Oldroyd and Moran 1983; Moritz 1984a). The genetic covariance among the workers of a colony headed by a naturally mated queen is composed as follows:

$$\begin{aligned}\mathrm{Cov}_{\mathrm{workers}} &= [(1/2 + 1/z)/2] V_A + (1/2)(1/z) V_D \qquad (7.25) \\ &= (1/4 + 1/2z) V_A + (1/2z) V_D,\end{aligned}$$

where z = number of drones which inseminated the queen;

if we use 10, as a common estimate for the degree of polyandry, we obtain

$$\mathrm{Cov_{workers}} = 0.27\mathrm{V_A} + 0.05\mathrm{V_D}, \qquad (7.26)$$

which is a helpful approximation if no other information concerning the mating of the queen and the breeding pedigree is available.

7.4.3.2 Genetic Covariance Among Drones

Characters of drones are rarely genetically analyzed. Owen (1989a) studied genetic variance components of drones in bumblebees. Similar information on drones of honeybees is lacking. The principle in decomposing the genetic covariance into non-linear and linear gene interactions is similar. Because the drones result from haploid eggs, there are no dominance effects. In diploid organisms the expected epistasis effects are small compared to dominance and additive gene effects. In drones, however, this is different and the expected covariance component is substantial.

Epistasis could cause similarity in two drones if alleles at two loci in both individuals are identical by descent (i.e. we have the same epistatic mechanism in both drones). This probability is 1/2 and we obtain for the covariance

$$\begin{aligned}\mathrm{Cov_{drones}} &= 1/2\mathrm{V_A} + 0\mathrm{V_D} + 1/2\mathrm{V_E} \qquad (7.27)\\ &= 0.5\mathrm{V_A} + 0.5\mathrm{V_E}\end{aligned}$$

where $\mathrm{V_E}$ = epistatic variance.

Under $\mathrm{V_A} = \mathrm{V_E}$, 50% of the total covariance in drones is due to epistatic effects, which is considerably more than in workers where we only find a very small contribution in similarities between workers due to epistasis. In half-sibs the component is zero, if the males are unrelated. For supersibs, the component with a value of 0.25 contributes only 20% to the total covariance. Using the average notation, we obtain for a ten-drone mating a coefficient of $0.025\mathrm{V_e}$ for the epistatic effect. Clearly, the drones provide a useful tool in the diagnosis of epistatic effects in the gene expression of honeybees.

7.4.3.3 Covariance Between Queen and Offspring

Finally, we have to address the problem of similarities between the queen and her offspring. They are usually estimated by a regression analysis of queen on offspring. This is not always possible because many characters are caste specific, and what is expressed in a queen is not necessarily expressed in her worker offspring, and certainly not in her drones. Rinderer (1977) pointed out the problems in using queen – offspring covariance in quantitative genetic analysis. One way to

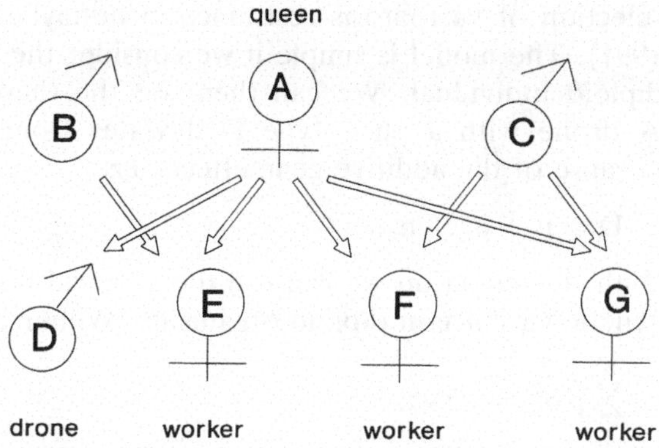

Fig. 7.9. Pedigree to explain the various covariances between the members of a honeybee colony

overcome the problem is to use the characters of the queen's sister workers and estimate a regression from the mean sister worker value of the offspring workers. Such an approach has been followed by Collins (1979), Milne (1985a,b,c), Milne and Friars (1984), and Moritz (1985a). Some characters, however, are expressed similarly in workers, queens and drones, and for those cases we will now derive the covariance composition using the pedigree in Fig. 7.9 and Eq. (20):

$$\mathrm{Cov}_{\text{queen-worker}} = 0.5V_A.$$

As long as the queen is not related to the inseminating drones, the only genetic similarities between the queen and her worker offspring are due to additive effects. For the queen-offspring drone covariance we obtain the same figure:

$$\mathrm{Cov}_{\text{queen-drone}} = 0.5V_A.$$

Again, in this case, only additive gene effects can cause similarities. Since the drones are haploid, there are no similarities due to dominance effects, and epistatic effects cannot cause similarities unless the queen is inbred and epistatic interactions between homozygous loci result in the same gene expression as between hemizygous loci.

7.4.3.4 Covariance Between Drone and Offspring

The genetic interpretation of the worker – drone covariance is only possible if the worker offspring of individual drones in a colony can be identified. The parameter recently gained significance for the selection of varroatosis resistance in honeybees (Moritz and Jordan 1991). The model is simple if we consider the drone as a homzygous diploid individual. We can then use the theory as outlined above. A drone with a pheontype D deviates from the population mean because of the additive gene effects α_a:

$$D = \mu + \alpha_a + \alpha_a. \quad (7.28)$$

With this equation we can use the classical notation for the additive genetic variance in diploid organisms (Weber 1978):

$$\sum_i^n p_i\alpha_i^2 = 1/2V_a. \quad (7.29)$$

A progeny worker, W, is diploid (A_bA_c) and deviates from the mean because of the additive gene effects α_b, α_c and a dominace deviation δ_{bc}. Thus

$$W = \mu + \alpha_b + \alpha_c + \delta_{bc}. \quad (7.30)$$

The covariance between these two individuals can be expressed as

$$\begin{aligned} cov_{wd} &= 2\alpha_a * (\alpha_b + \alpha_c + \delta_{bc}) \quad (7.31)\\ &= 2\alpha_a\alpha_b + \alpha_a\alpha_c + \alpha_a\delta_{bc}. \end{aligned}$$

If we consider allele c to be inherited from an unrelated queen, and a lacking covariance for additive and dominance gene effects, this reduces to

$$cov_{wd} = 2\alpha_a\alpha_b. \quad (7.32)$$

We can now extend this equation for all n alleles in the population and obtain

$$cov_{wd} = 2\sum_i^n p_i\alpha_i^2 = V_a. \quad (7.33)$$

Thus, because of the male haploidy, the covariance between drone and offspring worker directly corresponds to the additive genetic variance.

7.4.3.5 Covariance Between Mid-Parent and Offspring

Also the offspring – mid-parent covariance has to be reinterpreted to fit the case of male haploidy. Let M be the weighted mid-parent phenotype of a queen and a drone.

$$M = \mu + (2\alpha_a + \alpha_b + \alpha_c + \delta_{bc})/2. \quad (7.34)$$

The progeny worker phenotype is as in Eq. (7.30)

$$W = \mu + \alpha_d + \alpha_e + \delta_{de}.$$

The covariance between worker and parent mean can be written as

$$\mathrm{cov}_{wm} = (\alpha_a + \alpha_b/2 + \alpha_c/2 + \delta_{bc}/2)(\alpha_d + \alpha_e + \delta_{de}). \tag{7.35}$$

If the drone and the queen are unrelated, and allele d is inherited by the drone, this reduces to

$$\mathrm{cov}_{wm} = \alpha_a\alpha_d + \alpha_b\alpha_e/2 + \alpha_c\alpha_e/2. \tag{7.36}$$

Extending this equation to the generalized format with n alleles in the population we obtain

$$\begin{aligned}\mathrm{cov}_{wm} &= P(i{=}j)\sum p_i\alpha_i + (1/2)P(i{=}j)\sum p_i\alpha_i + (1/2)P(c{=}e)\sum p_i\alpha_i \\ &= 1/2V_a + 1/8V_a + 1/8V_a = 3/4V_a.\end{aligned} \tag{7.37}$$

This covariance is exclusively used in offspring on parent regression anslysis. Because the total phenotypic variance of the parent mean is expected to be only half of the individual variance (Falconer 1989), we obtain

$$b_{wm} = \frac{\mathrm{cov}_{wm}}{1/2V_p} = \frac{3/4V_a}{1/2V_p} = 3/2h^2 \tag{7.38}$$

$$h^2 = 2/3b_{wm}$$

where b_{wm} = regression slope (worker – mid-parent).

Thus both offspring of drone, and offspring on mid-parent regressions have to be interpreted very differently than in diploid species.

7.4.4 Similarities Between Related Colonies

For both natural and artificial selection, characters of individual bees seem to be less important than characters expressed by whole colonies. This becomes apparent if we look at similarities between colonies survived winter. As long as sufficient honey for winter is stored, the colony will be likely to survive well, irrespective of whether the foraging bees live three or only two weeks. Overwintering is a classical threshold problem, if you do not store enough you die, but if you store too much you may not be very efficient but still survive and therefore be very similar to other surviving colonies (see

Chap. 3). Colony phenotypes are particularly important in commercial breeding programmes where honey production usually is the focus of interest. Characters of individual honeybees are considered to be completely unimportant by practical bee breeders (just like characters of individual cells in cows that are selected for milk production). This leaves us with the question: what are the reasons for similarities between honeybee colonies and can we dissect such similarities in genetic terms?

7.4.4.1 Worker Effects

Certainly, colonies can be similar because the workers they are composed of are similar. We have already discussed the relatedness between worker groups and can directly translate those relationships into genetic covariances by using ϕ^* values for groups and individual ϕ values as in Eq. (7.22).

$$\text{Cov} = 1/2(\phi^* + \phi'^*)V_A + \phi\phi' V_D + (\phi\phi')^n V_{GI}, \qquad (7.39)$$

where V_{GI} = sum of non-linear genotype interactions,
n = group size,
ϕ^* = group relationship, female genes,
ϕ^* = group relationship, male genes,
ϕ = average individual probability for descent identity of female gene,
ϕ' = average individual probability for descent identity of male gene.

A theoretical example for gene interactions may show the meaning of the above equation explicitly. Let us assume we have a locus for hoarding with two alleles *a* (no hoarding) and *A* (1 unit hoarding) and let us further look at the basic two-workers-group for simplicity. Allele *A* is additive to *a* so that *Aa* individuals hoard 1 unit. Under the additive model, groups with the same number of *A* and *a* alleles (*Aa*/*Aa* and *AA*/*aa*) have the same phenotype (Table 7.7). The probability for this to occur is $(\phi'^* + \phi^*)/2$ as we have seen in Eq. (7.39).

If *A* is partially dominant over *a* the picture is completely different. Let the heterozygous workers hoard 1.5 units. Each combination of genotypes now has a specific phenotype. Table 7.7 shows that groups are only identical if the same genotype combination is in the group, even if the interactions between the genotypes are additive. Under a dominance model the group phenotype is determined by genotypic composition rather than by single alleles.

Table 7.7. Phenotypic values (units hoarding) of two-worker groups in a hypothetical hoarding experiment with additivity (left) and partial dominance (right) of *A* (1 unit hoarding) over *a* no hoarding. Note that the interactions between genotypes are considered to be additive in both examples

	Additive			Dominant		
Genotype	AA	Aa	aa	AA	Aa	aa
AA	4	3	2*	4	3.5	2
Aa	–	2*	1	–	3	1.5
aa	–	–	0	–	–	0

* = groups with different genotypes but identical phenotypic value.

Dominance can cause similarity between two groups whenever the identical allele combination appears in one individual each of both groups. This happens in PhiPhi′, the average of the individual dominance relationship, of all cases. Furthermore, non-additive genotype interactions can cause similarities, due to identical genotype combinations in the two groups. Consequently, in the above two-worker group example, we have to estimate the probability for identical pairs of genotypes in both groups $(\text{PhiPhi}')^2$ to evaluate similarities from non-additive genotype interactions. With increasing group size, the probability of finding a matching pair of individuals in each group will increase, and this parameter may become significant if the intragroup genotype interactions are additive. However, interactions in large groups involve many group members. The more complex the interactions between a large number of interacting individuals, the smaller its potential to cause similarities through non-linear interactions on the group phenotype. The probabilities for identical genotype combinations in two groups exponentially descend to zero with increasing group size, and this parameter is less significant in large groups.

This technique can be used for estimating genetic variance components is small worker groups (Moritz 1986b). In natural selection, however, complete colonies are selected which reflect the combined phenotype of queen and workers (and perhaps also drones). The above system therefore is only useful for true characters of worker groups or queenless colonies.

7.4.4.2 Combined Effects

The analysis of colony phenotypes which are composed of a queen effect and a worker effect has been studied by Chevalet and Cornuet

(1982a) and reviewed by Cornuet and Moritz (1989) in detail. Empirical evidence that the queen also has substantial impact on the prima vista "typical" worker trait "honey production" has been repeatedly documented (Bienefeld et al. 1989; Oldroyd 1988; Oldroyd and Goodman 1988, 1989). The principal model to analyze such queen – worker interactions is similar to the model we know from quantitative genetics for individuals. A colony's phenotype, P, is composed of the effect of a queen, Q, the average effect of the workers, W, and non-genetic effects, E:

$$P = Q + W + E. \qquad (7.40)$$

Because of this phenotype composition, we have to consider gene effects of the queen as well as the workers on the colony. Thus, for example there is not only one additive genetic variance component, but there are three of them: one for the queen (V_{Aq}), one for the workers (V_{Aw}), and, because some genes in workers and queen are identical, one component for the correlation between queen and workers (V_{Aqw}). The same is true for the dominance variance components. Let us look at a typical example in honeybee breeding, in which this theory can be applied.

Interpretation of Colony Similarities. Let us assume several unrelated groups of sister queens (queen mother in common) which mate with unrelated drones. If we use these queens in a colony performance test we will have several "families", i.e. groups of colonies which are headed by sister queens. If there is genetic variance for the character tested, sister colonies should be more similar to each other than to the rest of the population. This similarity can be estimated in an analysis of variance via the intraclass correlation. The question is, what does this similarity mean genetically? Here we need the relationships between queens and workers of the corresponding colonies which have been derived above. The pedigree in Fig. 7.10 may help to understand the derivation of the important relationships in a "family" of colonies given in Table 7.8. The expected intra-family covariance, if we assume a ten-drone mating of each queen, is composed as follows [using Eq. (7.39)]:

$$Cov_{family} = 0.6V_{Aq} + 0.16V_{Aw} + 0.6V_{Aqw} + 0.1V_{Dq}. \qquad (7.41)$$

Mother–Daughter Colony Relationship. This system is another design frequently used in quantitiative genetics of honeybees. Let us consider a group of colonies ("mothers") with unrelated naturally mated queens and another group of colonies ("daughters") with naturally mated daughter queens of the first set of queens.

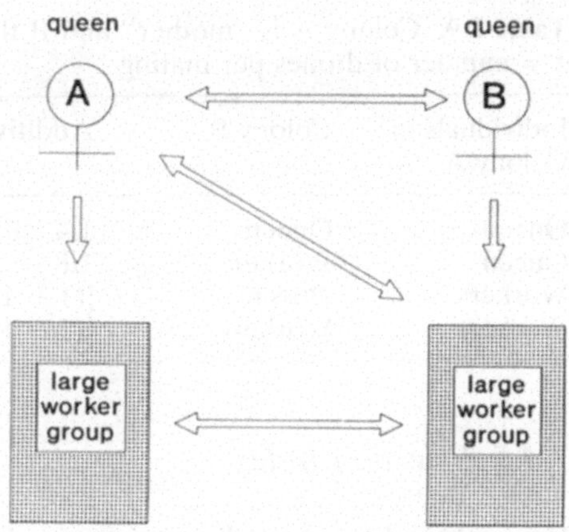

Fig. 7.10. Pedigree to explain the relationships between two colonies headed by queen *A* and *B* respectively)

Table 7.8. Important relationships for the evaluation of the intraclass correlation in a family of colonies headed by sister queens. Note that the queen – workers relationship is not the average individual relationship but the relationship of an individual to a group. The relationship between workers is an approximation of the group relationship between two very large groups (actually infinite group size). z = Number of drones per mating. Further details, see text

Individuals in Colony A	Colony B	Additive relationship	Dominance relationship
Queen	Queen	$[(1 - 1/z + 3/z]/4$	$1/z$
Queen	Workers	$[(1 - 1/z) + 3/z]/8$	0
Workers	Queen	$[(1 - 1/z) + 3/z]/8$	0
Workers	Workers	$[(1 - 1/z) + 3/z]/16$	0

Similarities between "mother" colonies and "daughter" colonies originate from common genes, as in the intra-family system before. We can apply the same notation as in the prior example, but, since the relationships are different, obtain other coefficients for the various variance components (Table 7.9).

If we assume an average mating of ten drones per queen, the similarity between "mother" and "daughter" colony can be decomposed as follows:

$$\mathrm{Cov}_{\text{mother-daughther}} = 0.5V_{Aq} + 0.55V_{Aqw} + 0.05V_{Dqw} + 0.15V_{Aw}. \quad (7.42)$$

Table 7.9. Colony A is "mother" and B the "daughter" colony. z = number of drones per mating

Individuals in Colony A	Colony B	Additive relationship	Dominance relationship
Queen	Queen	1/4	1/z
Queen	Workers	1/8	0
Workers	Queen	[(1 − 1/z) + 3/z]/8	1/2z
Workers	Workers	[(1 − 1/z) + 3/z]/16	0

7.4.5 Selection Index

For practical bee breeding it is essential to have an accurate estimate of the breeding value of a colony. Using the terminology from classical animal breeding, we can define a selection index for a colony, which is slightly more complex than that for individuals. The selection index of a colony is defined as the expectation of the performance of its "daughter" colony (Cornuet and Moritz 1989). Using the sake assumption as those for the estimation of the genetic variance components (natural mating, no relationship between families and drones), a selection index can be computed similarly as in Eqs. (7.28) and (7.29). For the sale of simplicity, let us consider only one character (e.g. honey production) which has be estimated from performance testing in one generation. The available information is restricted to an average family value ($m_{i.}$ for the i-th family) which is composed of several individual colony values (m_{ij}, for the j-th colony in the i-th family). If we standardize the data by the corresponding standard deviation, we obtain a simple index J_{ij} for each colony:

$$J_{ij} = \frac{Cov_{md} - Cov_{an}}{1 - Cov_{sist}}(m_{ij} - m_{i.}) + \frac{Cov_{md} + 11Cov_{an}}{1 - 11Cov_{sist}} m_{i.} \qquad (7.43)$$

where Cov_{md} = Covariance between mother and daughter colony,
Cov_{an} = Covariance between aunt and niece colony,
Cov_{sist} = Covariance between sister colonies.

It becomes clear that we need data from at least two generations to obtain the necessary information for the calculation of this index, since we have to evaluate similarities between two different generations. Once we have an estimate of this index, the genetic progress, δG_{female}, via female selection is simply the product of selection intensity, *i*, and the standard deviation of the selection index.

$$\delta G_{female} = i\sigma_J. \tag{7.44}$$

The selection intensity can be estimated according to Hill (1976):

$$i \sim i. - (n - 1)/[2i.(ns - nst + st + 1)], \tag{7.45}$$

where n = number of colonies per family,
s = number of selected colonies per generation,
t = correlation between indexes of sister colonies.

By using Smith's (1969) equation

$$i. \sim 0.8 + 0.41 \ln(n - 1). \tag{7.46}$$

Following Cornuet and Moritz's (1989) notation, t can be computed as follows:

$$t = \frac{Cov_{index\ sister}}{V_J}. \tag{7.47}$$

The variance of the selection V_J is

$$V_j = \frac{(n - 1)(Cov_{md} - Cov_{an})^2}{n(1 - Cov_{sist})} + \frac{[Cov_{md} + (n - 1) Cov_{an}]^2}{n[1 + (n - 1) Cov_{sist}}. \tag{7.48}$$

The covariance between the indexes is equal to:

$$Cov_{index\ sist} = \frac{Cov_{sist}(Cov_{md} - Cov_{an})^2}{(1 - Cov_{sist})^2} + \frac{(Cov_{an} - Cov_{md}Cov_{sist})[2Cov_{md} + (n-2) \times (Cov_{an} + Cov_{md}Cov_{sis}t) - (2n-2) Cov_{an}Cov_{sist}]}{(1 - Cov_{sist})^2 [1 + (n - 1) Cov_{sist}]}. \tag{7.49}$$

The male genetic progress in natural mating depends on mating control and on β, the proportion of drones in the population coming from selected colonies. Furthermore, the regression of the worker contribution on the selection index $b_{w/J}$ is needed to obtain the genetic progress

$$\delta G_d = \beta i b_{w/J} \sigma_J. \tag{7.50}$$

However, this entity is extremely difficult to obtain in natural breeding schemes, because β has to be determined with elaborate field experiments and $b_{w/J}$ is not available from the breeding scheme itself unless three successive generations are available. A detailed derivation of $b_{w/J}$ for standardized data ($\sigma = 1$) is given by Cornuet and Chevalet (1987, p. 260).

$$
\begin{aligned}
b_{w/J} = 1/V_J \Bigg\{ & (V_{Aq}/2 + V_{Aw}/4) \Bigg[\frac{Cov_{md}(1 - W)}{1 - z} \\
& + \left(\frac{Cov_{md}(W - Z)}{Z - Z^2} - \frac{Cov_{md}(ZCov_{gg} - WCov_{md}{}^2)}{Z(Z - Cov_{md}{}^2)} \right) \\
& \times (1/n + (n - 1)\,\phi/n) \Bigg] + (V_{Aq}/2 + \phi V_{Aw}) \\
& \times (ZCov_{gg} - WCov_{md}{}^2)/(Z - Cov_{md}{}^2) \Bigg\},
\end{aligned} \tag{7.51}
$$

where n = number of colonies per family.

Cov_{gg} = covariance grandmother − grandchild colony

$$
W = \frac{Cov_{md}/n + (n - 1)Cov_{an}/n}{Cov_{md}} \tag{7.52}
$$

$$
z = \frac{1 + (n - 1)Cov_{sis}}{n} \tag{7.53}
$$

In practical bee breeding with only marginal mating control, genetic progress through the males is low anyway. However, with instrumental insemination or isolated mating stations (e.g. on islands) the progress through males can be increased substantially.

7.4.6 Empirical Examples

After these theoretical reflections, the non-geneticist reader might ask: "what do I do with all of this?" or "How does this relate to anything in the real world?" It actually does relate to the real world, and therefore we will give some numerical examples for the application of the theory to reveal the importance of the quantitative theoretical framework for the understanding of selection in populations of superorganisms.

7.4.6.1 Analysis of Individual Characters

Characters of individuals can be assessed with classical quantitative genetics theory. It is necessary that the character can be studied at the individual level and is not expressed only in a group context. Morphological characters certainly fall into this category. We will use hypothetical data on a frequently studied character, the number of

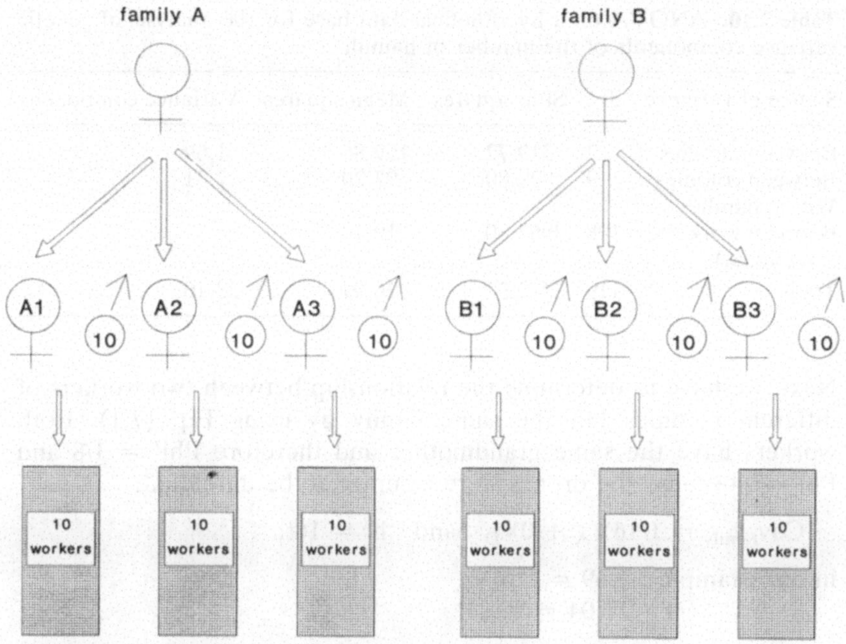

Fig. 7.11. Pedigree in a hypothetical experiment to determine genetic variance component for the number of hamuli. Four queens (only three shown) per family (two of three shown) are inseminated with ten unrelated drones each. Ten workers per colony are analyzed

hamuli on the forewing of a worker (e.g. Drescher 1971; Gonçalves 1972; Gonçalves and Stort 1978; Oldroyd and Moran 1983). Let us assume we have three honeybee families, each composed of four colonies (thus, in total, we have 12 colonies). The queens have naturally mated with ten drones, which are assumed to be unrelated. From our hypothetical experiment we have data on ten workers per colony which enables us to compute an analysis of variance on the data set (Fig. 7.11). Our hypothetical results are shown in Table 7.10.

We now have to know what the various variance components actually mean in genetic terms. The first step is the estimation of similarities in the form of covariances in order to evaluate genetic variance components. In an ANOVA, similarities can be expressed with the intraclass correlation, t, within a family and within a colony:

$$t_{family} = 1.69/28.11 = 0.06, \quad t_{colony} = 7.31/28.11 = 0.26.$$

Table 7.10. ANOVA on a hypothetical data base for the analysis of genetic variance components of the number of hamuli

Source of variance	df	Sum squares	Mean squares	Variance component
Between families	2	319.72	159.86	1.69
Between colonies Within families	9	829.80	92.20	7.31
Between workers (residual)	108	2062.80	19.10	
Total	119	3212.32	26.99	28.10

Next, we have to determine the relationship between two workers of different colonies but the same family by using Eq. (7.1). Both workers have the same grandmother and therefore Phi′ = 1/8 and Phi = 0 because the drones were assumed to be unrelated:

$$\mathrm{Cov}_{\mathrm{family}} = 1/16\mathrm{V_A} + 0\mathrm{V_D} \quad \text{and} \quad \mathrm{h}^2 = 16\mathrm{t},$$

in our example $1.69 = 1/16\mathrm{V_A}$
$27.04 = \mathrm{V_A}$,

from this we get a first estimate for the selectability

$$\mathrm{h}^2 = \mathrm{V_A}/\mathrm{V_P} = 27.04/28.11 = 0.96.$$

For sister workers in the same colony, we have to use the average individual sister relationship Phi′ = 1/2 and Phi′ = 1/z (z = number of drones) which, if z = 10, according to Eq. (7.23) gives

$$\mathrm{Cov}_{\mathrm{colony}} = 0.27\mathrm{V_A} + 0.05\mathrm{V_D}.$$

We obtain for our numerical example:

$$7.31 = 0.27\mathrm{V_A} + 0.05\mathrm{V_D}$$

$$7.31 = 0.27 \times 27.04 + 0.05\mathrm{V_D}$$

$$0.18 = \mathrm{V_D}.$$

In some cases we might obtain a negative estimate for the dominance effects which is, of course, biologically impossible and an artifact from the estimation procedure and a too small sample size. In our hypothetical example, we see that dominance may be less important for the phenotypic variation, whereas 96% is due to selectable additive gene effects. This example shows that no highly sophisticated breeding is necessary for the estimation of the basic parameters for characters of individual workers.

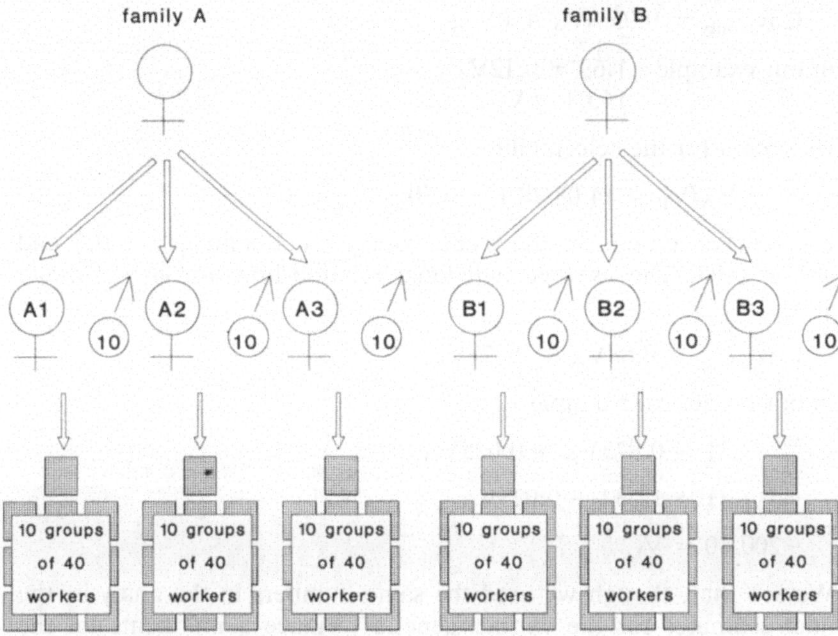

Fig. 7.12. Same pedigree as in Fig. 7.11 but characters of groups of 40 workers (ten groups per colony) are analyzed

7.4.6.2 Analysis of Social Traits of Small Worker Groups

Let us again work with a hypothetical, but typical example for this case. The response to alarm pheromone has been shown to be truly a group trait. Individual bees show no response to the pheromone whereas in groups the workers get highly agitated. Southwick and Moritz (1985) studied this behaviour and it was genetically analyzed by Moritz (1986b). We approach this group problem by using the theory from Section 7.4.3. Let us take the same pedigree as in the prior example with the hamuli: three families with four colonies each. Let us further take the same hypothetical data from the ANOVA table, being aware that ten groups of 40 workers instead of ten single workers per colony produced the data (Fig. 7.12). This does not affect the statistical outcome of the ANOVA but it does change the genetic interpretation of the data. For the relationship between two 40-worker groups from sister colonies we obtain:

$\phi'^* = 0.95/4$ and $\phi^* = 0$, thus

$$\mathrm{Cov}_{\mathrm{family}} = 0.95/8V_A + 0V_D;$$

in our example $1.69 = 0.12V_A$
$14.08 = V_A$,

we obtain for the selectability

$$h^2 = V_A/V_P = 14.08/28.1 = 0.50.$$

For worker groups in the same colony we obtain $\phi^* = 0.8$ and $\phi'^* = 0.95$. The average individual relationships are $\phi = 1/z$ and $\phi' = 1/4$,

$$\mathrm{Cov}_{\mathrm{colony}} = 0.88V_A + 0.025V_D,$$

in our numerical example

$$7.31 = 0.875V_A + 0.025V_D$$

$$7.31 = 12.32 + 0.025V_D$$

$$-200.40 = V_D.$$

We see that, though we used the same numbers in the analysis, the final estimates for the various genetic variance components for the social character are quite different than in the individual case.

7.5 Behavioural Genetics

Although honeybees play a central role in classical and modern ethological studies, the genetic analysis of honeybee behaviour is only a sideline in behavioural genetics. This is amazing, because honeybees provide all the necessary characteristics of a genetic model system: controlled mating, rapid generation cycle, mass production, and low costs for maintenance. The male haploid system allows elegant approaches whenever single gene expressions have to be studied. As soon as the genetics of social behaviour have to be analyzed, honeybees are the only readily available test system.

7.5.1 Analysis of Individual Behaviour

7.5.1.1 Learning Behaviour

The social structure of the honeybee colony cannot survive without communication. Communication, however, is only helpful if the members of the colony have two fundamental abilities. They need to

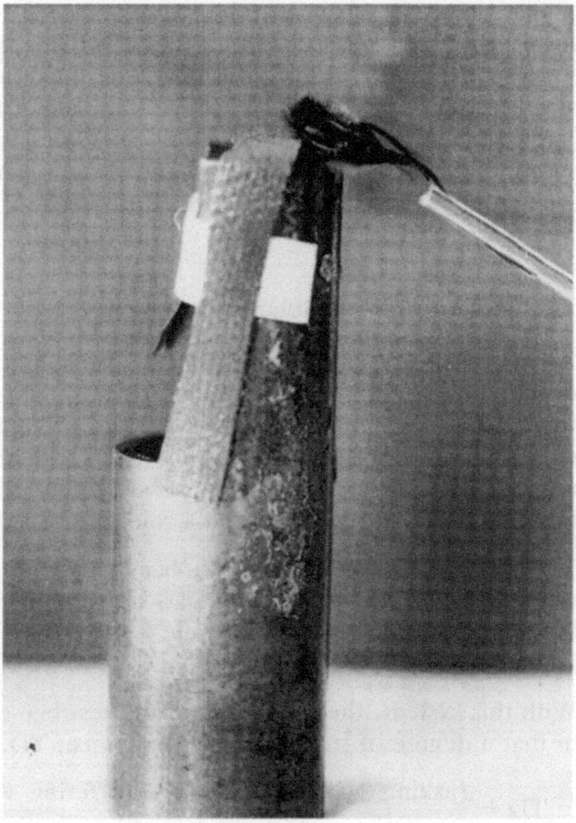

Fig. 7.13. Honeybee in brass tube. (Photo Menzel)

learn a certain message from communication and they have to *remember* a message to communicate it to nest members at a later time. The genetics of learning and memory has been studied in some detail by Brandes and co-workers (Brandes and Moritz 1983; Moritz and Brandes 1987; Brandes 1988, 1990, 1991; Brandes et al. 1988). They used the proboscis extension reflex designed by Kuwabara (1957) as the basic learning paradigm (Fig. 7.13).

Honeybee workers are individually constrained in small brass tubes with only the heads exposed. If the antennae are touched with a droplet of sugar syrup, the worker extends its proboscis in a reflex response. Workers which are exposed to odours while tasting the syrup, learn and remember these odours. In subsequent tests they expose the proboscis just to the presentation of the odour without the sugar syrup. Not all bees are equally good at learning. Some workers learn with high precision after only one conditioning phase. Others have extreme difficulties in associating odour and syrup. Brandes (1987, 1988, 1990a,b) could select for lines with good and poor learning ability and obtained a large genetic variance for this character in honeybee populations which resulted in h^2 estimates ranging from 0.38–0.44.

7.5.1.2 Trophallactic Behaviour

A bio-assay to study trophallactic behaviour quantitatively was developed by Korst and Velthuis (1982). A few workers (two or three) were kept in small cages and their trophallactic interactions monitored. Workers could either offer food or ask for food. This assay allows for the following possible interactions per encounter between a bee to be tested and a reference bee:

1 = both bees ask; 2 = both bees offer; 3 = test bee offers – reference bee asks; 4 = test bee asks – reference bee offers; 5 = test bee offers – reference bee ignores; 6 = test bee asks – reference bee ignores

With this system, the behaviour of the test bee can be easily ranked, in that a degree of trophallactic dominance, TD, is calculated

$$\mathrm{TD} = \frac{(\text{asking} + \text{being offered}) - (\text{offering} + \text{being asked})}{\text{total number of interactions}}.$$

Moritz and Hillesheim (1985) used this assay to evaluate the genetic variance of trophallactic behaviour and found a remarkable estimate for trophallactic dominance ($h^2 = 0.32$) in a population of the Cape honeybee (*Apis mellifera capensis*). The behavioural classification ranged from 0.92 (highly dominant = mostly fed by others) to −0.35 (subordinate = mostly offers food). The importance of these data becomes apparent, if we remember that this is not just phenotypic variation but up to 32% heritable and selectable variation. The well-fed dominant workers apparently mainly invest in their ovaries. The trophallactic dominant workers are more likely to become laying workers in queenless colonies than others (Korst and Velthuis 1982), which is an important issue in individual intracolonial selection among workers.

7.5.1.3 Reproductive Dominance

The trophallactic behaviour in social insects often reflects the hierarchy status of individuals in the colony (Wilson 1971; Franks and Scovell 1983). This seems to be valid for honeybee workers also, as we have seen above. Well-fed, dominant workers in queenless colonies of the Cape honeybee produce the queen substance (Hemmling et al. 1979; Crewe and Velthuis 1980; Crewe 1982; Velthuis et al. 1990). Its main compound, 9-oxo-2-decenoic acid (9-ODA), has been shown to suppress ovary development in workers (Butler and Fairey 1964), which is an extremely efficient way to maintain a hierarchy in the colony. Moritz and Hillesheim (1985) studied the genetic variance for the production of 9-ODA, the developmental stage of worker ovaries, and number of eggs laid in a given experimental period. They used a sib analysis on the worker offspring of laying capensis workers and found high estimates for the selectability of these characters (ranging from 0.29 for number of eggs laid to 0.89 for 9-ODA production). The high values for the selectability are particularly surprising because these characters are evidently tightly linked to direct individual fitness. Workers that produce offspring certainly are fitter than those remaining sterile, assuming the indirect fitness of dominant and subordinate workers in queenright colonies is equal. Since it was natural populations that were studied, forces other than direct individual fitness must be strong to maintain the large genetic variability. We will readdress this problem again when we discuss the modes of natural selection in honeybee populations (see Chap. 8).

7.5.2 Analysis of Social Behaviour

With honeybees genetic phenomena in social groups ranging from 2 to 80000 individuals can be studied with a scrutiny and precision not available in any other animal system. In the following, we will focus on three classical examples for the genetic analysis of group behaviour.

7.5.2.1 Hoarding Behaviour

When honeybee workers are caged in small groups of 50 workers and fed with sugar syrup, they will store some of this syrup into cells of a piece of comb, a trait coined as "hoarding behaviour" by Free and Williams (1972). Initially, a correlation between hoarding in the laboratory and honey crop in the field was expected. Disruptive selection for fast and slow hoarding workers gave promising results, in that there was considerable selectable genetic variance for this character (Rothenbuhler et al. 1979). However, the selection results were asymmetrical and the response in the slow line was much stronger than in the fast (economically desirable) line. The initial enthusiasm was dampened after Rinderer and Collins (1986) pointed out that the correlation between hoarding in the lab and honey production is ambiguous. Field tests do not reveal significant differences between the honey production of slow- and fast-hoarding bees after four generations of selection for hoarding behaviour.

Although hoarding behaviour seems to have only limited predictive power for honey production of colonies, it is a very good bioassay to analyze group behaviour genetically. For example, Moritz and Hillesheim (1989) found that genotypic variance can have substantial impact on hoarding. They composed experimental groups of workers from one, two, three or four patrilines from the same colony (the queens were inseminated with four drones, each carrying a different combination of genetic markers). Non-mixed worker groups of each of the four patrilines had the same hoarding ability. Genotypically mixed groups, however, hoarded significantly less syrup and there was a negative correlation between the number of patrilines in the group and the amount of hoarded syrup. Apparently, strong nonlinear genotype interactions affected the expression of this social behaviour. The mixed group consumed more sugar and stored less. Does this indicate that there is more conflict in such groups and less coordination? It could be, but it also might not be, and we certainly need more experiments to draw meaningful conclusions at the behavioural, ecological, or evolutionary level. Nevertheless, we see that soical behaviour is much more than something like an average worker behaviour. Here, we have the experimental evidence that the phenotype of a superorganism is strongly affected by genotype interactions which do not have to be additive at all.

7.5.2.2 Hygienic Behaviour

The textbook classic, and probably the best-known example of honeybee genetics, is the inheritance of hygienic behaviour. Rothenbuhler (1964a), following observations of Park (1937), showed that hygienic behaviour is one of the major factors in resistance to infections of American foul brood (*Bacillus larvae*). Hygienic honeybee workers detect diseased brood, uncap the wax capping of the brood cells and remove the infected larvae. From a classical back-crossing experiment of F_1 queens and drones of parental hygienic and non-hygienic colonies, Rothenbuhler (1964b) concluded that a simple two-locus model suffices to explain the genetics of hygienic behaviour. One recessive allele *u* for uncapping and one allele *r* for removing. It was indeed exciting to find a complex social behaviour to be determined by a very basic genetic mechanism. However, very soon Rothenbuhler and co-workers (Thompson 1964; Trump et al. 1967; Palmquist-Momot and Rothenbuhler 1971) indicated that the phenomenon might be more complex. The major problem is to decide, which colony is "hygienic" and which is not. The differences between behavioural types are not so clear-cut in Rothenbuhler's (1964b) back-crossing experiment. Since full-size colonies were studied, it actually would be extremely difficult to make precise estimates of single gene actions, because interactions between the workers may easily override certain individual phenotypes. A good example came from a Trump et al. (1967) study, which showed that mixed groups of hygienic and non-hygienic workers always expressed the hygienic phenotype. The hygienic worker subgroup was large enough to take care of all the uncapping and removing. The central problem in Rothenbuhler's (1964) original analysis is to assign behavioural phenotypes to distinct classes of individuals with corresponding genotypes. Instead of using a simple classification of "hygienic" and "non-hygienic", Rothenbuhler's (1964b) data can be analyzed with statistical methods. After searching for clusters, one can specify a mean and a variance for behavioural types (Moritz 1988b). With this approach it becomes apparent that the original data do not support the two-locus hypothesis. The data for removing behaviour significantly deviate from a bimodal distribution, which is expected if there are only two types of workers, removers and non-removers. At least two loci form the genetic basis for removing, but in light of possible epistatic interactions any other genetic model is equally likely. Quantitative genetic studies using a laboratory bio assay (Milne 1982, 1983), showed only moderate estimates for genetic variance (Milne 1985d,e) with $h^2 = 0.14$ for uncapping and $h^2 = 0.02$ for removing. This is another indication that either non-linear interactions or environmental effects strongly affect the expression of hygienic behaviour.

7.5.2.3 Defensive Behaviour

Although honeybees are widely regarded as beneficial insects, they certainly have undesirable traits. The worst character which conflicts most dramatically with the interests of human society is defensive behaviour (see Chap. 4). The disturbance threshold for a colony defence varies from subspecies to subspecies and from colony to colony. Also the extent of the colony defence can vary substantially. A large part of this variation is due to environmental effects (Brandeburgo et al. 1982; Collins 1981; Collins and Rinderer 1985; Drum and Rothenbuhler 1984; Schua 1952; Southwick and Moritz 1986), but repeated studies in the field and in the laboratory have shown that a large amount of the phenotypic variation is due to selectable genetic variation (Collins 1979; Collins et al. 1984; Moritz 1986b; Moritz et al. 1987a,b; Rinderer et al. 1983). Defensive behaviour has been the focus of interest in applied honeybee biology recently, especially because of the rapid dispersion of Africanized honeybees in South and Central America. In a vast number of studies, defensive behaviour has been quantitatively analyzed after the initial field tests of Stort (1974). The various field tests follow the same pattern: the colony is alarmed mechanically or by pheromone application, and the number of the attacking bees is determined either via the number of stings in a leather target or by simply counting the bees of the defence force. Laboratory tests try to eliminate the environmental variation and to improve the selectability. Collins (1979) tested caged group of bees which were briefly exposed to the alarm pheromone. She behaviourally evaluated the response of the bees by the time they needed to react and the intensity of the response. A true quantitative test was used by Southwick and Moritz (1985) and Moritz et al. (1985). They determined the increase in metabolic activity after pheromone exposure as a measure of reactivity (Fig. 7.14). The estimates for the genetic variance are given in Table 7.11.

In a series of papers Stort (1975a,b,c, 1976, 1978) presents studies on various characters related to defensive behaviour and tries a genetic analysis. Stort claims that several sets of two-locus systems control defence behaviour. Rinderer and Collins (1986) correctly argue that there is only poor evidence for a two-locus system controlling the number of stings in the glove and another two-locus system for the number of stings in the leather ball, as proposed by Stort. For the practical purpose of selecting docile honeybees and for the understanding of the importance of defensive behaviour in natural selection, the quantitative genetic approaches should give better information than classical genetic analyses. However, at present the quantitative h^2 estimates range from 0.03 to 1.28 and precludes any sensible conclusion. Clearly, even more work needs to be done to elucidate the picture of the genetics of colony defence.

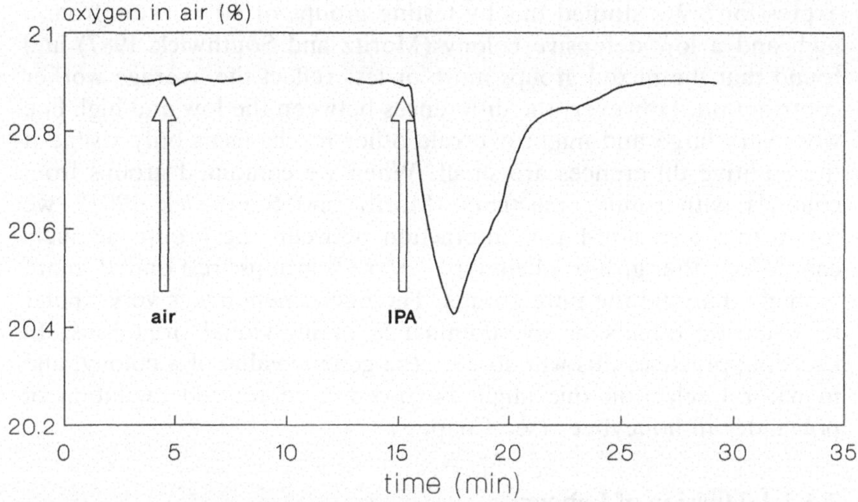

Fig. 7.14. A group of 40 bees was exposed to isopentyl acetate (IPA, major compound of alarm pheromone) in a flow-through system for 5 s. Exposure to plain air did not cause any reaction. IPA increased the metabolic rate by about 20 µl O_2/min per bee

Table 7.11. Estimates for the selectability, h^2, of various behavioural traits related to the colony defence

Behaviour	Assay	h^2 Estimate	Author
Time to react to alarm	Field	0.31–1.28	Collins et al. (1984)
pheromone	Field	0.03	Rinderer et al. (1983)
	Lab	0.68	Collins (1979)
Time to react to target	Field	0.31–0.69	Collins et al. (1984)
Number of stings in	Field	0.57	Collins et al. (1984)
target	Field	0.30–0.57	Moritz et al. (1987a)
Metabolism in alarm	Lab	0.36	Moritz et al. (1987b)
reaction	Lab	0.98[a]	Moritz (1986)

[a] Based on pooled data of different subspecies; does not reflect genetic variance of a natural population.

The complexity of defence behaviour becomes obvious in social groups composed of mixed genotypes, a common situation in the honeybee colony where several patrilines coexist in the colony. What happens if we combine a group of docile workers with highly defensive ones? Will it express an average group phenotype or will non-linear interactions among the group members affect the group phenotype expression? We studied this by testing groups of 40 workers from a high and a low defensive colony (Moritz and Southwick 1987) and found that the mixed groups more or less reflect the average worker contribution. However, the differences between the low and high line were very large and might overrule other mechanisms only visible if the additive differences are small. When we combined groups from colonies with similar reactivity (Moritz and Southwick 1987), we found that also non-linear interaction between the group members can affect the group phenotype. Mixed groups responded more strongly than did the pure groups. The phenomenon was very similar to heterotic effects or overdominance in individual organisms. In breeding practice, this will obscure the genetic value of a colony, and in natural selection this might be one reason for the evolution of polyandry in honeybees (see Chap. 8).

7.5.2.4 Division of Labour

Environmental effects, such as age of workers, are major determinants for polyethism (Rösch 1927, 1930; Lindauer 1952). However, the system is not fixed and workers show considerable variance for performing certain tasks (Seeley 1982a). Some of this variability seems to be of genetic origin. We already discussed the genetic variance for laying and subordinate workers in queenless colonies (Moritz and Hillesheim 1985). Actually, this is nothing more than genetically determined division of labour, such that there are workers which lay the eggs and those that take care of them. Similar results were recently found by Page and co-workers (Page and Robinson 1991; Page et al. 1989a,b; Calderone and Page 1987; Rothenbuhler and Page 1989). They observed that workers of particular patrilines, which they could identify by isozymes, specialized in certain tasks (Fig. 7.15).

Kolmes et al. (1989) and Frumhoff and Baker (1988) came to the same conclusion using phenotypic mutant strains of honeybees. For example, in a given colony, workers of one patriline can be represented in the guard force at a significantly higher frequency than others. Yet another patriline could be more frequently represented in the "undertaker" worker group that removes debris and dead bees. These studies, revealing genetic variance for division of labour, do not falsify the whole concept of age-dependent polyethism. They just show that it is not completely unimportant which genotype a honeybee has when it is performing a certain task. The patrilines are not distributed at random over the various behavioural categories found

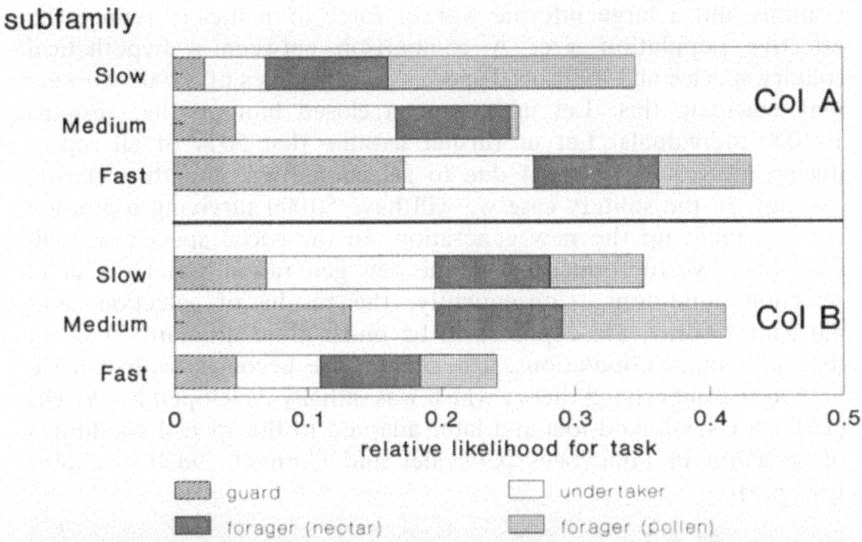

Fig. 7.15. Labour allocation of the various patrilines in two colonies (*A* and *B*). The patrilines were identified by isozyme markers (*slow*, *medium*, *fast*) at the MDH locus. Clearly, there is strong specialization for each of the subfamilies. For example, in colony *A*, workers of the "slow" patriline are more likely to become foragers than the "medium" workers which most likly will join the undertaker force (removal of debris and dead bees). (Data of Robinson and Page 1989a)

in a honeybee colony. The population genetic and evolutionary consequences are not yet completely understood, but it would certainly be of high selective advantage at the superorganismic level if the queen could select a good mix of drones on her mating flight in order to provide the necessary genetic variability for the colony (see Chap. 8).

7.6 Population Genetics

7.6.1 The Effective Population Size of Superorganisms

Male haploid populations have a smaller effective population size, N_e, than equally large diploid populations, if the sex ratio is not strongly female-biased (Crozier 1976). In honeybee populations, it is not only male haploidy that affects N_e. The superorganism system itself has a much stronger impact on N_e than the male haploid population structure. Eusociality with colonies of overlapping generations and a large infertile worker force dramatically reduce the effective population size. A comparison between a hypothetical solitary species and a eusocial species with colonies of 10000 workers may elucidate this. Let us assume a closed biotope that sustains 100000 individuals. Let us further assume that 50% of all reproducing individuals are lost due to selection (e.g. predation during mating). In the solitary case we will have 50000 surviving reproductives to build up the new generation. In the social species we will have only five reproductives for the new generation under the same selection conditions. Consequently, the results of selection, and particularly drift, are expected to be much more dramatic in social than in solitary populations. The significance becomes evident if we look at the underlying theory which was initially developed by Wright (1933) for sex-linked loci and later adapted to the special conditions of selection in honeybees [Chevalet and Cornuet 1982b, see also Eq. (6.3)]:

$$1/N_e = \frac{4z + 2}{9zN_q}, \tag{7.54}$$

where z = number of drones mating one queen,
N_q = number of selected colonies.

Obviously, the larger the number of colonies, the larger is N_e. Furthermore, it becomes clear that multiple mating does increase N_e, but not as much as one might assume at first. This is because the genetic variability of the drones is limited by the number of queens in the population. Even with a very large z, N_e will never exceed

$$\lim_{z \to \infty} N_e = 2.25N_q \tag{7.55}$$

whereas with an increasing N_q, N_e grows proportionally. Thus, again we see, that the colony is the critical unit in yet another genetical discipline. The number of colonies sets the limit for the effective size, and the number of individual drones and queens has a much smaller impact.

7.6.2 Polymorphisms in Nuclear Genome

7.6.2.1 Isozyme Polymorphism

The limitation of the effective population size has been suggested as having consequences for the maintenance of polymorphic loci. The low variability found in isozyme studies on honeybees (Table

7.7) may be one phenomenon caused by the small N_e. The average decrease of heterozygosity per generation equals $1/(2N_e)$ and a small effective population size should result in a rapid loss of heterozygosity. On the other hand, the wide mating range of honeybees may counteract this effect (Chap. 6). Nevertheless, only few enzyme systems proved useful in population genetic studies (Table 7.12). Considerable variability is found for malate dehydrogenase (MDH), alcohol dehydrogenase (ADH), malic enzyme (ME) and several esterase loci (Est). Recently, the esterase six locus (Est-6) was also shown to be polymorphic in Italian populations (Basiolo and Comparini 1990). Pamilo et al. (1978) found that genetic variability was particularly reduced in eusocial Hymenoptera. Besides male-haploidy also the controlled environment in a colony of eusocial insects, similar to the homoeostasis of individual organisms, may be a further reason for a low genetic variability. However, although a heterogeneous environment generally is expected to favour genetic variation (e.g. Levene 1953; Dempster 1955; Haldane and Jayakar 1963), many conditions are possible under which a variable environment reduces the genetic variance below the variability level expected under neutrality (Pamilo 1988). Robertson (1962) showed this for overdominance in finite populations, and Hedrick (1986) found similar results in a more general model for multiple niche polymorphisms. If there is habitat selection the chances for stable polymorphisms are even smaller (Hoekstra et al. 1985). Nevertheless, there is one common feature in the above models. The smaller the N_e, the faster the elimination of polymorphisms, and this certainly affects social populations.

7.6.2.2 Molecular DNA Markers

A very powerful tool to reveal polymorphisms in honeybee populations is given through molecular DNA markers. Hall (1986) used diagnostic DNA probes that revealed characteristic differences between Africanized and European honeybees. The use of synthetic oligonucleotides in DNA fingerprinting seems to be even more powerful to reveal genetic variability in honeybee populations. Moritz et al. (1991c) could discriminate between super- and half-sisters using the DNA fingerprint technique. Such probes prove extremely helpful in the analysis of genetic intra-and intercolonial variation. So far, only preliminary work has been done, but DNA fingerprinting will gain a central role in the future research of the genetics of superorganisms.

Table 7.12. Enzyme systems and allele frequencies in honeybee populations. Mdh = malate dehydrogenase; Adh = Alcohol dehydrogenase; Est = Esterase; Me = Malic enzyme

Author	Mdh1			Adh1			Est1			Est-3		Me		
	S	M	F	S	M	F	S	M	F	S	F	S	M	F
Sylvester (1976)	0.70	0.11	0.20	–	–	–	–	–	–	–	–	–	–	–
	0.01	0.16	0.84	–	–	–	–	–	–	–	–	–	–	–
	0.39	0.37	0.24	–	–	–	–	–	–	–	–	–	–	–
	0.50	0.13	0.37	–	–	–	–	–	–	–	–	–	–	–
Contel et al. (1977)	0.71	0.15	0.14	–	–	–	–	–	–	–	–	–	–	–
	0.03	0.20	0.77	–	0.44	0.56	–	–	–	0.95	0.05	–	–	–
Gartside (1980)	0.31	0.32	0.37	–	0.20	0.80	–	–	–	1.0	–	–	–	–
	0.70	0.10	0.20	–	0.30	0.70	–	–	–	1.0	–	–	–	–
	0.20	0.50	0.30	–	0.30	0.70	–	–	–	0.98	0.02	–	–	–
Nunamaker (1980)	0.10	0.20	0.70	–	–	–	–	–	–	–	–	–	–	–
	–	0.01	0.99	–	–	–	–	–	–	–	–	–	–	–
	–	–	1.0	–	–	–	–	–	–	–	–	–	–	–
Nunamaker and Wilson	0.04	0.03	0.93	–	–	–	–	–	–	–	–	–	–	–
(1981)	–	–	1.0	–	–	–	–	–	–	–	–	–	–	–
Cornuet and Louveaux	0.75	–	0.25	–	–	–	–	–	–	–	–	–	–	–
(1981)	–	1.0	–	–	–	–	–	–	–	–	–	–	–	–
Badino et al. (1983)	0.77	0.03	0.21	–	–	–	–	–	–	–	–	–	–	–
	–	1.0	–	–	–	–	–	–	–	–	–	–	–	–
Martins et al. (1977)	–	–	–	0.05	0.70	0.25	–	–	–	–	–	–	–	–
	–	–	–	–	0.10	0.90	–	–	–	–	–	–	–	–
Mestriner and Contel (1972)	–	–	–	–	–	–	–	–	–	0.91	0.09	–	–	–
	–	–	–	–	–	–	–	–	–	0.98	0.02	–	–	–
Bitondi and Mestriner (1983)	–	–	–	–	–	–	0.02	–	0.98	0.98	0.02	–	–	–
Badino et al. (1982, 1984,		Clines		–	–	–		Clines		–	–	–	–	–
1985, 1988)	0–0.9	0–1	0–1	–	–	–	0–1	0–1	0–0.6	–	–	–	–	–
Sheppard and Berlocher	0.04	0.85	0.11	–	–	–	–	–	–	–	–	0.14	0.86	–
(1984, 1985)	0.64	0.06[a]	0.30	–	–	–	–	–	–	0.94	0.06	–	0.94	0.06
Lobo et al. (1989)		Clines for all enzyme systems in South American samples												
Del Lama et al. (1990)		Clines for all enzyme systems in South American samples												

[a] Mdh-1 M allele 87.

7.6.3 *Inbreeding*

7.6.3.1 Inbreeding and Practical Breeding Schemes

The constraints set by the male haploid structure of honeybee populations have not only consequences for natural selection but also for artificial selection by man. In small populations related drones and queens will mate and inbred individuals will appear. Inbreeding in diploid individuals is quantified by the inbreeding coefficient, *F*, which is the probability of finding two alleles identical by descent at the same locus. Inbred organisms often show a reduced vigour due to inbreeding depression, and breeding schemes can quickly suffer from serious inbreeding depressions. For example, thermoregulation, hoarding and recruiting ability are substantially impaired in inbred honeybee workers (Brückner 1976, 1979, 1980). The same is true for brood rearing, hygienic behaviour and honey production of colonies with inbred workers (Plass 1953; Bienefeld 1988; Bienefeld et al. 1989). The increase of the inbreeding coefficient has been studied in open and closed population breeding schemes (Chevalet and Conuet 1982b, Moritz 1984b, Cornuet and Moritz 1989).

7.6.3.2 Inbred Superorganisms

When we deal with colonies as superorganisms, we should focus on inbreeding not only from the individual but also from the colonial point of view. In individuals, the inbreeding coefficient reflected the probability for descent of identical alleles at a randomly chosen locus. A close transformation of this definition to the superorganism concept is given by the probability to find descent-identical alleles at the same locus in two randomly chosen alleles. This definition, however, is close to what we have derived in Eq. (7.12) in Section 7.3.4.1, the average individual relationship ϕ between the workers of the colony. In addition, we have to look at the probability that both chosen alleles are in the same individual. Thus, the inbreeding coefficient, F^* of a superorganism can be represented by

$$F^* = (1 - 1/n)\phi + F/n, \qquad (7.56)$$

where ϕ = average individual relationship between two group members,
F = average inbreeding coefficient of a group member,
n = group size.

Such a measure could become very meaningful for natural and artificial selection, if the initial evidence of the importance of genotypic variance for social behaviour proves to be important for fitness or for economic productivity of honeybee colonies.

7.6.4 *Variation of Mitochondrial Genes*

Genetic variation not only occurs in the chromosomal genes but also in mitochondrially housed genes. Because mitochondria are expected to be maternally inherited, they offer good opportunities to dissect selection on the male and the female sex. However, are mitochondria exclusively maternally inherited in honeybees? Honeybees have a polyspermic fertilization, i.e. many sperms enter the egg including their mitochondria-rich tail. Observations of Nachtsheim (1914) indicate that only one sperm participates in the zygote formation and the other accessory sperms somehow degenerate. Occasionally these accessory sperms gain their own mitotic activity, form male tissue, and finally gynandromorphs develop (Woyke and Hillesheim 1989). This might imply that there is also a fair chance for male mitochondria to survive in the egg after fertilization.

7.6.4.1 Mitochondrial DNA of Honeybees

MtDNA of honeybees has been analyzed in several recent studies (Moritz et al. 1986; Smith 1988; Smith and Brown 1988; Smith et al. 1989; Hall and Muralidharan 1989; Crozier et al. 1989; Hall 1990). Several authors (Hall and Muralidharan 1989; Smith et al. 1989; Moritz and Meusel 1991; Sheppard et al. 1991) used this technique in combination to analyze the Africanized bee problem (see Chap. 8). The 18 000 base-pair cyclic DNA have been characterized by restriction site maps (Fig. 7.16) and several regions have been sequenced. The cytochrome oxidase genes as well as several RNA coding regions have been identified (Crozier et al. 1989; Vlasak et al. 1987). There is considerable variation among the various species of honeybees. Particularly the cytochrome oxidase region I and II (CO-I, CO-II) have been studied in detail. Crozier et al. (1989) sequenced this region for *Apis mellifera ligustica* and Garnery et al. (1991) the 5′ end of the CO-II region for *A. dorsata*, *A. florea* and *A. cerana*. Cornuet et al. (1991) studied the CO-I and CO-II region in more detail and found inserts in form of 54 bp fragment and 196 bp repeats in the intergenic region. They argue that this region, in fact, might contain

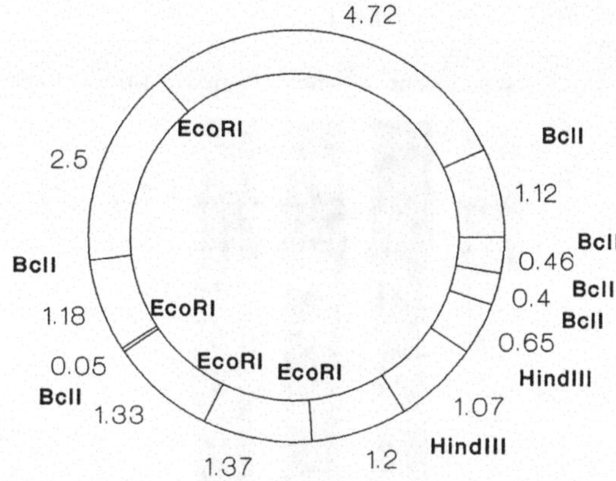

Fig. 7.16. Restriction site map of honeybee mtDNA based on three enzymes (*BclI*, *EcoRI*, *HindIII*) (according to Crozier et al. 1989). The fragment sizes are given in 1000 base pairs

an origin of replication. The high AT content and a secondary structure strongly favour this hypothesis.

Because of these repeated duplications, subspecies of *Apis mellifera* can readily be distinguished on the basis of restriction fragment length polymorphisms (RFLP). This feature was used to study the maternal mode of mitochondrial inheritance (Meusel and Moritz 1991). *Apis mellifera capensis* and *Apis mellifera carnica* have very distinct differences in the restriction site map (Fig. 7.17). Although there is substantial paternal mtDNA transfer (up to 30% of the total mtDNA) in freshly fertilized eggs, adult hybrid workers invariably revealed the maternal type. This indicates that in spite of the polyspermic fertilization in honeybees, the female determines the mitotype of the offspring. Paternal mtDNA is detectable throughout the embryonic development but degrades rapidly in the early larval stage (L1). In L2-stage larvae no paternal mtDNA was found (Meusel and Moritz 1991).

7.6.4.2 Cyto-Nuclear Disequilibrium

Because of the maternal mode of inheritance, mtDNA can be a very helpful tool in the assessment of sexual selection and swarming in natural selection. Particularly the dynamics of the disequilibrium, D, between genotype and mitotype reveal the impact of male and female fitness on natural selection. The disequilibrium can be estimated by

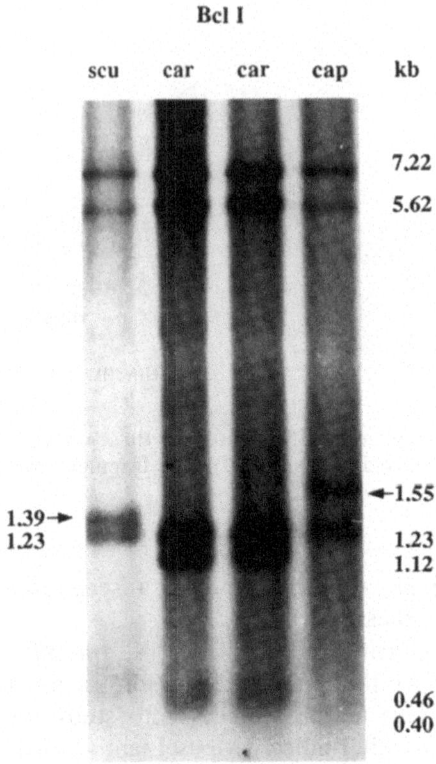

Fig. 7.17. MtDNA differences between *Apis mellifera capensis* (cap) from South Africa and *A.m. carnica* (car) from Central Europe. scu = A.m. scutellata from South Africa

using the method of Asmussen et al. (1987) which can be easily adapted to the male haploid situation in honeybees (Moritz and Meusel 1991). The cyto-nuclear disequilibrium is particularly useful to monitor the introgression of one population into another. Under the assumption that the fitness differences between mitotypes are much smaller than between genotypes, and are essentially neutral, the major force in the propagation of mitochondrial genes must be hitchhiking. Depending on the female fitness, the mitotype associated to the more fit genotype will increase its frequencies until equilibrium is reached. Under additivity of the fitness-determining genes, the fitness of drones has a significant impact on the equilibrium frequency. The more fit the superior genotype, the smaller will be the equilibrium frequency of the initially associated mitotype. It may look counter-intuitive that the drones, though they do not inherit mtDNA themselves, have a strong impact on the population genetics of mitochondria. However, if paternal (fit drone × unfit queen) and maternal (fit queen × unfit drone) hybrids are equally fit, it is obvious that the mitotype, which is initially associated with the inferior genotype, will be more strongly propagated the higher the frequency of paternal hybrids in the population (Moritz and Meusel 1992). Arnold et al. (1988) showed that permanent disequilibria are possible in hybrid zones if mating preference is determined by epistatic interactions between nuclear and cytoplasmic loci. Disequilibria phases can be extended under assortative mating in very recent hybrid zones (Asmussen et al. 1987). Reports on the hybridization of African and European honeybees in South and Central America might be a symptom of such disequilibria in honeybee hybrid zones (Smith et al. 1989; Hall and Muralidharan 1989; Moritz and Meusel 1991; Sheppard et al. 1991; for a discussion of the Africanized honeybee problem, see Chaps. 5 and 8).

Chapter 8 Natural Selection

It is certain that whatever changes may overtake biology in the future, we must henceforth grapple with the organism as a dynamic agency in a very complex and unstable environment. (William Morton Wheeler 1911)

As we have seen in the chapter on genetics, selection can be applied to both the individual and the colony. Thus we do not follow Buss' (1987) concept for the evolution of individuality in that selection "*will act solely upon the higher unit . . . if the lower unit is physically contained within the higher unit* (Buss 1987, p. 184). In fact, this axiom may only hold for true organisms but not for superoganisms (higher unit), in which the lower units are individuals themselves. Clearly, selection can act upon individual workers, queens and drones within the unit of a honeybee colony. This does not mean that it is easy to discriminate between selection pressures on the different levels. Most often individual and colony-level selection work synergistically. For example, an individually fit queen may produce many worker offspring, which will have consequences at the colony level and may result in a strong and fit colony. Usually, both selection mechanisms will be tightly interwoven and hard to dissect with experimental biological techniques.

8.1 Selection at the Individual Level

Classical Darwinian individual selection obviously operates on the sexual reproductives in honeybee populations. If a queen is a poor flyer, she certainly has a reduced fitness, because she may for example never reach the drone congregation area for mating. During the phase of the mating and sexual reproduction, natural selection operates undeniably at the individual level. The drones compete directly for access to the queen, and queens have to mate with sufficient drones to fill the spermatheca with semen, and clearly sexual selection is exclusively individual selection in the honeybee (see Chap. 6). However, direct individual selection operates on the workers as well as on the sexual reproductives. Intracolonial competition among laying workers has dramatic consequences for natural selection. Nevertheless, we will see that there are only a few examples in which we can clearly separate individual fitness from colony fitness. In all cases, selection at the individual level interacts very closely with selection at the colony level, causing very particular phenomena that can only be found in superorganismic species.

8.1.1 Selection of Sexual Reproductives

Natural selection of individual queens has been studied most intensively in connection with mating strategies. Since a single drone produces enough semen to fill the spermatheca, it seems puzzling, at first, why queens mate with many drones. Polyandry increases the time of the mating flight and increases the risk of queen losses. This becomes very obvious when we observe repeated mating flights of queens that are insufficiently inseminated in prior matings. The queens take substantial risks on each of these flights which leaves us with the question: where is the adaptive value of this behaviour? Can this phenomenon be explained exclusively by individual selection theory?

Crozier and Page (1985) reviewed a variety of explanations for polyandry in social insects. Only a few explanations rely strictly on individual selection theory. Ratnieks (1990a) offers an additional model in which the removal of diploid drone brood and the limited comb area in the nest are highlighted. He also needs to invoke the constraints of the bee colony structure to explain the evolution of polyandry. Cole (1983) pointed out that polyandry is more frequent in social insects with large colonies than species with small ones. Therefore, it is not surprising that the evolution of polyandry is so difficult to explain outside the context of sociality. Apparently, polyandry goes hand in hand with the social organization of the colony. Because of this correlation, it seems difficult to deny that selection at the colony level or at the superorganismic level is important for the evolution of polyandry. Furthermore, the few individualistic theories for the evolution of polyandry are more or less irrelevant for the case of honeybees. WestEberhard (1975) suggested that polyandry in social insects evolves because of insufficient semen transfer in single matings. This hypothesis, which was also stressed by others (Trivers and Hare 1976; Cole 1983) does not apply to the honeybee because single drones do have sufficient semen to inseminate the queen. This was the very reason why for a long time, until the reports of Roberts (1944), honeybee queens were believed to be monandrous. Moreover, the theory seems to be generally weak because there is no reason why natural selection would not favour males that produce sufficient semen for single matings. Actually, in both social bees and wasps, males are almost always able to provide sufficient spermatozoa for complete inseminations (Crozier and Page 1985).

Some other theories have been cited as individual selection models but are embedded in the biology of the superorganism. A particularly attractive theory has been put forward by Page (1980) and Page and Metcalf (1982). In honeybees, homozygosity at the sex locus causes diploid drones that are functionally lethal (see Chap. 7). If the queens mate with single drones, then most colonies would have a regular brood production and only a few would suffer from 50% lethal eggs whenever the queen has a sex allele in common with the drone. If all queens mate with a large number of drones, all colonies would be affected by diploid drones, but only very weakly. The average proportion of diploid drones produced in the population is not affected by polyandry. It is the variance of diploid male production between colonies that is reduced by polyandry. Page (1980) suggests a non-linear threshold model for colony reproduction that explains polyandry nicely. His basic assumption is that the colony needs to exceed a certain number of workers in spring to produce a successful swarm. Colonies with a low level of diploid drone load will easily reach this threshold and are not substantially less fit than colonies with no diploid drones. However, colonies with 50% diploid drones will fail to produce any swarms (Fig. 8.1). Thus, though the average frequency of diploid drones is not changed, the number of swarming colonies is.

In this model, natural selection will favour polyandrous queens, but is this really individual selection? The actual selection takes place at the colony level. Those colonies with polyandrous queens are more frequently selected by nature to produce offspring swarms than those with monandrous queens. Just the use of the number of produced swarms (clearly superorganismic groups) as major fitness parameter unequivocally reveals the colonial-base system of the model. Under strict individual selection the monandrous queens should have a higher fitness than polyandrous females because of the reduced

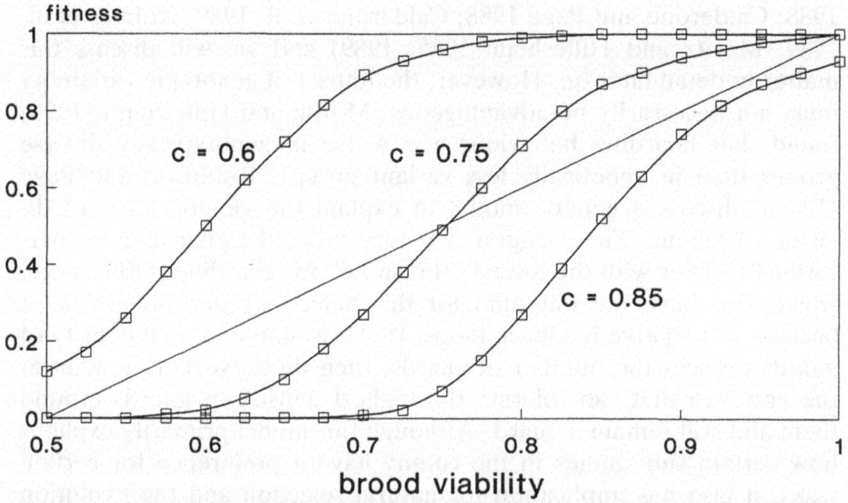

Fig. 8.1. Colony growth can be expressed with the aid of a logistic growth function. The relationship between brood viability and fitness depending on the inflection point (c in viability units) of the growth function is shown in the figure. If there is a linear relationship between brood viability and fitness, then there is neither selection for polyandry nor for monandry. If $c = 0.6$ then selection will favour multiple matings irrespective of the number of sex alleles in the population. For $c = 0.75$ natural selection favours polyandry if there are five or more alleles, but selects for monandry in populations with less than three alleles. If the growth curve is characterized by $c < 0.85$, then natural selection will favour fewer matings regardless of the number of alleles in the population

mating risk. The decision whether a polyandry or monandry is advantageous depends on colony growth which is without any doubt a group character composed of the joint action of queen and workers. The sex-locus-polyandry theory actually supplies the first example for a balance between group and individual selection.

Other models invoke colony selection more directly. Polyandry certainly enhances the genetic variance in a colony. One plausible argument is that colonies with an increased genotypic variability are better able to cope with environmental variance. This is a theory analogous to that used to explain the evolution of sexuality (Maynard Smith 1978). Polyandry could also be selected for, if division of labour depends on genetic variance. Recently, several studies have indicated that genotypic variance has a strong impact on division of labour in honeybee colonies (Page et al. 1989a,b; Robinson and Page 1988, 1989a,b; Rothenbuhler and Page 1989; Frumhoff and Baker

1988; Calderone and Page 1988; Calderone et al. 1989; Kolmes et al. 1989; Moritz and Hillesheim 1985, 1989) and we will discuss this matter in detail later on. However, the impact of genotypic variability must not necessarily be advantageous. Moritz and Hillesheim (1990) found that hoarding behaviour was worse in genotypically diverse groups than in genetically less variant groups. Robinson and Page (1989a) discuss a genetic model to explain the genetic basis of division of labour. They suggest that rare tasks like guarding are performed by bees with the lowest "threshold" for guarding. Other bees, which also have the potential for this behaviour just do not do it because of negative feedback loops. If, for example, a high density of guards reduces the number of guards, then those workers remain at the entrance that can tolerate the highest density of guards around them and still remain a guard. Although this model primarily explains how certain subfamilies in the colony have a preference for certain tasks, it also has implication for natural selection and the evolution of polyandry. Singly mated queens at the average would produce colonies with much higher behavioural thresholds than colonies with multiply mated queens. Also the threshold variance between colonies would be larger under single mating. In a highly polyandrous system every colony is expected to have a subfamily with a low behavioural threshold for any given task and there will be no "specialist colonies". Consequently, the genetically variable colony would be expected to be more fit under the Robinson and Page (1989a) model, and multiply mated queens should produce more swarms and drones than monandrous queens.

Sherman et al. (1988) suggested that resistance to certain parasites or pathogenes can evolve polyandry if (a) sperm is mixed in the spermatheca, (b) various patrilines genetically differ in their resistance, and (c) there is a concave relationship between individual worker survival and colonial survival similar to that proposed by Page (1980), discussed above in detail. Indeed, Shykoff and Schmid-Hempel (1991a,b) could recently show that parasite-mediated frequency-dependent selection can play an important role in counteracting kin selection for high intracolonial relatedness in bumblebees. In colonies with closely related individuals the susceptibility of *Bombus terrestris* colonies to a trypanosome parasite, *Crithidia bombi*, was higher than in genetically mixed colonies.

Another way to select for polyandry was suggested by Starr (1984) and Moritz (1985b). They pointed out that the optimal sex investment ratios in colonies with multiply mated queens becomes equal for queen and workers. Under single mating, the workers should prefer to invest three times as much into female offspring as into the male offspring (Trivers and Hare 1976). The queen should prefer a 1:1 investment ratio, which results in potential conflict between queen and worker. If colonies that avoid such conflict are more productive than those with conflict, then multiple mating of the queen should be selected for. Moritz (1985b) argued that the queens should optimize the fitness gain over the mating risk. Data collected by Woyke (1956) show that the mating flight is a linear function of the mating frequency. A regression analysis on his data set reveals a mean time of 2.04 min per mating plus an additional 10 min to and from the mating site. Assuming the total flight time to be the major determinant of the risk of a queen loss, we can treat the 10-min flight to the mating site as 5 "mating time units" and obtain a probability for surviving queens of

$$W_q = (1 - c)^z(1 - c)^5, \qquad (8.1)$$

where c = probability of loosing a queen in a single mating,
z = number of matings,
W_q = survival of queen.

If conflict is affecting colony fitness W_c, then a low conflict with equal investments interests for both queen and workers should result in a high fitness. Since the optimal investment ratio for the queen is $SR_q = 1$ and for the workers $SR_w = 1 + 2/z$, W_c can be expressed as

$$W_c = z/(2 + z). \qquad (8.2)$$

Hence, there is an optimal mating frequency for any given mating risk to maximize the combined fitness W_cW_q:

$$(z/(z + 2))(1 - c)^{z+5} = \text{maximum}. \qquad (8.3)$$

With a decreasing mating risk, the optimal mating frequency increases (Fig. 8.2). Under the mating conditions sketched by Adams et al. (1978) (17 drones per mating) or Koeniger and Ruttner (1989) (8 drones) the total mating risk should range between 8 and 20%. This seems to be well within the range of the natural mating risk. Tiesler (1972) reported on an average of 26.4% losses in a 10-year study and a data base of several thousand queens. Even though this figure seems high, it does not include losses due to introduction of the queens in mating nuclei, as suspected by Ratnieks (1990a). Ratnieks found in a similar study with 52 queens a loss of only 2 queens which seems a very low and optimistic figure, probably achieved under optimal climatic conditions. The number of queen losses is even

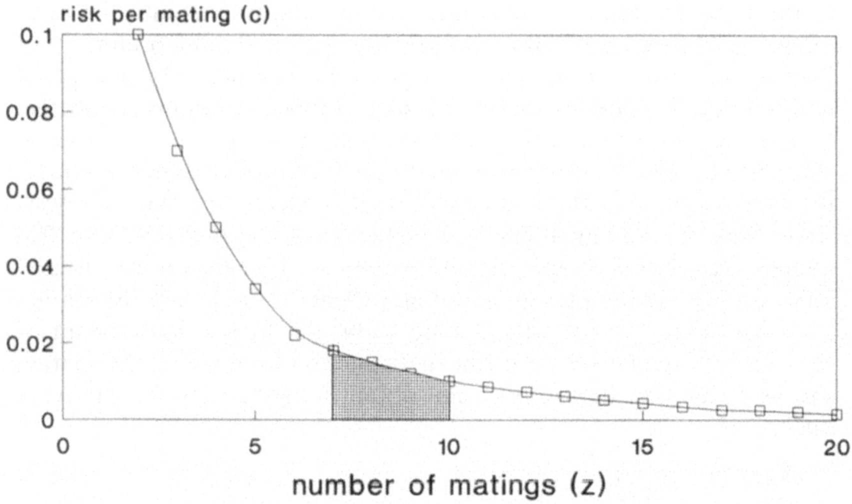

Fig. 8.2. Optimal number of matings (z) for queens at a given mating risk (c) to minimize the sex allocation conflict. The *shaded area* represents typical numbers of matings of queens in temperate climate (7–10 drones, G. Koeniger 1986). The expected mating risk fits well with empirical observations of queen losses during mating

higher under instrumental insemination, where the queen encounters only minimal risks because she performs no mating flight (Moritz and Kühnert 1984, in a study of 3440 queens). The risk of mating should not be underestimated and it certainly causes substantial evolutionary consequences for the multiple mating strategy of the queen.

Crozier and Page (1985) proposed that, especially in large colonies, polyandry may be the only way to avoid the conflict situation, because the queen cannot physically control the investment of workers. This was to rescue the model accounting for Cole's (1983) observation that polyandry correlates with colony size. However, in honeybees, such control could be easily achieved via pheromonal regulation, and the minimal-conflict hypothesis (Starr 1984; Moritz 1985b) is neither strongly supported nor contradicted by Cole's (1983) arguments. The pheromonal queen control could be a trait that has evolved after polyandry was established.

Ratnieks (1990a) offered an alternative model for the evolution of polyandry. He focussed on the frequency of diploid drones and the efficiency of cell use in the brood nest. Clearly, the workers should avoid investing in diploid drones if they are unable to reproduce. Any brood space occupied by diploid drone larvae slows down the colony development. Ratnieks (1990a) rationale is as follows: A certain proportion, p, of all eggs is not reared to the imaginal stage because of homozygosity at the sex locus. These p diploid drones are detected and removed at time k of the larval life, and after another delay, d, the queen will lay a new egg into the cleaned cell. The brood chamber efficiency (BCE) can now be expressed as

$$\mathrm{BCE} = 1/(1 + \sum p^n(1 - p)nk + \sum p^n(1 - p)nd + d), \tag{8.4}$$

where n = total number of diploid males,
p = probability for diploid drones,
k = delay for removal,
d = delay for oviposition in cleaned cell.

Equation (8.4) reduces to

$$\mathrm{BCE} = 1/(1 + (k + d)(p/(1 - p)) + d). \tag{8.5}$$

The brood chamber efficiency can vary between zero and one. BCE = 1 indicates that all cells contain a diploid female that will be reared to maturity. Polyandry affects the brood chamber efficiency. With increasing number of drones per mating BCE improves with a point of minimal returns at a queen mating frequency of four drones per queen.

It is difficult to decide which of the above theories and models (if any) is the correct one to explain the evolution of polyandry. Most likely all of them somehow will be effective. The one with the least assumptions, and therefore a very attractive one, is certainly the brood-chamber-efficiency model, but just the simplicity of the model alone does not tell us anything about its biological significance.

In looking at the problem of polyandry, not only are the queen and colony aspect important, but obviously the drones also take part in the procedure. What could be the selective advantage of drones to multiply inseminate a queen? Why does a drone inseminate a queen that has already mated with 15 others? A weak effect of a "late male" advantage has been found in artifically inseminated queens (Moritz 1986a). Such a late male advantage may explain why queens carrying a mating sign are more attractive to drones than unmated virgin queens (Koeniger 1986, 1990). However, since the advantage of the last male over the prior males is relatively small (only 15–20% of all offspring stems from the last of eight semen portions) this does not seem to be a major selective force. Laidlaw and Page (1984) could not find such phenomena in an experiment where they used four instead of eight different semen portions, and Woyke (1963b) failed to find a relation between insemination sequence and sperm usage. However, in both studies the number of sperms injected was not controlled and just sperm-number variability between drones might have overridden the effect of the insemination sequence.

The simplest explanation for polyandry from the male point of view seems to be that a successful mating of a drone is an extremely infrequent event and therefore the male should mate whenever possible. Irrespective of how little the contribution of a male in an insemination is, it is still better than waiting for an extremely unlikely second chance in mating. Thornhill and Alcock (1983) argue that in view of this, the obligatory death of the drone after mating once may also make sense. If the opportunity for mating a queen is very small, all the investment should be focussed on the achievement of at least one mating. Also the increase of the semen load or other physiological and anatomical adaptations to allow for multiple matings are unlikely to be adaptive if the chance for an encounter with a second virgin queen is virtually nil. In general, it seems that selection at the drone level operates at extreme selection intensities. Only very few are selected by nature to reproduce. In spite of this strong individual selection, the mating strategies reflect more important selection at other levels. It certainly would be in the drones' individual interest to be the only mating male. It is obvious that in spite of the mating sign he leaves in the queen after mating, he is not very successful in preventing other drones to mate the queen. Therefore, selection at other levels than at the individual drone seems to have larger adaptive value. These levels could be either the queen or the colony as discussed above. Polyandry in honeybees is unlikely to be a result from direct individual selection on drones.

8.1.2 Intracolonial Selection Among Laying Workers

Individual selection is not restricted to the sexual reproductives. In honeybee colonies, the queen, the drones, and also the workers are clearly individuals and their direct individual fitness is by no means zero. Whenever a virgin queen is lost, the colony is usually hopelessly queenless because it cannot rear a new queen from the remaining brood. In these cases, laying workers will develop which are not insignificant for the production of drones in the population. The reproductive capacity of colonies with laying workers can be of considerable importance for natural selection (Page and Erickson 1988). Reproductive workers have been particularly well studied in the Cape honeybee, *Apis mellifera capensis*. Laying workers of this subspecies behave as "false queens" (Anderson 1963) and parthenogenetically produce female instead of male offspring (see Chap. 7). Laying *A.m. capensis* workers produce a queen-like signal in their mandibular glands (Ruttner et al. 1967; Hemmling et al. 1979; Crewe and Velthuis 1980; Crewe 1982; Velthuis et al. 1990) that suppresses the ovary development in other non-laying workers. False queens apparently have a superior individual fitness compared to sterile workers. Moritz and Hillesheim (1985) found that those behavioural and physiological characters of capensis workers that are closely related to reproductive success have a surprisingly high genetic variance. This was surprising and caused Crozier (1987) to formulate this quendary in his essay on sociogenetics in insects "*the discovery of genuinely high heritabilities for several traits is interesting but also suggests that there has not been much selection on the characteristics involved, which is slightly puzzling*" (Crozier 1987, p. 326). Indeed, the data did not agree with classical understanding of genetic variance of characters closely related to fitness. Fisher (1930) and Falconer (1989) clearly show that the genetic variance of fitness should be zero in populations at equilibrium. This seems quite plausible, because fitness-related characteristics are very strongly selected for and should be at fixation after due time.

A possible solution to the problem was given by Hillesheim et al. (1989). They showed that both the phenotype of the individual worker and the colony's phenotype are important for natural selection (see Chap. 2). Hillesheim et al. (1989) selected lines with dominant and subordinate behaviour. They then established colonies that were exclusively made up from dominant, subordinate or 1:1 mixed colonies. These experimental colonies were then tested in various bioassays characterizing social behaviour of honeybees (hoarding behaviour, comb construction, brood care). The higher the frequency of dominant workers, the poorer was the social cooperation in the colony. Colonies composed of dominant workers only neglected brood care, never constructed comb, and were poor in hoarding behaviour (Fig. 8.3). Obviously the increased individual fitness of the dominant workers had negative impact on group phenotypes. A population genetic model showed that stable polymorphisms are possible between a hypothetical locus with one allele each for dominance and for subordination (Moritz 1989). Figure 8.4 shows the frequency of the dominance allele *d* under various conditions. Obviously, the more frequently virgin queens are lost, the higher is the equilibrium frequency for the *d* allele. If all queens survive, the *d*

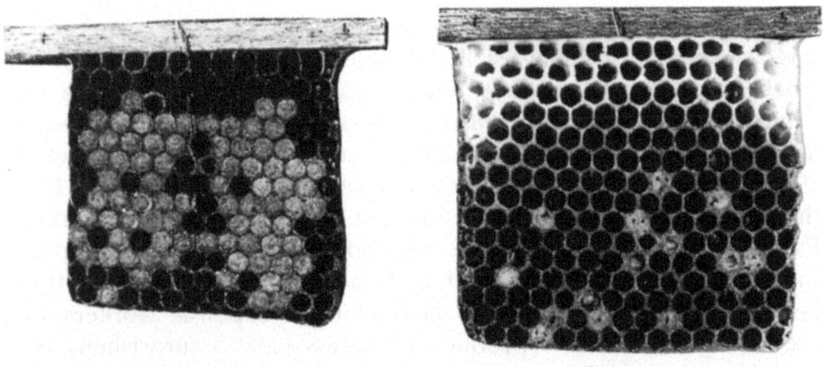

Fig. 8.3. Comb with eggs reared to pupae by subordinate workers (*left*) and non-attended brood in colonies with dominant workers (*right*). (From Hillesheim et al. 1990)

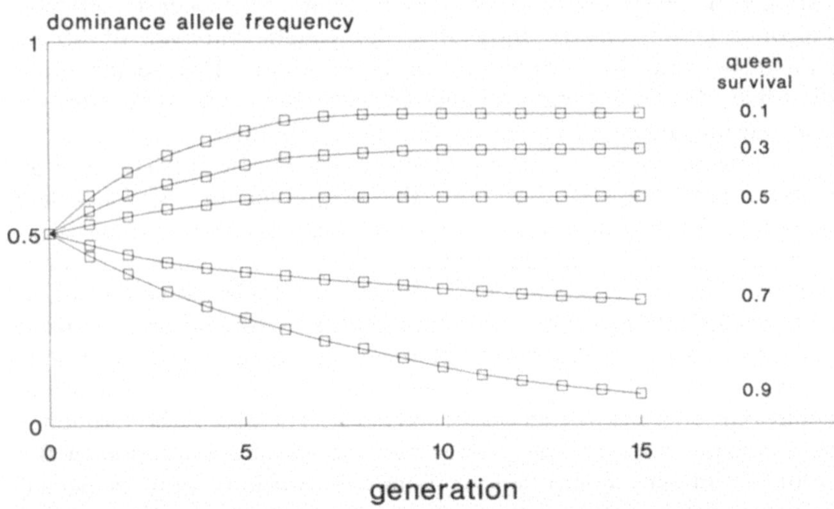

Fig. 8.4. Change of the worker-dominance allele frequency over time (*generations*) with various rates of queen survival. The queens are assumed to mate with ten drones. (Redrawn after Moritz 1990)

allele has no selective advantage. This seems plausible because queenright colonies with dominant workers are less fit in the model than those with a higher frequency of subordinate workers. Not only is the risk of losing a queen important, but also the degree of polyandry is significant for the fate of the *d* allele in the population. In an extension of this analytic model, a simulation shows that the equilibrium frequency of *d* rises with increasing numbers of drones mating with the queen (Moritz 1989). Interestingly, the most polymorphic situation under a queen risk of 70% is given at 17 drones per mating, the figure claimed by Adams et al. (1977) for the average number of drones per naturally mated queen in South American populations. The data we know for the reproductive behaviour of honeybees generally predict a polymorphic situation under the above model. Other explanations based exclusively on classical or neoclassical individual Darwinian selection fail to explain the phenomenon of a high genetic variance for reproductive dominance. The superorganism itself seems to be the crucial unit in natural selection, balanced by selection at the worker level within the colony. It seems very unlikely that the inheritance of reproductive dominance is determined by a single locus with two alleles. Nevertheless, the above model shows that in principle there are genetic mechanisms that maintain the large genetic variance for direct individual worker fitness. It also shows that selection in honeybees can operate within the superorganism, adding an additional level to selection on the sexual reproductives and the colony.

8.2 Kin Selection

Due to the close relationship of the colony members, kin selection as defined by Hamilton (1964a,b) is expected to be a substantial factor in evolutionary processes of honeybees (see Chap. 2). For the nonreproductive workers, especially, kin selection seems to be important because they only have indirect individual fitness via the reproductive success of their sister queens. Obviously, a worker would be particularly genetically successful if a supersister queen became head of the colony instead of a half-sister from a different patriline. The indirect individual fitness of a supersister worker of the new queen is three times as high as that of the half-sister worker. Consequently, those workers with an ability to discriminate between super- and half-sisters should have a large fitness advantage in honeybee colonies. Therefore, the presence of kin recognition would be nice evidence to support the significance of kin selection in honeybees. But is selection at this level significant for honeybee colonies? It could easily be that selection pressures at other levels (e.g. the colony) overrule potential advantageous effects of kin recognition on individual fitness of workers.

Certainly, the ability of kin recognition would form the basis for strong conflict between the workers of the various patrilines in the colony. Colonies with conflict, however, are not necessarily better off than those that are good at cooperation. Various scenarios can be constructed that lead to the fact that kin recognition is detrimental to the overall colony fitness in spite of its potential benefit to individual workers. Furthermore, even without kin recognition, the inclusive fitness concept is still viable in honeybees as long as Hamilton's rule $k > 1/r$ is valid. In other words: if a worker is still closely enough related to a half-sister so that the costs of an investment in a half-sister queen are less than the quotient between benefit to the recipient and relationship, then eusociality and the superorganism provide an evolutionary stable strategy (ESS).

The controversy on the significance of kin recognition in animals was recently fuelled by Grafen (1990) and Blaustein et al. (1991) who both shed doubts on the selective importance of kin recognition. Also, its adaptive significance in honeybees seems to be less convincing than initially thought (Moritz 1988a, 1991; Moritz and Hillesheim 1990; Oldroyd et al. 1990). The major problem is the difficulty to experimentally reveal the evolutionary significance of discriminatory behaviour of worker bees (Blaustein 1983). Breed et al. (1988) for example could show that behavioural potential for intracolonial kin discrimination is completely overriden by effects at the colony level that govern nestmate recognition.

8.2.1 Discrimination Among Queens

Discrimination between super- and half-sisters would be most effective for the sterile workers if performed at the level of queens. This becomes especially important during the process of queen rearing before swarming of the colony. The workers decide which larvae become gyne and which ones remain as workers by feeding them differential diets. Clearly, under strong kin selection, workers should show nepotism and preferentially feed their supersisters. Indeed, several experimental approaches seem to indicate such discriminatory mechanisms for young larvae and eggs (Visscher 1986; Noonan 1986; Page et al. 1989a). However, Oldroyd et al. (1990) pointed out that none of these reports really shows that kin recognition and nepotism have any significance for natural selection in honeybees. They argue that the observed phenomena are equally plausibly explained by genetic task specialization, nestmate recognition and stochastic effects. Recognition systems apparently are very difficult to demonstrate. Breed et al. (1984) found no kin discrimination ability for rearing queens. Completely unrelated larvae were reared at frequencies similar to the related ones. Visscher (1986) proposed that the royal jelly masked the recognition labels of the larvae and only very young stages can be discriminated. But also in his experiments the discrimination ability was not overwhelming and statistically just significant.

Kin discrimination has likewise been claimed to be found at the adult queen level. Getz et al. (1982) report on workers joining their supersister queen during swarming. However, this experiment could not be replicated (Getz and Winston, in Wintson 1987) and the data are not convincing in that only two swarms were tested. In both cases, the same patriline revealed a stronger tendency for swarming which is the only conclusive evidence of the primary data. Invoking aggregation due to kin recognition seems like unnecessary speculation on the data set which neither supports nor rejects any hypothesis on kin recognition. In other experiments, queens were discriminated on the basis of genetic relationship (Breed 1981; Getz and Smith 1983; Page and Erickson 1986a,b). Moritz and Crewe (1988a) could identify the secretions of the tergal glands as possible sources for the individual queen labels. They definitely could exclude mandibular gland secretions as potential discrimination labels because the composition of the signals showed only limited genetical variability and workers were unable to discriminate between signals of related and unrelated queens. In spite of the potential tergal discrimination labels, it remains unclear whether the discrimination reflects true intracolonial patriline recognition or just a very refined nestmate recognition. Bees learn very quickly and learned templates may easily overrule possible innate templates (Page and Erickson 1986b; Fletcher 1987; Moritz 1988a). Although, at present, the documentation of recognition of queens is still fragmentary, it seems to be possible, at least in theory, for three reasons:

1. There are only a few queens at most in a colony, and it might be possible for the workers to associate the volatile signals to the corresponding queens.
2. There is sufficient genetic variability in the tergal gland secretions among queens so that workers potentially could use them as discrimination labels.
3. There is as yet no experimental evidence that shows that worker queen kin recognition is insignificant for natural selection.

8.2.2 *Kin Discrimination Among Workers*

A potential adaptive significance of kin recognition at the queen level seems obvious. At the worker level this becomes less clear. Workers make up the sterile caste and therefore one might assume it should be fairly unimportant for a non-reproductive worker in which of her worker nestmates she invests. However, workers are not always sterile as we have seen above. Laying workers appear in all hopelessly queenless colonies and produce male or, in the case of the Cape honeybee, female offspring. Therefore the importance of laying workers in a population of honeybees is far from being insignificant.

Korst and Velthuis (1982) showed that laying workers are particularly characterized by their trophallactic behaviour. Dominant reproductive workers are fed more frequently by other workers and feeding is believed to be crucial for the development of the laying workers. Similarly to the queen-recognition case, those workers that preferentially feed their super sister workers should gain the most indirect fitness. Workers with the ability to discriminate between super and half sister workers should have a selective advantage.

A similar large body of studies seems to indicate that kin recognition among workers is a usual characteristic of the honeybee worker. This view somehow collides with the superorganism model, where we actually would expect selection at the colony level to be more efficient. For the superorganism, cooperation would be more profitable than the unavoidable intracolonial competition resulting from kin recognition. This being so, do the studies on worker – worker recognition truly reflect kin recognition, or are the above reservations on queen recognition also valid? In the following paragraphs we will see that the same reservations apply here, too. Actually, there is not a single study on worker kin recognition that could not be explained on the basis of colony-level selection. In fact, there are several studies that show that worker–worker recognition, though it is biologically possible, is almost irrelevant at the intracolonial level (Oldroyd et al. 1990).

The theory of patriline recognition amongst workers was most strongly emphasized by the group of Getz. Similarly to Breed (1981), Getz and Smith (1983) found that super sisters were less frequently bitten in worker–worker interactions than half-sisters or unrelated workers. Though in Breed's (1981) experiments nestmate recognition might have been involved, Getz and Smith (1983) avoided this problem by rearing super- and half-sisters on the same comb in the same colony. However, they still allowed the workers to learn "their" potential patriline in a 6-day period of confinement in cages of 100 to 200 bees of the same subfamily. As Getz (1991) points out correctly, in the colony bees have no chance to experience the pure odour of their own patriline, and Moritz (1988a) actually shows that the discrimination phenomenon found in these studies may reflect just nestmate recognition, because workers in the colony lack the patriline recognition ability. Just because there is a genetic component to an odour label by no means tells us whether the discrimination ability is kin or nestmate recognition. Even the ability of workers to learn genetically determined patriline specific odours and use these in conditioning experiments (Getz et al. 1986, 1988) just shows that these odours could form a potential label basis for recognition. It does not tell us whether the bees really use those odours in the colony for patriline recognition. Because of the extremely well-developed olfactory receptive mechanism of bees, one can condition bees to many more or less uniform odours of no apparent biological significance.

Another complication arises in those studies analyzing worker interactions in observation hives with a limited number of patrilines (usually two). Evers and Seeley (1986) reported on a higher degree of aggression among half- than super sisters and interpreted this as kin recognition. However, in the light of genetic task specialization (Frumhoff and Baker 1988; Calderone and Page 1988; Robinson and Page 1988, 1989a,b; Page et al. 1989a,b) another explanation becomes equally plausible. Oldroyd et al. (1990) argue that if some subfamilies had a stronger predisposition for biting, then the identical phenomenon would result. The same holds true for Frumhoff and Schneider's (1987) study. They found that super sisters have more trophallactic interactions than half sisters. It seems one can safely assume that workers performing the same task are more likely to meet in a colony than workers executing very different tasks. Unfortunately, Frumhof and Schneider (1987), though aware of this problem, did not test for differences in patriline frequencies at different localities and deviations from frequencies found in emerging brood. From Frumhof and Baker's (1988) and the other studies cited above, we can see that such non-random distributions of patrilines are common in honeybee colonies.

The polyandrous mating system poses a major problem for potential patriline recognition. Hogendoorn and Velthuis (1988) concluded from the frequency of trophallactic encounters that the discrimination ability virtually disappeared in highly polyandrous colonies. Page et al. (1990) questioned the statistical analysis, but

Moritz and Heisler (1991) found similar results in a quantitative trophallactic bioassay testing groups with two to eight patrilines (Fig. 8.5). The recognition modes seemed to result from the experimental setup and did not reflect adaptive behaviour in the biological context of a colony.

Another argument against kin recognition at the worker level is the high genetic variance found for this character. Moritz and Hillesheim (1990) found that 39% of the total variability is due to selectable genetic variance ($h^2 = 0.39$). This is much higher than selectability estimates of characters related to individual fitness in other organisms (Mousseau and Roff 1986). Empirical studies show that characters related to fitness vary around $h^2 = 0.27$ (Gustafsson 1986; Mousseau and Roff 1986) and behavioural characters less strongly related to fitness have average estimates of $h^2 = 0.37$. From this point of view one could argue that intracolonial patriline recognition at the worker level is less significant in comparison to other mechanisms in natural selection. On the other hand, the genetic task specialization model (Robinson and Page 1989a,b) pre-

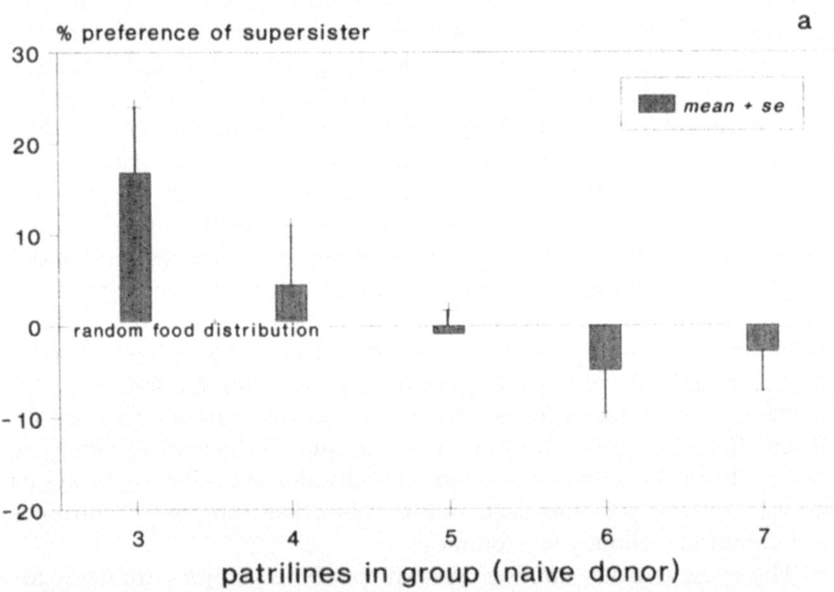

Fig. 8.5. The amount of transferred food was determined after trophallactic interactions in small experimental groups. With an increasing number of subfamilies in the group, the preferential feeding of super sisters and the ability of kin recognition decreased

dicts a high genetic variance for discrimination. This becomes clear if we look in more detail at the data of Moritz and Hillesheim (1990). In spite of the high genetic variance (or maybe because of this) in all tested colonies at least one patriline was present with a wellexpressed discrimination ability. This fits nicely with the task specialization model if the discriminators were guard bees. Guard bees ought to preclude foreign intruders to enter the hive and perform precise nestmate recognition. It would be highly adaptive if they can detect friend or foe in time before they enter the hive. The high genetic variance could therefore just be the consequence of genetic task specialization, and intracolonial frequency-dependent selection, i.e. colonies with the right frequency of "genetic" guards, undertakers, foragers, and other task specialists are fitter than others and genetic variability is maintained via polyandry.

Ratnieks and Reeve (1991) argue in a similar fashion that nepotism is a very sensitive evolutionary strategy towards reinvasion on non-nepotists if there is a cost to discriminating kin from less related nestmates. Particularly the high degree of polyandry in the honeybee promotes the loss of nepotistic behaviour, even if the cost of discrimination is very low. Polymorphic systems between nepotists and non-nepotists will result if colony efficiency suffers from kin discrimination by a few percent only.

The task specialization or colony efficiency model is not necessarily simpler than a plain individualistic kin selection model, but it seems better supported by the empirical data currently available. In terms of a genetic model, kin selection is actually a simpler model than the nestmate recognition/task specialization model because it only invokes individual selection and avoids additional assumptions on colony composition and colony fitness. On the other hand, looking at a honeybee colony as a superorganism, plain individual selection models seem to leave out the most interesting part the honeybee system provides: the effect of eusociality on processes in natural selection.

8.3 Selection on Phenotypes of Superorganisms

Clearly, superorganisms have phenotypes (i.e. colony characteristics) which are exposed to natural selection. This becomes most obvious in human-performed selection for economically important characters of honeybee colonies. Selection is almost exclusively based on colonial characteristics like honey production, colony defence, swarming behaviour and overwintering. All of these characters of commercial interest are also critical for the processes in natural selection. Colonies with small honey stores are prone to winter death. A lack of colony defence allows predation, and a suboptimal swarming activity directly relates to a reduced reproductive rate. Though it seems plausible to acknowledge the importance of selection at the superorganismic level, the actual genetic mechanisms forming the basis for such a selection are less transparent.

8.3.1 Genetic Analysis of Division of Labour

A rapidly increasing body of papers indicates that for most tasks, genetic variance is an important factor (Calderone and Page 1988; Frumhoff and Baker 1988; Kolmes et al. 1989; Moritz and Hillesheim 1985, 1989; Moritz and Southwick 1987; Page et al. 1989a; Page and Robinson 1991; Robinson and Page 1988, 1989a,b; Robinson et al. 1990; Rothenbuhler and Page 1989). Specific worker tasks in the colony, such as guarding, defence, undertaking, foraging or reproductive dominance, were shown to be strongly determined by gene effects. The gene effects were either analyzed with quantitative genetic techniques (Moritz and Hillesheim 1985) or genetic markers (e.g. isozymes in Robinson and Page 1988, 1989a,b). An example for guarding behaviour presented by Robinson and Page (1989) neatly shows how the system operates. In a colony composed of three patrilines (characterized by an *S*(low), *M*(edium) and *F*(ast) allele for malate dehydrogenase, MDH), workers of the *S* and *F* patriline were more frequently found, and workers of the *M* patriline more rarely found in the guard force than expected under the overall gene frequency in the colony.

The significance of genetic effects for division of labour does not mean that the age polyethism concept is without merit. It just shows that workers of a specific patriline are more likely to become, for example, guards than workers of the same age of another subfamily. Robinson and Page (1989a) developed a negative feedback loop model for task regulation in the colony. Workers of different patrilines have different response thresholds to perform certain tasks. An example may illustrate how this model works. Let us think of a colony with two patrilines, *A* and *B*. Workers of patriline *A* will engage in guarding the hive entrance if they are of the appropriate age and there are less than 20 guards on duty. Workers of patriline *B* have a higher threshold for guarding and feel sufficiently safe as soon as ten workers guard the entrance. If we start with zero guards, workers of both patrilines will initially engage in guarding. However, as soon as ten guards are active, no more workers of patriline *B* will become guard bees. Furthermore, whenever a guard bee quits service, it will be replaced by a worker from patriline *A* with the 20-worker guard force threshold. Eventually all guards will be workers of the *A* patriline. Obviously, this model oversimplifies what is actually going on in a bee colony, nevertheless it provides a plausible explanation for how genotypic task specialization may operate in principle.

8.3.2 Genetic Variability and Task Specialization

Page et al. (1989b) proposed a model for genetically determined task specialization in a honeybee colony. Their model is based on the assumption that the phenotype of a bee deviates from the mean by an age-dependent effect, *A*, a genetic effect, *S*, and environmental effects, *E*. Condensed to an equation this is:

$$P(T)_k = \mu + S_i + A_{ij} + E_{ijk}, \quad (8.6)$$

where $P(T)_k$ = probability for worker k to perform task T,
μ = mean probability for performing task T in the population,
S_i = effect of the subfamily i on the probability to perform task T,
A_{ij} = effect of the age group j in subfamily i,
E_{ijk} = special environmental effects on worker k.

A colony phenotype could then be formally calculated by summing up all possible tasks (including non-tasks) over all workers. This, of course, represents an extremely simplified mathematical model for genetic task specialization. Interactions between workers and subfamilies strongly affect social behaviour (Moritz and Hillesheim 1989), and threshold phenomena (Robinson and Page 1989a) will interfere with the above linear model. The significance of the genetic task specialization model becomes more obvious in the two-locus model designed by Page et al. (1989b). Consider two loci, *A* for foraging on pollen and *B* for foraging on nectar. The phenotypes of the workers are given in Table 8.1.

Let us assume the colony would be most fit if it allocates equal numbers of pollen and nectar foragers. Obviously there are several ways to achieve equal forager allocation. The colony could either

Table 8.1. Percentage of workers performing either pollen or nectar foraging depending on the genotype at the two loci

		AB	Ab	aB	ab
AB	Pollen %	50	75	25	50
	Nectar %	50	25	75	50
Ab	Pollen %	–	100	50	25
	Nectar %	–	0	50	75
aB	Pollen %	–	–	0	25
	Nectar %	–	–	100	75
ab	Pollen %	–	–	–	50
	Nectar %	–	–	–	50

have omnipotent bees (50:50, pollen:nectar) or could follow a strategy to be composed of equal numbers of specialists (100:0, 0:100). Both result in equal labour allocation and an equal fitness $W_a = 1$ due to the allocation ratio. Specialists are workers that either forage nectar or pollen. They have a degree of specialization of $S = 1$. Workers that perform both tasks at equal ratios are not specialized and their degree of specialization is $S = 0$. If it is advantageous to have specialists because they are somehow more efficient by a proportion c = coefficient of specialization, then the fitness of the colony, can be written as

$$W_c = W_a(1 + cS), \tag{8.7}$$

where W_c = colony fitness,
W_a = fitness due to allocation ratio,
S = average degree of specialization,
c = coefficient of specialization.

Thus, if a colony entirely composed of specialists (S = 1) is twice as fit as a colony of omnipotent workers (S = 0), the coefficient of specialization equals $c = 1$ ($W_c = 2W_a$). Page et al. (1989b) modelled various scenarios and found that if the queen mates twice, the minimum value of $c = 1.1$ is sufficient to result in stable polymorphic systems. Furthermore, the average fitness of populations with $c > 1.1$ increases and may be favoured by selection. Thus, particularly under high polyandry, it may pay to have specialists at very low thresholds of c, the coefficient of specialization.

8.3.3 Non-Linear Worker Interactions

The phenotype expression of groups in the above model of Robinson and Page (1989a) is based on individual genotypes that are expressed or not expressed. However, interactions between the members of the group may also cause overexpression, as we have seen in Chap. 7. Just the genotypic variability itself may affect the group phenotype, overriding the phenotypes of the individual group members. This becomes clear if we compare the behaviour of three hypothetical experimental groups of bees from two colonies and we look at intragroup conflict. Two groups are composed of bees from one colony each and the third one is a mixed group with bees from both colonies. The pure groups will cooperate without much conflict, whereas the workers in the mixed group are mainly involved in killing each other, a typical reaction of unfamiliar worker bees due to intruder rejection. The interaction between the group members is overriding the phenotypes of the non-mixed groups and intragroup conflict is overexpressed. Although this example is not really suitable to explain non-linear group-member interactions in honeybee colonies, because under natural conditions bees from different nests rarely combine (however see Hepburn and Whiffler 1988), it certainly shows the principal mode of operation of potential interactions. Just the combination of bees and not the individual genetic value may determine a group phenotype. Whenever such traits appear they are, of course, selectable neither by man nor by nature, because it is not the genes but just the mixing of genotypes that determines the phenotype. If such characters are important for the fitness of the colony, then they can only be selected indirectly on the individual level via the mating strategy of the queen. Thus, if genotypic diversity is advantageous for the colony, then polyandrous queens have a greater reproductive success than singly mated ones.

Both models, the task specialization and the genotypic diversity model actually are very closely related and just highlight different special cases (overexpression and threshold) of the same base theory. The superorganismic phenotype is composed of individual and group characters that are not necessarily linearly related. This makes a priori expectations of natural selection very uncertain, and for each trait of interest the mode of inheritance will have to be studied separately. Only based on experimental and empirical evidence can the response to selection by nature and/or man be predicted.

8.4 Africanized Honeybees: A Large-Scale Evolutionary Experiment

8.4.1 The History of Africanization

In the past two decades the public became excited about evolutionary biology of honeybees and its consequences for man. The Africanized honeybee got strong attention by the public media because of its high potential for defensive behaviour which caused several lethal incidents. The problem arose in 1956 when honeybees from Africa were introduced into Brazil to improve the honeybee stock of European origin mainly used by beekeepers. The goal was to increase the productivity via a bee that was better adapted to the tropical habitat of South America. Twenty-six swarms of African *Apis mellifera scutellata* were introduced into Brazil in 1956 (Kerr 1967). Some of these escaped and revealed the good adaptation of the African type to the tropics. Within 30 years they covered completely all of Central America and northern South America from Mexico in the north, to Argentina in the south. Further spread northward has been predicted by various authors (Dietz et al. 1985; Taylor 1977, 1988; Taylor and Spivak 1984; Rinderer et al. 1987; Southwick et al. 1990) and the near future will show where the ecological limits of the Africanized honeybees are situated (see Chap. 5).

8.4.2 Reproductive Advantage of Africanized Honeybees

Morphologically the feral bees of South America are very similar to the African *Apis mellifera scutellata* (Daly and Belling 1978). However, there is clear hybridization between the introduced African and the European races (Kerr and Bueno 1970). Lobo et al. (1989) and Del Lama et al. (1990) showed that there is substantial racial admixture in the Africanized honeybee and that is why the terminology "Africanized" instead of "African" honeybee genetically makes much sense. The surprising thing in the Africanized bee problem is the speed of the spread and the repeated reports of the complete extinction of European honeybees after the Africanized honeybee arrived (Hall and Muralidharan 1989; Hall 1990; Roubik 1989; Southwick et al. 1990). If there are repeated hybridizations, why does the African genotype not dilute? The potential selective advantages of the African type over the European have been discussed in detail in Chap. 5. Yet the question remains: which are the most important selective forces in the process of Africanization? Is the increased swarming the most critical factor or is the sexual reproductive advantage of queens and drones most significant? The population genetical tools to study this problem are available with the analysis of nuclear DNA and mitochondrial (mt) DNA. Because mtDNA is exclusively maternally inherited in honeybees (Meusel and Moritz 1992), whereas nuclear DNA is transmitted via both sexes, the role of swarming and sexual reproduction can be analyzed in empirical studies. Hall (1990) used this approach in his study on European, African and Africanized honeybees and showed its powerful potential. He found that managed colonies of European maternal lineage revealed more often African nuclear markers than in the reciprocal case. Hall (1990) interpreted this as indication for a limited introgression of European genes into Africanized populations and ready introgression of African alleles into European lineages. However, a simple fitness advantage of the African genes explains his results equally well. The crucial population genetic parameters to understand the process of Africanization in the field are still to be determined. A theoretical model of Moritz and Meusel (1992) showed why it is a time-consuming procedure to get reasonable samples and a clear genetic picture of the Africanized bee phenomenon. They found that not only do the queens and the colony, but also the drones, have an impact on the mitotype equilibrium frequency. At first glance this is counter-intuitive because drones do not transfer mitochondria. Nevertheless, it is a highly plausible phenomenon. With increasing fitness of African drones, a growing number of European queens will be mated with African drones. Consequently, the fraction of hybrid queens in the population with the European mitotype will expand and the equilibrium frequency for the African mitotype will drop. The newly introduced mitotype is unlikely to reach fixation unless the fitness of the African drones is very small compared to the resident European type.

8.4.3 Genetic Evidence from Field Studies

The theoretical expectation has implications for using mtDNA as a diagnostic tool for the identification of Africanized (not African!) honeybees. In contrast to claims based on field data from Mexico (Hall and Muralidharan 1989; Smith et al. 1989; Hall 1990) it seems extremely unlikely that mtDNA can be used in such a way. Recent empirical evidence reveals a fair amount of European mitotypes (20%) in Africanized honeybees in Brazil (Moritz and Meusel 1992). Sheppard et al. (1991) studied a cline from Brazil to Argentina and found similar data indicating hybridization between both Africanized and European types. Both results are plausibly supported by the ecological data (increased swarming capacity and drone fitness of Africanized bees). The fitness advantage of the Africanized drones is also increased through "reproductive parasitism" (Rinderer et al. 1987). Africanized drones drift into European host colonies which in response reduce their own drone production. Apparently the colony can only maintain a certain maximum "drone titre" which prohibits any further drone rearing. With an increased proportion of Africanized drones in the population, one would expect European queens to be mated more frequently by Africanized drones, thereby producing hybrids maintaining the European mitotype. At the present stage of evidence there seems to be no reason to invoke special evolutionary mechanisms, such as nuclear-mitochondrial interactions (Hall and Muralidharan 1989) or extensive assortative mating (Taylor 1985) to explain the Africanized honeybee phenomenon. Also, Lobo et al. (1989) and Del Lama et al. (1990) found no evidence for a competitive displacement mechanism between African and European populations. In their study of South American populations in Brazil and Uruguay based on isozyme analysis, racial admixture was the rule and they found no signs of sympatric reproductive isolation between Africanized and European populations.

If we accept that defensive behaviour and the high swarming tendency are determined by a nuclear polygenic system which is subject to genetic variation as defined in quantitative genetic theory, then it may become difficult to use molecular markers as indicators for the Africanized type. For example, most isozyme markers are believed to be selectively neutral or at least under very weak selection in South American populations (Lobo et al. 1989), although substantial selection on the MDH – locus cannot be excluded in European populations (Badino et al. 1982). Only if we can firmly correlate the behavioural traits in question with certain DNA regions, will the molecular techniques make sense. Hall (1986, 1988) used diagnostic probes for the identification of Africanized honeybees. However, only if they prove to be correlated with the significant fitness characters of the Africanized bees, will they be helpful in the analysis of the actual evolutionary mode of Africanization. As long as neutral markers are tested, we can only reveal hitchhiking phenomena similar to those illustrated in the above model for mitochondrial genes.

8.5 Selection in Superorganisms: Too Complex to Study?

In this chapter we have seen how multifaceted selection on superorganisms can be. Classical individual, colonical and kin selection interact simultaneously and obscure clear-cut mechanisms in natural selection we know from populations of solitary individuals. Shall we therefore restrict ourselves to the study of individuals to get clearer answers on the questions we pose? First of all, the true stimulus for science and research is to reveal the riddles given by complexity, and just because of this philosophical aspect it seems to be justified to study the mode of natural selection in superorganismic populations. Population geneticists often regard molecular geneticists as "gene jockeys" and reductionists who are ignorant of what actually happens in organisms. The same argument, however, also holds for many population geneticists who are ignorant of what happens in superorganisms. This ignorance led to a complete rejection of group and colony selection models in social insect populations for almost four decades in evolutionary biology (Starr 1979). Only recently have sociogeneticists (Crozier 1987; Owen 1986, 1989b; Wilson and Sober 1989) pointed out the importance of the social group for natural selection. Superorganisms allow for the analysis of natural selection at the gene, the individual, and the superorganismic level and therefore offer just one more degree of freedom in evolutionary studies than classical individual organisms.

The study of the modes of natural selection in superorganisms is most intriguing because of their tremendous success in nature. For example, eusocial insects make up three-quarters of the total insect biomass in the Amazonian rain forest (Beck 1971; Fittkau and Klinge 1973). The majority of vertebrates are socially organized and solitary individuals are the rare exception. Since sociality is so widely abundant and successful, it is actually surprising that evolutionary biologists do not more intensely tackle the problem of natural selection in superorganisms. Certainly, the basic principles of inheritance are more easily studied in solitary organisms like *Drosophila*, yeasts or bacteria, but students of evolutionary biology are likely to miss a large array of significant problems. Evolutionary studies relevant for the bulk of the fauna abundant on this planet should certainly address the problem of social groups and superorganisms. It is absolutely necessary that we continue to increase our knowledge in understanding gene action at the DNA level, but it seems equally important not to lose scope for the overall (and certainly more complex) problem of the consequences of these gene actions for natural selection, not only at the organismic but also at the superorganismic level.

References

Abdellatif MA (1965) Comb cell size and its effect on the body weight of the worker bee, *Apis mellifera* L. Am Bee J 105:86–87

Adams J, Rothman ED, Kerr WE, Paulino ZL (1977) Estimation of sex alleles and queen matings from diploid male frequencies in a population of *Apis mellifera*. Genetics 86:583–596

Adelman RC (1978) Endocrine regulation of enzyme activity in aging animals of different genotypes. In: Bergsma D, Harrison D (eds) Genetics effects of aging. Natl Found, New York, pp 355–387

Akahira Y, Sakagami SF, Zucchi R (1970) Die Nähreier von den Arbeiterinnen einer stachellosen Biene, *Trigona* (*Scaptotrigona*) *postica*, die von der Königin kurz vor der eigenen Eiablage gefressen werden. Zool Anz 185:85–93

Alber M, Jordan R, Ruttner F, Ruttner H (1955) Von der Paarung der Honigbiene. Z Bienenforsch 3:1–28

Alexander RD (1974) The evolution of social behavior. Annu Rev Ecol Syst 5:325–383

Alexander RD, Borgia (1978) Group selection, altruism, and the levels of or ganization of life. Annu Rev Ecol Syst 9:449–474

Alexander RD, Sherman PW (1977) Local mate competition and parental investment in social insects. Science 196:494–500

Allee WC, Emerson AE, Park O, Park T, Schmidt KP (1949) Principles of animal ecology. Saunders, Philadelphia

Allen MD (1956) The behaviour of honeybees preparing to swarm. Br J Anim Behav 4:14–22

Allen MD (1957) Observations on honeybees examining and licking their queen. Br J Anim Behav 5:81–84

Allen MD (1959) Respiration of worker honeybees of different ages at different temperatures. J Exp Biol 36:92–101

Allen MD (1965a) The production of queen cups and queen cells in relation to the general development of honeybee colonies and its connection with swarming and supersedure. J Apic Res 4:121–141

Allen MD (1965b) The effect of a plentiful supply of drone comb on colonies of honeybees. J Apic Res 4:109–119

Allen TJ (1966) Performance of information channels in the transfer of technology. Ind Manage 8:87–98

Allen TU (1978) Managing the flow of technology. MIT Press, Cambridge, MA

Almond Board of California (1987) Statistical tables. California Almonds, Sacramento, CA

Altmann G, Gontarski H (1963) Über den Wasserhaushalt der Winterbienen. Symp Genet Biol Ital 12:308–328

Ambrose JT, Morse RA, Boch R (1979) Queen discrimination by honey bee swarms. Ann Entomol Soc Am 72:673–675
Anderson RH (1963) The laying worker in the Cape honeybee, *Apis mellifera capensis*. J Apic Res 2:85–92
Andersson M (1984) The evolution of eusociality. Annu Rev Ecol Syst 15:165–189
Anon (1985) The mathematics of the honeycomb. Sci Dig 93:74–77
Aristotle (1949) Tierkunde. P Gohlke, Paderborn
Armbruster L (1913) Chromosomenverhältnisse bei der Spermatogenese solitärer Apiden (*Osmia cornuata* Latr.). Arch Zellforsch 11:242–326
Armbruster L (1938) Versteinerte Honigbienen aus dem obermiozänen Randecker Maar. Arch Bienenkd 19:1–48, 97–133
Arnold J, Asmussen MA, Avise J (1988) An epistatic mating system model can produce permanent cytonuclear disequilibria in a hybrid zone. Proc Natl Acad Sci USA 85:1893–1896
Aschoff J (1989) Temporal orientation: circadian clocks in animals and humans. Anim Behav 37:881–889
Asencot M, Lensky Y (1976) The effect of sugars and juvenile hormone on the differentiation of the female honey bee larvae (*Apis mellifera* L.). Life Sci 18:693–700
Asencot M, Lensky Y (1984) Juvenile hormone induction of "queenliness" on female honey bee (*Apis mellifera* L.) larvae reared on worker jelly and stored on royal jelly. Comp Biochem Physiol 78 B:109–117
Asencot M, Lensky Y (1988) The effect of soluble sugars in stored royal jelly on the differentiation of female honey bee (*Apis mellifera* L.) larvae to queens. Insect Biochem 18:127–133
Asmussen MA, Arnold J, Avise JC (1987) Definition and properties of disequilibrium statistics for associations between nuclear and cytoplasmic genotypes. Genetics 115:755–768
Autrum H, Zwehl V von (1962) Zur spektralen Empfindlichkeit einzelner Sehzellen der Drohne (*Apis mellifica*). Z Vgl Physiol 46:8–12
Aviles L (1986) Sex ratio bias and possible group selection in the social spider *Anelosimus eximius*. Am Nat 128:1–12
Aviles L (1987) Female biased sex ratios in social spiders: a result of group selection? In: Eder J, Rembold H (eds) Chemistry and biology of social insects. Peperny, Munich, p 349
Avitabile A (1978) Brood rearing in honey bee colonies from late autumn to early spring. J Apic Res 17:69–73
Avitabile A, Kasinskas JP (1977) The drone population of natural honeybee swarms. J Apic Res 16:145–149
Avitabile A, Morse RA, Boch R (1975) Swarming honeybees guided by pheromones. Ann Entomol Soc Am 68:1079–1082
Avitabile A, Stafstrom DP, Donovan KJ (1978) Natural nest sites of honey bee colonies in trees in Connecticut, USA. J Apic Res 17:222–226
Badino G, Celebrano G, Manino A (1982) Genetic variability of *Apis mellifera ligustica* Spin. in a marginal area of its geographical distribution. Experientia 38:540–541
Badino G, Celebrano G, Manino A (1983) Population structure and Mdh-1 locus variation in *Apis mellifera ligustica*. J Hered 74:443–446

Badino G, Celebrano G, Manino A (1984) Population genetics of Italian honey bee (*Apis mellifera ligustica* Spin.) and its relationships with neighbouring subspecies. Boll Mus Reg Sci Nat Torino 2:571–584

Badino G, Celebrano G, Manino A, Longo S (1985) Enzyme polymorphism in the Sicilian honeybee. Experientia 41:752–754

Badino G, Celebrano G, Manino A, Ifantidis MD (1988) Allozyme variability in Greek honey bees (*Apis mellifera* L.). Apidologie 19:377–386

Bailey L (1954) The respiratory currents in the tracheal system of the adult honeybee. J Exp Biol 31:589–593

Bailey L (1981) Honey bee pathology. Academic Press, New York

Bak P, Chen K, Creutz M (1989) Self-organized criticality in the "game of life." Nature (Lond) 342:780–782

Baker HG, Baker I (1973) Amino acids and their evolutionary significance. Nature (Lond) 241:543–545

Barbier M, Hügel MF (1961) Synthese de l'acide ceto-9-decen-2-cis-oique, isomere de la "Substance Royale." Bull Soc Chim Fr 202:1324–1326

Barbier M, Lederer E (1960) Structure chemique de la "substance royale" de la reine d'abeille (*Apis mellifica*). C R Acad Sci Paris 250:4467–4469

Barbier M, Lederer E, Nomura T (1960) Synthese de l'acide ceto-9-decene-2-trans-oique (substance royale) et de l'acide ceto-8-nonene-2-trans-oique. C R Acad Sci Paris 251:1133–1135

Barbour MG, Burk JH, Pitts WD (1976) Terrestrial plant ecology. Benjamin Cummings, Reading, MA

Barker RG, Jay SC (1974) A comparison of foraging activity of honey bee colonies with large and small populations. Manit Entomol 8:48–5

Baroni Urbani C (1979) Territoriality in social insects. In: Hermann H (ed) Social insects, vol I. Academic Press, New York, pp 91–120

Basalingappa S (1970) Environmental hazards to reproductives of *Odontotermes assmuthi* Holmgren. Indian Zool 1:45–50

Basiolo A, Comparini A (1990) Esterase-6-locus, a new enzyme polymorphism in *Apis mellifera*. Apidologie 21:123–126

Bastian J, Esch H (1970) The nervous control of the indirect flight muscles of the honey bee. Z Vgl Physiol 67:307–324

Batra SWT (1964) Behavior of the social bee, *Lasioglossum zephyrum*, within the nest. Insect Soc 11:159–186

Batra SWT (1966) Life cycle and behavior of the primitively social bee, *Lasioglossum zephyrum*. Univ Kans Sci Bull 46:359–423

Baumgartner D, Roubik D (1989) Ecology of necrophilous and filth-gathering stingless bees (Apidae: Meliponinae) of Peru. J Kans Entomol Soc 62:11–22

Beck L (1971) Bodenzoologische Gliederung und Charakterisierung des amazonischen Regenwaldes. Amazoniana 3:69–132

Beck SD (1980) Insect photoperiodism. Academic Press, New York

Beck WS (1971) Human design: molecular, cellular, and systematic physiology. Harcourt Brace Jovanovich, New York

Beeling I (1929) Über das Zeitgedächtnis der Bienen. Z Vgl Physiol 9: 259–338

Beenakkers AMT, Van Der Horst DJ, Van Marrewijk WJA (1984) Insect flight metabolism. Insect Biochem 14:243–260

Beetsma J (1979) The process of queen-worker differentiation in the honeybee. Bee World 60:24–39

Bego LR, Simoes D, Zucchi R (1983) On some structures in *Nannotrigona* (*Scaptotrigona*) *postica* Latreille (Hymenoptera, Meliponinae). Bol Zool Univ Sao Paulo 6:79–88

Beig D (1972) The production of males in queenright colonies of *Trigona* (*Scaptotrigona*) *postica*. J Apic Res 11:33–39

Berg S (1988) Grössenabhängige Flugdauer beim Paarungsflug der Drohne (*Apis mellifera* L.). In: Nachtigall W (ed) The flying honeybee. G Fischer, Stuttgart, pp 43–50

Bertholf LM (1925) The moults of the honeybee. J Econ Entomol 18: 380–384

Betts AD (1943) Temperature and food consumption of wintering bees. Bee World 24:60–62

Beutler R (1953) Nectar. Bee World 34:106–116, 123–136, 156–162

Bevan E (1827) The honey-bee; its natural history, physiology, and management. Baldwin, Cradock, Joy, London

Bienefeld K (1988) Vererbung von Leistungseigenschaften bei der Honigbiene (*Apis mellifica* L.). Dissertation, Tech Univ Munich, pp 159

Bienefeld K, Pirchner F (1990) Heritabilities for several colony traits in the honeybee (*Apis mellifera carnica*). Apidologie 21:175–183

Bienefeld K, Reinhardt F, Pirchner F (1989) Inbreeding effects of queens and drones on colony traits in the honeybee. Apidologie 20:439–450

Bitondi MMG, Mestriner MA (1983) Esterase isozymes of *Apis mellifera*: substrate and inhibition characteristics, developmental ontogeny, and electrophoretic variability. Biochem Genet 21:985–1002

Blau PM (1969) Objectives of sociology. In: Bierstedt R (ed) A design for sociology: scope, objectives, and methods. Am Acad Pol Soc Sci, Philadelphia, pp 345–389

Blau PM (1977) Inequality and heterogeneity: a primitive theory of social structure. Free Press, New York

Blaustein AR (1983) Kin recognition mechanisms: phenotype matching or recognition alleles. Am Nat 121:749–753

Blaustein AR, Bekoff M, Byers JA, Daniels TJ (1991) Kin recognition in vertebrates: what do we really know about adaptive value? Anim Behav 41:1079–1084

Blows MW, Schwarz MP (1991) Spatial distribution of a primitively social bee: does genetic population structure facilitate altruism? Evolution 45:680–693

Blum MS (1974) Deciphering the communicative rosetta stone. Bull Entomol Soc Am 20:30

Blum MS (ed) (1985) Fundamentals of insect physiology. John Wiley, New York

Boch R, Morse RA (1974) Discrimination of familiar and foreign queens by honeybee swarms. Ann Entomol Soc Am 67:709–711

Boch R, Morse RA (1979) Individual recognition of queens by honeybee swarms. Ann Entomol Soc Am 72:51–53

Boch R, Morse RA (1981) Effects of artificial odors and pheromones on queen discrimination by honey bees (*Apis mellifera* L.) Ann Entomol Soc Am 74:66–67

Boch R, Morse RA (1982) Genetic factor in queen recognition odors of honey bees. Ann Entomol Soc Am 75:654–656

Boch R, Rothenbuhler WC (1974) Defensive behaviour and production of alarm pheromone in honeybees. J Apic Res 13:217–221

Boch R, Shearer DA (1962) Identification of geraniol as the active component in Nassanoff pheromone of the honey bee. Nature (Lond) 194:704–706

Boch R, Shearer DA (1964) Identification of nerolic and geranic acids in the Nassanoff pheromone of the honey bee. Nature (Lond) 202:320–321

Boch R, Shearer DA (1966) Iso-pentyl acetate in stings of honeybees of different ages. J Apic Res 5:65–70

Boch R, Shearer DA, Stone BC (1962) Identification of iso-amyl acetate as an active component in the sting pheromone of the honey bee. Nature (Lond) 195:1018–1020

Bodenheimer FS (1937) Studies in animal populations, II. Seasonal population trends of the honey-bee. Q Rev Biol 12:406–425

Bohart GE (1972) Management of wild bees for the pollination of crops. Annu Rev Entomol 17:287–312

Boomsma JJ (1987) The empirical analysis of sex allocation in monogynous ants. In: Eder J, Rembold H (eds) Chemistry and biology of social insects. Peperny, Munich, pp 355–357

Boomsma JJ, Grafen A (1990) Intraspecific variation in ant sex ratios and the Trivers-Hare hypothesis. Evolution 44:1026–1034

Boomsma JJ, Grafen A (1991) Colony-level sex ratio selection in the eusocial Hymenoptera. J Evol Biol 4:383–407

Boorman SA, Levitt PR (1980) The genetics of altruism. Academic Press, New York

Boreham MM, Roubik DW (1987) Population change and control of Africanized honeybees in the Panama canal area. Bull Entomol Soc Am 33:34–39

Böttcher FK (1975) Beiträge zur Kenntnis des Paarungsfluges der Honigbiene. Apidologie 6:233–281

Boyd R, Richerson PJ (1980) Effects of phenotypic variation on kin selection. Proc Natl Acad Sci USA 77:7806–7810

Brandeburgo MM, Conçalves LS, Kerr WE (1982) Effects of Brazilian climatic conditions upon the aggressiveness of Africanized colonies of honeybees. In: Jaisson P (ed) Social insects in the tropics, vol 1. Presse de l'Universite Paris Nord, Paris, pp 256–280

Brandes C (1987) Effects of bidirectional selection on learning behaviour in honeybees (*Apis mellifera capensis*). In: Eder J, Rembold H (eds) Chemistry and biology of social insects. Peperny, Munich, pp 192–193

Brandes C (1988) Estimation of heritability of learning behaviour in honeybees (*Apis mellifera capensis*). Behav Genet 18:119–132

Brandes C (1990) Characterization of bidirectionally selected lines in honeybees shows that inheritance of learning is mainly due to additive genetic factors. In: Veeresh GK, Mallik B, Viraktamath (eds) Social insects and the environment. Oxford & IBH, New Delhi, pp 505–506

Brandes C (1991) Genetic differences in learning behavior in honeybees (*Apis mellifera capensis*). Behav Genet 21:271–294

Brandes C, Moritz RFA (1983) Evaluation of the heritability of learning behaviour of *Apis mellifera* using the proboscis extension reflex. Verh Dtsch Zool Ges 1983, p 258
Brandes C, Frisch B, Menzel R (1988) Time course of memory formation differs in honey bee lines selected for good and poor learning. Anim Behav 36:981–985
Breed MD (1981) Individual recognition and learnign of queen odors by worker honeybees. Proc Natl Acad Sci USA 78:2635–2537
Breed MD (1983) Nestmate recognition in honeybees. Anim Behav 31:86–91
Breed MD (1989) Introduction: theory and empirical results in sociogenetics. In: Breed MD, Page RE (eds) The genetics of social evolution. Westview Press, Boulder, CO, pp 1–13
Breed MD, Bennet B (1987) Kin recognition in highly eusocial insects. In: Fletcher DJC, Michener CD (eds) Kin recognition in animals. Wiley, Chichester, pp 243–285
Breed MD, Gamboa GJ (1977) Behavioral control of workers by queens in primitively eusocial bees. Science 195:694–696
Breed MD, Page RE (1989) The genetics of social evolution. Westview Press, Boulder
Breed MD, Velthuis HHW, Robinson GE (1984) Do worker honey bees discriminate among unrelated and related larval phenotypes. Ann Entomol Soc Am 77:737–739
Breed MD, Stiller TM, Moor MJ (1988) The ontogeny of kin discrimination cues in the honey bee, *Apis mellifera*. Behav Genet 18:439–448
Brett RA (1990) The population structure of naked mole-rat colonies. In: Sherman PW, Jarvis JVM, Alexander RD (eds) The biology of the naked mole-rat. Princeton Univ Press, Princeton, NJ, pp 97–136
Brain MV (1965) Social insect populations. Academic Press, London
Bridges CB (1925) Sex in relation to chromosomes and genes. Am Nat 59:127–137
Brody S (1945) Bioenergetics and growth. Reinhold, New York
Brouwers EVM (1984) The glucose-fructose ratio in the food of the honey bee larvae. J Apic Res 23:94–101
Brückner D (1976) Nachweis von Heterosis für das Merkmal Temperatur regulierung bei der Honigbiene (*Apis mellifera*). Apidologie 7:243–249
Brückner D (1979) Effects of inbreeding on worker honeybees. Bee World 60:137–140
Brückner D (1980) Hoarding behaviour and life span of inbred, non-inbred and hybrid honeybees. J Apic Res 19:35–41
Buchmann SL (1989) Long-term pollen records from honey bee debris middens in the Sonoran Desert. Eastern Branch, Entomol Soc Am 61st Annu Meet, Providence, Rhode Island
Buchmann SL, Thoenes SC (1990a) Scouting and foraging behavior by bees in natural swarms. Am Bee J 130:798–799
Buchmann SL, Thoenes SC (1990b) The electronic scale honey bee colony: a management and research tool. Bee Sci 1:40–47
Buckle GR, Greenberg L (1981) Nestmate recognition in sweat bees (*Lasioglossum zephyrum*): does an individual recognize its own odour or only odours of its nestmates? Anim Behav 29:802–809

Bühlmann G (1985) Assessing population dynamics in a honeybee colony. Mitt Dtsch Ges Allg Angew Entomol 312–316

Bühlmann G, Imdorf A, Wille H (1987) Pollen harvest and colony development of *Apis mellifera* in Switzerland. In: Eder J, Rembold H (eds) Chemistry and biology of social insects. Peperny, Munich, pp 159–160

Bulmer MA (1983) Sex ratio theory in social insects with swarming. J Theor Biol 100:329–339

Bunyard P, Goldsmith E (1988) Gaia, the thesis, the mechanisms and the implications. Quintrall, Wadebridge, UK

Burrill RM, Dietz A (1981) The response of honeybees to variations in solar radiation and temperature. Apidologie 12:319–328

Burt WH (1943) Territoriality and home range concepts as applied to mammals. J Mammal 24:346–352

Buss LW (1987) the evolution of individuality. Princeton Univ Press, Princeton, NJ

Butler CG (1954) The method and importance of the recognition by a colony of honeybees (*A. mellifera*) of the presence of its queen. Trans R Entomol Soc Lond 105:11–29

Butler CG (1957) The process of queen supersedure in colonies of honeybees (*Apis mellifera* L.). Insect Soc 4:211–223

Butler CG (1959) Queen substance. Bee World 40:269–275

Butler CG (1960) The significance of queen substance in swarming and supersedure in honey-bee (*Apis mellifera* L.) colonies. Proc R Entomol Soc Lond A 35:129–132

Butler CG (1961) The scent of queen honey bees (*Apis mellifera*) that causes partial inhibition of queen rearing. J Insect Physiol 7:258–264

Butler CG (1973) The queen and the "spirit of the hive." Proc R Entomol Soc Lond A 48:59–65

Butler CG, Fairey EM (1964) Pheromones of the honeybee: biological studies of the mandibular gland secretion of the queen. J Apic Res 3:65–76

Butler CG, Free JB (1952) The behaviour of worker honeybees at the hive entrance. Behaviour 4:262–292

Butler CG, Simpson J (1967) Pheromones of the queen honey bee (*Apis mellifera* L.) which enable her workers to follow her when swarming. Proc R Entomol Soc Lond A42:149–154

Butler CG, Callow RK, Johnston NC (1961) The isolation and synthesis of queen substance, 9-oxodec-trans-2-enoic acid, a honeybee pheromone. Proc R Soc Lond B155:417–432

Butler CG, Callow RK, Chapman JR (1964) 9-Hydroxydec-trans-2-enoic acid, a pheromone stabilizing honeybee swarms. Nature (Lond) 201:733

Butler CG, Callow RK, Greenway AR, Simpson J (1974) Movement of the pheromone, 9-oxodec-2-enoic acid, applied to the body surfaces of honeybees (*Apis mellifera*). Ext Exp Appl 17:112–116

Buttel-Reepen H (1906) Apistica. Beiträge zur Systematik, Biologie, sowie zur geschichtlichen und geographischen Verbreitung der Honigbiene (*Apis mellifica* L.), ihrer Varietäten und der übrigen *Apis*-Arten. Veroeff Zool Mus Berlin, pp 118–120

Calderone NW, Page RE (1988) Genotypic variability in age polyethism and task specialization in the honey bee, *Apis mellifera* (Hymenoptera: Apidae). Behav Ecol Sociobiol 22:17–25

Calderone NM, Robinson GE, Page RE (1989) Genetic structure and division of labour in honeybee societies. Experientia 45:765–767

Cale GH Jr, Rothenbuhler WC (1975) Genetics and breeding of the honey bee. In: Dadant P (ed) The hive and the honey bee. Dadant & Sons, Hamilton, IL, pp 157–184

Cale GH, Banker R, Powers J (1946) Management for honey production. In: Dadant P (ed) The hive and the honey bee. Dadant & Sons, Hamilton, IL, pp 355–412

Cale GH, Banker R, Powers J (1968) Pollen gathering relationship to honey collection and egglaying in honey bees. Am Bee J 108:8–9

Callow RK, Johnston NC (1960) The chemical constitution and synthesis of queen substance of honeybees (*Apis mellifera*). Bee World 41:152–153

Camazine S (1991) Self-organizing pattern formation on the combs of honey bee colonies. Behav Ecol Sociobiol 28:61–76

Camazine S, Morse RA (1988) The Africanized honey bee. Am Sci 76: 465–471

Camazine S, Sneyd J (1991) A model of collective nectar source selection by honey bees: self-organization through simple individual rules. J Theor Biol 149:547–571

Camazine S, Sneyd J, Jenkins J, Murray JD (1990) A mathematical model of self-organized pattern formation on the combs of honeybee colonies. J Theor Biol 147:553–571

Cameron SA, Robinson GE (1990) Juvenile hormone does not affect division of labor in bumble bee colonies (Hymenoptera: Apidae). Ann Entomol Soc Am 83:626–631

Caron DM, Greve GW (1979) Destruction of queen cells placed in queenright *Apis mellifera* colonies. Ann Ent Soc Am 72:405–407

Casey TM (1981) Insect flight energetics. In: Herreid CF, Fourtner CR (eds) Locomotion and energetics in arthropods. Plenum Press, New York pp 419–452

Casey TM, May ML, Morgan KR (1984) Flight energetics of euglossine bees in relation to morphology and wing stroke frequency. J Exp Biol 116: 271–289

Cerpet WL, Taylor DW (1985) The diversification of the Leguminosae: first fossil evidence of Mimosiodeae and Papilionodeae. Science 228: 1087–1089

Chance MM (1983) Honeybees observed feeding on the blood of a bear. Bee World 64:177

Charnov EL (1978) Sex-ratio selection in eusocial Hymenoptera. Am Nat 112:317–326

Chevalet C, Cornuet JM (1982a) Etude theorique sur la selection du caractere production du miel chez l'Abeille. 1. Modele genetique et statistique. Apidologie 13:39–65

Chevalet C, Cornuet JM (1982b) Evolution de la conanguinite dans une population d'abeilles. Apidologie 13:157–168

Cheverud JM (1985) A quantitative genetic model of altruistic selection. Behav Ecol Sociobiol 16:239–243

Chippendale GM (1978) The functions of carbohydrates in insect life processes. In: Rockstein M (ed) Biochemistry of insects. Academic Press, New York, pp 2–55
Clark AB (1978) Sex ratio and local resource competition in a prosimian primate. Science 210:163–165
Cockerell TDA (1907) A fossile honeybee. Entomologist (Lond) 40:227–229
Coelho JR, Mitton JB (1988) Oxygen consumption during hovering is associated with genetic variation of enzymes in honey-bees. Funct Ecol 2:141–146
Coenen-Stass DB, Schaarschmidt B, Lamprecht I (1980) Temperature distribution and calorimetric determination of heat production in the nest of the wood ant, *Formica polyctema* (Hymenoptera, Formicidae). Ecology 61:238–244
Cole BJ (1983) Multiple mating and the evolution of social behavior in the Hymenoptera. Behav Ecol Sociobiol 12:191–201
Collins AM (1979) Genetics of the response of the honeybee to an alarm chemical, isopentyl acetate. J Apic Res 18:285–291
Collins AM (1980) Effect of age on the response to alarm phermones by caged honey bees. Ann Entomol Soc Am 73:307–309
Collins AM (1981) Effects of temperature and humidity on honeybees response to alarm pheromones. J Apic Res 20:13–18
Collins AM, Rinderer TE (1985) Effect of empty comb on the defensive behaviour of honeybees. J Chem Ecol 11:333–338
Collins AM, Rinderer TE, Tucker KW, Sylvester HA, Lackett JJ (1980) A model of honeybee defensive behaviour. J Apic Res 19:224–231
Collins AM, Rinderer TE, Harbo JR, Brown MA (1984) Heritabilities and correlations for several characteristics in the honey bee. J Hered 75: 135–140
Contel EPB, Mestriner MA (1974) Esterase polymorphisms at two loci in the social bee. J Hered 65:349–352
Contel EPB, Mestriner MA, Martins E (1977) Genetic control and developmental expression of malate dehydrogenase in *Apis mellifera*. Biochem Genet 15:859–876
Cooper PD, Schaffer WM, Buchmann SL (1985) Temperature regulation of honey bees (*Apis mellifera*) foraging in the Sonoran Desert. J Exp Biol 114:1–15
Corkins CL (1930) The metabolism of the honeybee colony during winter. Bull Wyo Agric Exp Stn 175:1–54
Corkins CL, Gilbert CS (1932) The metabolism of honeybees in winter cluster. Univ Wyo Agric Exp Stn Bull 187:1–30
Cornuet JM (1983) Reproduction génétique et sélection de l'abeille. Bull Tech Apic 10:13–36
Cornuet JM, Chevalet C (1987) Etude theoretique sur la sélection du caractère "production du miel chez l'abeille II. Plan de sélection combinée de reines en fécondation naturelle. Apidologie 18:253–266
Cornuet JM, Louveaux J (1981) Aspects of genetic variability in *Apis mellifera* L. In: Howse PE, Clement JL (eds) Biosystematics of social insects. Academic Press, London, pp 85–94
Cornuet JM, Moritz RFA (1989) Selection theory and selection programmes.

In: Moritz RFA (ed) The instrumental insemination of the queen bee. Apimondia, Bucharest, pp 125–141
Cornuet JM, Garnery L, Solignac M (1991) Putative origin and function of the intergenic region between COI and COII of *Apis mellifera* L. mitochondrial DNA. Genetics 128:393–403
Correvon H, Pouyanne M (1916) Un curieux cas de mimetisme chez les Ophrydées. J Soc Nat Hort 29:23–84
Crabtree B, Newsholme EA (1975) Comparative aspects of fuel utilization and metabolism by muscle. In: Usherwood PNR (ed) Insect muscle. Academic Press, New York, pp 405–500
Craig R (1980) Sex investment ratios in social Hymenoptera. Am Nat 116:311–323
Crailsheim K (1985) Distribution of haemolymph in the honeybee (*Apis mellifera*) in relation to season, age and temperature. J Insect Physiol 31:707–713
Crailsheim K (1990) The protein balance of the honey bee worker. Apidologie 21:412–429
Crane E, Walker P (1984) Important honeydew sources and their honeys. Bee World 65:105–112
Crane PR, Friis EM, Pederson KR (1986) Lower cretaceous angiosperm flowers. Fossil evidence on early radiation of dicotyledons. Science 228:852–854
Crewe RM (1976) Aggressiveness of honey bees and their pheromone production. S Afr J Sci 72:209–212
Crewe RM (1982) Compositional variability: the key to social signals produced by honeybee mandibular glands. In: Breed MD, Michener CD, Evans HH (eds) The biology of social insects. Westview Press, Boulder, CO, pp 318–322
Crewe RM (1985) Bees observed feeding on an *Impala* carcass. Bee World 66:8
Crewe RM (1988) Natural history of honey-bee mandibular gland secretions: development of analytical techniques and the emergence of complexity. In: Needham GR, Page RE, Delfinado-Baker M, Bowman CE (eds) Africanized honeybees and bee mites. Ellis Horwood, Chichester, pp 149–158
Crewe RM, Velthuis HHW (1980) False queens: a consequence of mandibular gland signals in worker honeybees. Naturwissenschaften 67:467–469
Croat TC (1978) Flora of Barro Colorado Island. Stanford Univ Press, Palo Alto, CA
Crow JF, Kimura M (1970) An introduction to population genetics theory. Harper & Row, New York
Crozier RH (1970) Coefficients of relationship and the identity of genes by descent in the Hymenoptera. Am Nat 104:216–217
Crozier RH (1975) Animal cytogenetics, vol 3. Insecta 7, Hymenoptera. Borntraeger, Berlin
Crozier RH (1976) Counter-intuitive property of effective population size. Nature (Lond) 262:384
Crozier RH (1977) Evolutionary genetics of the Hymenoptera. Annu Rev Entomol 22:263–288

Crozier RH (1979) Genetics of sociality. In: Hermann HR (ed) Social insects. vol 1. Academic Press, New York, pp 223–286
Crozier RH (1982) On insects and insects: twists and turns in our under standing of the evolution of eusociality. In: Breed MD, Michener CD, Evans HE (eds) The biology of social insects. Westview Press, Boulder Co, pp 4–9
Crozier RH (1987) Towards a sociogenetics of social insects. In: Eder J, Rembold H (eds) Chemistry and biology of social insects. Peperny, Munich, pp 325–328
Crozier RH, Brückner D (1981) Sperm clumping and the population genetics of Hymenoptera. Am Nat 117:561–563
Crozier RH, Luykx P (1985) The evolution of termite eusociality is unlikely to have been based on a male-haploid analogy. Am Nat 126:867–869
Crozier RH, Page RE (1985) On being the right size: male contributions and multiple mating in social Hymenoptera. Behav Ecol Sociobiol 18:105–115
Crozier RH, Pamilo P (1980) Asymmetry in relatedness: who is related to whom? Nature (Lond) 283:604
Crozier RH, Smith BH, Crozier YC (1987) Relatedness and population structure of the primitively eusocial bee *Lasioglossum zephyrum* (Hymenoptera: Halictidae) in Kansas. Evolution 41:902–910
Crozier RH, Crozier YC, Mackinlay AG (1989) The CO-I and CO-II region of honeybee mitochondrial DNA: evidence for variation in insect mitochondrial evolutionary rates. Mol Biol Evol 6:399–411
Cruz-Landim C (1967) Estudo comparativo de algumas glandulas das abelhas (Hymenoptera, Apidae) e respectivas implicacoes evolutivas. Arq Zool (Sao Paulo) 15:177–290
Dadant CC (1975) Beekeeping equipment. In: Dadant CP (ed) The hive and the honey bee. Dadant & Sons, Hamilton, IL, pp 303–328
Daly H, Balling SS (1978) Identification of Africanized honeybees in the western hemisphere by discriminant analysis. J Kans Entomol Soc 51: 857–869
Darchen R (1962) Observation directe du dévelopment d'un rayon de cire. Le rôle des chaines d'abeilles. Insect Soc 9:103–120
Darchen R (1968) Le travail de la cire et la construction dans la ruche. In: Chauvin R (ed). Traité de biologie de l'abeille, vol II. Masson, Paris, pp 240–331
Darchen R, Pain J (1966) Le nid de *Trigona* (*Dactylurina*) *staudingeri Gribidoi* (sic). Biol Gabonica 2:25–35
Darwin C (1859) On the origin of species. J Murray, London
Davis K (1991) Forest. In: Lorimer LT (ed) Encyclopedia Americana, vol 11. Grolier, Danbury, CT, pp 582–590
Dawkins R (1976) The selfish gene. Cambridge Univ Press, Cambridge, UK
Dawkins R (1982) The extended phenotype. Oxford Univ Press, Oxford, UK
DeJong D (1982) Orientation of comb building by honeybees. J Comp Physiol A 147:495–501
Deligne J, Quennedey A, Blum MS (1981) The enemies of and defense mechanisms of termites. In: Hermann HR (ed) Social insects, vol 2. Academic Press, New York, pp 1–76
Del Lama MA, Lobo JA, Soares AEE, Del Lama SN (1990) Genetic differentiation estimated by isozymic analysis of Africanized honeybee

populations from Brazil and from Central America. Apidologie 21: 271–280
Dempster ER (1955) Maintenance of genetic heterogeneity. Cold Spring Harbor Symp Quant Biol 20:25–31
Deneubourg JL, Aron S, Goss S, Pasteels JM (1989) The self-organizing exploratory pattern of the Argentine ant, *Iridomyrmex humilis*. J Insect Behav 3:159–168
Deodikar GB, Thakar CV (1966) Cytogenetics of Indian honeybees and bearing on taxonomic and breeding problems. Indian J Genet 26A: 286–393
Deodikar GB, Tahkar CV, Pushpa NS (1959) Cytogenetic studies in Indian honeybees I. Somatic chromosome complement in *Apis indica* and its bearing on evolution and phylogeny. Proc Indian Acad Sci 49:196–207
Dietz A, Lambremont E (1970) Caste determination in honeybees, II. Food consumption of individual honey bee larvae, determined with 32P-labelled royal jelly. Ann Entomol Soc Am 63:1342–1345
Dietz A, Krell R, Eischen FA (1985) Preliminary investigation on the distribution of Africanized honey bees in Argentina. Apidologie 16: 99–108
Dixon AFG (1985) Aphid ecology. Chapman and Hall, New York
Doncaster L (1906) Spermatogenesis of the hive bee (Apis mellifica). Anat Anz 29:490–491
Doncaster L (1907) Spermatogenesis of the hive bee (Apis mellifica). Correction. Anat Anz 31:168–169
Drescher W (1968) Die Flugaktivität von Drohnen der Rassen *Apis mellifera* und *Apis mellifera ligustica*. Z Bienenforsch 9:390–409
Drescher W (1971) Selektionsversuche an einem Merkmal mit polygener Basis bei *Apis mellifera* L. 23rd Int Congr Apimondia, Moscow
Drescher W, Rothenbuhler WC (1964) Sex determination in the honeybee. J Hered 55:195–201
Drum NH, Rothenbuhler WC (1984) Effect of temperature on nonstinging aggressive response of worker honeybees to diseased and healthy bees. J Apic Res 23:82–87
Duchateau MJ (1989) Agonistic behaviours in colonies of the bumblebee, *Bombus terrestris*. J Ethol 7:141–151
Duchateau MJ, Velthuis HHW (1988) Development and reproductive strategies in *Bombus terrestris* colonies. Behaviour 107:186–207
Duchateau MJ, Velthuis HHW (1989) Ovarian development and egg laying in workers of *Bombus terrestris*. Entomol Exp Appl 51:199–213
Dyce EJ, Morse RA (1960) Wintering honeybees in New York State. Cornell Univ Ext Bull 1054
Dyer FC (1985a) Mechanisms of dance orientation in the Asian honeybee *Apis florea* L. J Comp Physiol A 157:183–198
Dyer FC (1985b) Nocturnal orientation of the Asian honey bee, *Apis dorsata*. Anim Behav 33:769–774
Dyer FC (1987) New perspectives on the dance orientation of the Asian honey bees. In: Menzel R, Mercer A (eds) The neurobiology and behaviour of honey bees. Springer, Berlin Heidelberg New York, pp 234–246

Dyer FC (1991) Bees acquire route based memories but not cognitive maps in a familiar landscape. Anim Behav 41:239–246
Dzierzon J (1845) Gutachten über die von Herrn Direktor Stöhr im ersten und zweiten Kapitel des General-Gutachtens aufgestellten Fragen. Eichstädter Bienenz 1:109–113, 119–121
Eckert JE (1933) The flight range of the honeybee. J Agri Res 47:257–285
Eckert R, Randall D (1983) Animal physiology. Freeman, New York
Engels E, Engels W (1988) Age-dependent queen attractiveness for drones and mating in the stingless bee, *Scaptotrigona postica*. J Apic Res 27:3–8
Engels W (1988) Reproductive strategies in bees. Verh Dtsch Zool Ges 81:155–167
Engels W (1990) Social insects: an evolutionary approach to caste and reproduction. Springer, Berlin Heidelberg New York
Engels W, Engels E (1977) Vitellogenin und Fertilität bei stachellosen Bienen. Insect Soc 24:71–94
Engels W, Imperatriz-Fonseca VL (1990) Caste development, reproductive strategies, and control of fertility in honeybees and stingless bees. In: Engels W (ed) Social insects. An evolutionary approach to castes and reproduction. Springer, Berlin Heidelberg New York, pp 167–230
Erber J (1981) Neural correlates of learning in the honeybee. Trends Neurobiol Sci 4:270–273
Esch H (1960) Über die Köpertemperaturen und den Wärmehaushalt von *Apis mellifica*. Z Vgl Physiol 43:305–355
Esch H (1961) Über die Schallerzeugung beim Werbetanz der Honigbiene. Z Vgl Physiol 45:1–11
Esch H (1964) Über den Zusammenhang zwischen Temperatur, Aktionspotentialen und Thoraxbewegungen bei der Honigbiene (*Apis mellifica*). Z Vgl Physiol 48:547–551
Esch H (1967a) The sound produced by swarming honey bees. Z Vgl Physiol 56:408–411
Esch H (1967b) The evolution of bee language. Sci Am 216:3–9
Esch H (1967c) The sounds produced by flies and bees. Z Vergl Physiol 54:256–267
Esch H (1976) Body temperature and flight performance of honey bees in a servo-mechanically controlled wind tunnel. J Comp Physiol 109:265–277
Esch H (1988) The effects of temperature on flight muscle potentials in honeybees and cuculiinid winter moths. J Exp Biol 135:109–117
Esch H, Bastian JA (1968) Mechanical and electrical activity in the indirect flight muscles of the honey bee. Z Vergl Physiol 58:429–440
Esch H, Esch I, Kerr WE (1965) Sound: an element common to communication of stingless bees and to dances of honeybees. Science 149:320–321
Esch H, Nachtigall W, Kogge SN (1975) Correlations between aerodynamic output, electrical activity in the indirect flight muscles and wind positions of bees in a servomechanically controlled wind tunnel. J Comp Physiol 100:147–159
Esch H, Goller F, Heinrich B (1991) How do bees shiver. Naturwissenschaften 78:325
Escherich K (1909) Die Termiten oder weissen Ameisen. W Klinkhardt, Leipzig

Evers CA, Seeley TD (1986) Kin discrimination and aggression in honey bee colonies with laying workers. Anim Behav 34:924–925
Faegri K, Pijl L van der (1979) The principles of pollination ecology, 3rd edn. Pergamon Press, Oxford
Fahrenholz L, Lamprecht I, Schricker B (1989) Thermal investigations of a honey bee colony: thermoregulation of the hive during summer and winter and heat production of members of different castes. J Comp Physiol B 159:551–560
Fahrenhorst H (1977a) Nachweis übereinstimmender Chromosomenzahlen (n = 16) bei allen 4 *Apis*-Arten. Apidologie 8:89–100
Fahrenhorst H (1977b) Chromosome number in the tropical honeybee species *Apis dorsata* and *Apis florea*. J Apic Res 16:56–58
Falconer DS (1981) Introduction to quantitative genetics, 2nd edn. Longman, New York
Falconer DS (1989) Introduction to quantitative genetics, 3rd edn. Longman Scientific & Technical, Essex, UK
Farrar CL (1952) Ecological studies on overwintered honey bee colonies in the northern states. US Dep Agric Circ 702
Farrar CL (1960) From need to plenty – through the cold of winter. Am Bee J 100:306–308
Fell RD, Ambrose JT, Burgett DM, DeJong D, Morse RA, Seeley TD (1977) Seasonal cycle of swarming in honey bees (*Apis mellifera* L.). J Apic Res 16:170–173
Fenton FA (1952) Field crop insects. Macmillan, New York
Ferguson AW, Free JB (1980) Queen pheromone transfer within honeybee colonies. Physiol Entomol 5:359–366
Festinger L (1949) The analysis of sociograms using matrix algebra. Hum Relat 2:153–158
Fieberg K (1912) Schlafende Insekten. Jena Z Naturwiss 48:315–364
Filmer (1932) Brood area and colony size as factors in activity of pollination units. J Econ Entomol 25:336–343
Fisher RA (1930) The genetical theory of natural selection. Clarendon, Oxford
Fittkau EJ, Klinge H (1973) On biomass and trophic structure of the central Amazonian rain forest ecosystem. Biotropica 5:2–14
Fleig R (1990) Engrailed expression and body segmentation in the honeybee *Apis mellifera*. Wilhelm Roux's Arch Dev Biol 198:467–473
Fleig R, Sander K (1986) Embryogenesis of the honeybee *Apis mellifera* L. (Hymenoptera: Apidae): an SEM study. Int J Insect Morphol Embryol 15:449–462
Fletcher DJC (1975) Significance of dorsoventral abdominal vibration among honey bees (*Apis mellifera* L.). Nature (Lond) 256:721–723
Fletcher DJC (1978) The African bee, *Apis mellifera adansonii*, in Africa. Annu Rev Entomol 23:151–171
Fletcher DJC (1987) Recognition of actual and potential female reproductives in colonies of social Hymenoptera. In: Pasteels JM, Deneubourg JL (eds) From individual to collective behavior in social insects. Birkhäuser, Basel, pp 365–378
Fletcher DJC, Blum M (1983a) Regulation of queen number by workers in colonies of social insects. Science 212:73–75

Fletcher DJC, Blum M (1983b) The inhibitory pheromone of queen fire ants: effects of disinhibition on dealation and oviposition by virgin queens. J Comp Physiol 153:467–475
Fletcher DJC, Crewe RM (1981) Nest structure and thermoregulation in the stingless bee *Trigona* (*Plebeina*) *denoiti* Vachal (Hymenoptera: Apidae). J Entomol Soc S Afr 44:183–196
Fluri P, Lüscher M, Wille H, Gerig L (1982) Changes in weight of the pharyngeal gland and hemolymph titres of juvenile hormone, protein and vitellogenin in worker honey bees. J Insect Physiol 28:61–68
Forel A (1910) Das Sinnesleben der Insekten. Reinhardt, Munich
Fowler HG (1981) On emigration of leaf-cutting ant colonies. Biotropica 13:316
Frank SA (1987) Variable sex ratio among colonies of ants. Behav Ecol Sociobiol 20:195–201
Franks NR (1989) Army ants: a collective intelligence. Am Sci 77:138–145
Franks NR, Scovell E (1983) Dominance and reproductive success among slave making ants. Nature (Lond) 304:724–725
Free JB (1959) The transfer of food between the adult members of a honeybee community. Bee World 40:193–201
Free JB (1960) The pollination of fruit trees. Bee World 41:141–151, 169–186
Free JB (1961) Hypopharyngeal gland development and division of labour in honeybee (*Apis mellifera* L.) colonies. Proc R Entomol Soc A 36:5–8
Free JB (1963) The flower constancy of honeybees. J Anim Ecol 32:119–131
Free JB (1967) The production of drone comb in honeybees. J Apic Res 6:29–36
Free JB (1968) The conditions under which foraging honeybees expose their Nasonov gland. J Apic Res 7:139–145
Free JB (1970) The insect pollination of crops. Academic Press, London
Free JB (1977) The social organization of the honeybees. Arnold, London
Free JB (1987) Pheromones of social bees. Cornell Univ Press, Ithaca, NY
Free JB, Butler CG (1955) An analysis of the factors involved in the formation of a cluster of honeybees (*Apis mellifera*). Behaviour 7:304–316
Free JB, Racey PA (1966) The pollination of *Freesia refracta* in glasshouses. J Apic Res 5:177–182
Free JB, Simpson J (1963) The respiratory metabolism of the honeybee colonies at low temperatures. Entomol Exp Appl 6:234–238
Free JB, Simpson J (1968) The alerting pheromones of the honeybee. Z Vgl Physiol 61:361–365
Free JB, Spencer-Booth Y (1960) Chill-coma and cold death temperatures of *Apis mellifera*. Entomol Exp Appl 3:222–230
Free JB, Williams IH (1970) Exposure of Nasonov gland by honeybees (*Apis mellifera*) collecting water. Behaviour 37:286–290
Free JB, Williams IH (1972) Hoarding by honey bees (*Apis mellifera* L.). Anim Behav 20:327–334
Free JB, Williams IH (1974) Factors determining food storage and brood rearing in honeybee (*Apis mellifera* L.) comb. J Entomol Am 49:47–63
Free JB, Williams IH (1975) Factors determining the rearing the rejection of drones by the honeybee colony. Anim Behav 23:650–675

Free JB, Williams IH (1979) Communication by pheromones and other means in *Apis florea* colonies. J Apic Res 18:16–25
Free JB, Williams IH (1983) Scent-marking of flowers by honeybees. J Apic Res 22:86–90
Frisch B (1984) Circadiane Aktivitätsrhythmik bei der Honigbiene. Apidologie 15:278–280
Frisch B, Aschoff J (1987) Circadian activity in honeybees: entrainment by feeding cycles. Physiol Entomol 12:41–49
Frisch B, Koeniger N (1991) Synchronization of circadian rhythms in honeybees. Behav Ecol Sociobiol (in press)
Frostic G (1978) The infinite destiny. Presscroft Papers, Benzonia, MI
Frostic G (1981) The evolving omnity. Presscroft Papers, Benzonia, MI
Frumhoff PC, Baker J (1988) A genetic component to division of labour within honey-bee colonies. Nature (Lond) 333:358–361
Frumhoff PC, Schneider S (1987) The social consequences of honey bee polyandry: the effects of kinship on worker interactions within colonies. Anim Behav 35:255–262
Fukuda H, Sekiguchi K (1966) Seasonal change of the honey bee worker longevity in Sapporo, North Japan, with notes on some factors affecting the life span. Jpn J Ecol 16:206–212
Furgala B (1975) Fall management and the wintering of productive colonies. In: Dadant CP (ed) The hive and the honeybee. Dadant & Sons, Hamilton, IL, pp 471–490
Gadagkar R (1985a) Evolution of insect sociality – a review of some attempts to test modern theories. Proc Indian Acad Sci 94:309–324
Gadagkar R (1985b) Kin recognition in social insects and other animals – a review of recent findings and a consideration of their relevance for the theory of kin selection. Proc Indian Acad Sci 94:587–621
Gadagkar R (1990) Evolution of eusociality: the advantage of assured fitness returns. Philos Trans R Soc Lond B 329:17–25
Gambino P, Hoelmer K, Daly HV (1990) Nest sites of feral honey bees in California, USA. Apidologie 21:35–45
Garnery L, Vatrin D, Cornuet JM, Solignac M (1991) A phylogenetic tree of four *Apis* species from mitochondrial DNA sequence data. Apidologie (in press)
Garofalo CA, Kerr WE (1975) Sex determination in bees. 1. Balance between femaleness and maleness genes in *Bombus atratus*. Genetica 45:203–209
Gartside DF (1980) Similar allozyme polymorphism in honeybees (*Apis mellifera*) from different continents. Experientia 36:649–650
Gary NE (1960) A trap to quantitatively recover dead and abnormal honeybees from the hive. J Econ Entomol 53:782–785
Gary NE (1961) Mandibular gland extirpation in living queen and worker honey bees (*Apis mellifera*). Ann Entomol Soc Am 54:529–531
Gary NE (1962) Chemical mating attractants in the queen honey bee. Science 136:773–774
Gary NE (1963) Observations of mating behaviour in the honeybee. J Apic Res 3:65–76
Gary NE (1966) Robbing behaviour in the honbey bee. Am Bee J 106: 446–448

Gary NE (1975) Activities and behavior of honey bees. In: Dadant CP (ed) The hive and the honey bee. Dadant & Sons, Hamilton, IL, pp 185–264
Gary NE, Marston J (1971) Mating behaviour of drone honeybees with queen models (*Apis mellifera* L.). Anim Behav 19:299–304
Gary NE, Witherell PC, Marston JM (1972) Foraging range and distribution of honey bees used for carrot and onion pollination. Environ Entomol 1:71–78
Gates BN (1914) The temperature of the bee colony. Bull US Dep Agric 96:1–29
Gehring WJ (1987) Homeotic genes, the homeobox, and the spatial organization of the embryo. Harvey Lect 81:153–172
Gerig L (1971) Wie Drohnen auf Königinnen-Attrappen reagieren. Schweiz Bienenztg 94:558–562
Gerling D, Velthuis HHW, Hefetz A (1989) Bionomics of the large carpenter bees of the genus *Xylocopa*. Annu Rev Entomol 34:163–190
Getz WM (1991) The honey bee as a model kin recognition system. In: Hepper PG (ed) Kin recognition. Cambridge Univ Press, Cambridge, UK (in press)
Getz WM, Smith KB (1983) Genetic kin recognition: honey bees discriminate between full and half sisters. Nature (Lond) 302:147–148
Getz WM, Brückner D, Parisian TR (1982) Kin structure and the swarming behavior of the honey bee, *Apis mellifera*. Behav Ecol Sociobiol 10: 265–270
Getz WM, Brückner D, Smith KB (1986) Conditioning honeybees to discriminate between heritable odors from full and half sisters. J Comp Physiol A 159:251–256
Getz WM, Brückner D, Smith KB (1988) Variability of chemosensory stimuli within honeybees (*Apis mellifera*) colonies: a differential conditioning assay for discrimination cues. J Chem Ecol 14:249–260
Ghent RL, Gary NE (1962) A chemical alarm releaser in the honey bee stings. Psyche 69:1–6
Ghisalberti EL (1979) Propolis: a review. Bee World 60:59–84
Ghiselin MT (1974) The economy of nature and sex. Univ California Press, Berkeley, CA
Gösswald K (1951) Die rote Waldameise in Dienste der Waldhygiene. Metta Kienau, Lueneburg, FRG
Goewie EA (1978a) Regulation of caste differentiation in the honey bee (*Apis mellifera* L.) after topical application of JH-III. Proc K Ned Akad Wet 79C:466–469
Goewie EA (1978b) Regulation of caste differentiation in the honey bee (*Apis mellifera* L.). Meded Landbouwhogesch Wageningen 78–15:1–75 pp 3–8
Goller F, Esch H (1990) Comparative study of chill-coma temperatures and muscle potentials in insect flight muscles. J Exp Biol 150:221–231
Goller F, Esch H (1991) Oxygen consumption and flight muscle activity during heating in workers and drones of *Apis mellifera*. J Comp Physiol B 161:61–67
Gonçalves LS (1972) Untersuchungen über das morphologische Merkmal "Zahl der Flügelhäckchen" bei *Apis mellifica*. In: Ruttner F, Ruttner H,

Harnaj V (eds) Paarungskontrolle und Selektion bei der Honigbiene. Apimondia, Bucharest, pp 83–86
Gonçalves LS, Stort AC (1978) Honey bee improvement through behavioral genetics. Annu Rev Entomol 31:197–213
Goodman LJ, Fisher RC (1991) Physiology and behaviour of honeybees. CAB International, Oxon, UK
Gordon MS, Bartholomew GA, Grinnell AD, Jorgensen CB, White FN (1982) Animal physiology, principles and adaptations, 4th edn. Macmillan, New York
Gould JL (1980) The case for magnetic sensitivity in birds and bees (such as it is). Am Sci 68:256–268
Gould JL (1982) Why do bees have dialects? Behav Ecol Sociobiol 10: 53–56
Gould JL (1984) The natural history of honey bee learning. In: Marler P, Terrace H (eds) The biology of learning. Springer, Berlin Heidelberg New York, pp 149–180
Gould JL (1985) How do bees remember flower shape? Science 227: 1492–1494
Gould JL (1986) The landmark map of honey bees: do insects have cognitive maps? Science 232:861–863
Gould JL (1987a) Landmark learning in honey bees. Anim Behav 34: 990–997
Gould JL (1987b) Honey bees store learned flower-landing behaviour according to time of day. Anim Behav 35:1579–1581
Gould JL (1988) Resolution of pattern learning in honey bees. J Insect Behav 1:225–233
Gould JL (1991) The ecology of honeybee learning. In: Goodman LJ, Fisher RC (eds) The behaviour and physiology of bees. CAB International, Oxon UK, pp 306–322
Gould JL, Gould CG (1988) The honey bee. Freeman, New York
Gould JL, Towne WT (1987) Honey bee learning. Adv Insect Physiol 20: 55–75
Gould JD, Dyer FC, Towne WF (1985) Recent progress in the study of the dance language. In: Hölldobler B, Lindauer M (eds) Experimental behavioral ecology and sociobiology. G Fischer, Stuttgart, pp 141–161
Grafen A (1985) A geometric view of relatedness. Oxford Surv Evol Biol 2:28–89
Grafen A (1986) Split sex ratios and the evolutionary origins of eusociality. J Theor Biol 122:95–121
Grafen A (1990) Do animals really recognize kin? Anim Behav 39:42–54
Graham GH, Balzer L (1982) The effects of the introduced honeybee (*Apis mellifera*) on Australian native bees. NSW Natl Parks Wildlife Serv Rep, Australian Museum, Sydney
Grassé PP, Noirot C (1951) La sociotomie: migration et fragmentation de la termitière chez les *Anoplotermes* et les *Trinervitermes*. Behaviour 3: 146–166
Grassi B, Sandias A (1893) Costituzione e sviluppo della societa dei Termitidi. Atti Accad Gioenia Sci Nat 6:1–75
Greenberg L (1979) Genetic component of bee odor in kin recognition. Science 206:1095–1097

Greenberg L (1982) Persistent habituation to female odor by male sweat bees (*Lasioglossum zephyrum*) Hymenoptera: Halictidae). J Kans Entomol Soc 55:525–531

Greenberg L, Buckle GR (1981) Inhibiting of worker mating by queens in a sweat bee (*Lasioglossum zepyrum*). Insect Soc 28:347–352

Gregor H-J (1982) Die jungtertiären Floren Süddeutschlands. Enke, Stuttgart

Griffing B (1967) Selection in reference to biological groups I. Individual and group selection applied to populations of unordered groups. Aust J Biol Sci 20:127–139

Griffing B (1968) Selection in reference to biological groups II. Consequences of selection in groups of one size when evaluated in groups of a different size. Aust J Biol Sci 21:1163–1170

Griffing B (1969) Selection in reference to biological groups IV. Application of selection index theory. Aust J Biol Sci 22:131–142

Griffing B (1982a) A theory of natural selection incorporating interaction among individuals IX. Use of groups consisting of a mating pair and sterile diploid caste members. J Theor Biol 95:181–198

Griffing B (1982b) A theory of natural selection incorporating interaction among individuals X. Use of groups consisting of a mating pair together with haploid and diploid caste members. J Theor Biol 95:199–223

Groot AP de (1951) Effect of a protein containing diet on the longevity of caged bees. K Ned Akad Wet 54:2–4

Gunn A, Gatehouse AG (1988) The development of enzymes involved in flight metabolism in *Spodoptera exempta* and *Mythimna separata*. J Comp Physiol B 91:315–324

Gustafsson L (1986) Lifetime reproductive success and heritability: empirical support of Fisher's fundamental theorem. Am Nat 128:761–764

Hachinoe Y, Oshini N (1952) On the meiosis of the drone honey bee (*Apis mellifica*). Bull Natl Inst Agric Sci G 3:83–87

Hagenguth H, Rembold H (1978) Kastenspezifische Modulation des Ecdysteroid-Titers bei der Honigbiene. Mitt Dtsch Ges Allg Angew Entomol 1:296–298

Haldane JBS, Jayakar SD (1963) Polymorphism due to selection in varying directons. J Genet 58:237–242

Hall HG (1986) DNA differences found between Africanized and European honeybees. Proc Natl Acad Sci USA 83:4874–4877

Hall HG (1988) Characterization of the African honey-bee genotype by DNA restriction fragments. In: Needham GR, Page RE, Delfinado-Baker M, Bowman CE (eds) Africanized honey bees and bee mites. Ellis Horwood, Chichester, pp 287–293

Hall HG (1990) Parental analysis of introgressive hybridization between African and European honeybees using nuclear DNA RFLPs. Genetics 125:611–621

Hall HG, Muralidharan K (1989) Evidence from mitochondrial DNA that African honey bees spread as continuous maternal lineages. Nature (Lond) 339:211–213

Hallund V (1956) Om biernes overvintring. Nord Bitidsskr 8:106–113

Hamilton WD (1964a) The genetical evolution of social behaviour. I. J Theor Biol 7:1–16

Hamilton WD (1964b) The genetical evolution of social behaviour, II. J Theor Biol 7:17–52
Hamilton WD (1967) Extraordinary sex ratios. Science 156:477–488
Hamilton WD (1972) Altruism and related phenomena, mainly in social insects. Annu Rev Ecol Syst 3:193–232
Hamilton WD (1975) Innate social aptitudes of man: an approach from evolutionary biology. In: Fox R (ed) Biosocial anthropology. Wiley, New York, pp 133–155
Hamilton WD (1987) Discriminating nepotism: expectable, common, overlooked. In: Fletcher DJC, Michener CD (eds) Kin recognition in animals. Wiley, Chichester, pp 417–437
Hammann E (1957) Wer hat die Initiative bei den Ausflügen der Königin, die Königin oder die Arbeitsbienen? Insect Soc 4:91–106
Hancocks D (1973) Master builders of the animal world. Harper & Row, New York
Harris WV (1971) Termites. Their recognition and control. Longman, London
Harrison JM (1985) Caste specific changes in honeybee flight capacity. Physiol Zool 59:175–187
Hartfelder K (1985) Composition of larval food in the stingless bee, *Scaptotrigona postica*. Mitt Dtsch Ges Allg Angew Entomol 4:339–341
Hartfelder K (1990) Regulatory steps in caste development of eusocial bees. In: Engels W (ed) Social insects. Springer, Berlin Heidelberg New York, pp 245–264
Hawkins AJS, Bayne BL, Day AJ (1986) Protein turnover, physiological energetics and heterozygosity in the blue mussel, *Mytilus edulis*: the basis of variable age specific growth. Proc R Soc Lond B 299:161–176
Haydak MH (1930) Division of labor in the colony. Cesky Vcelar 64:166–168
Haydak MH (1935) Brood rearing by honeybees confined to a pure carbohydrate diet. J Econ Entomol 28:657–660
Haydak MH (1970) Honey bee nutrition. Annu Rev Entomol 15:143–156
Haydak MH, Dietz A (1965) Influence of the diet on the development and brood rearing of honeybees. Proc Int Beekeeping Congr, Bucharest. Apimondia 20:158–162
Hazelhoff EH (1941) De luchtverversching van een bijenkast gedurende den zomer. Maandschr Bijenteelt 44:16
Heath JE, Adams PA (1967) Regulation of heat production by large moths. J Exp Biol 47:21–33
Hedrick PW (1986) Genetic polymorphism in heterogenous environments: a decade later. Annu Rev Ecol Syst 17:535–566
Heinrich B (1974a) Thermoregulation in endothermic insects. Science 185:747–756
Heinrich B (1974b) Thermoregulation in bumblebees. I. Brood incubation by *Bombus vosnesenskii* queens. J Comp Physiol 88:129–140
Heinrich B (1975) Thermoregulation in bumblebees. II. Energetics of warm up and free flight. J Comp Physiol B 95:155–166
Heinrich B (1976) The foraging specializations of individual bumblebees. Ecol Monogr 46:105–128
Heinrich B (1979a) "Majoring" and "minoring" in bumblebees: an experimental analysis. Ecology 60:245–255

Heinrich B (1979b) Bumblebee economics. Harvard Univ Press, Cambridge, MA

Heinrich B (1980a) Mechanisms of body-termperature regulation in honeybees, *Apis mellifera*. I. Regulation of head temperature. J Exp Biol 85:61–72

Heinrich B (1980b) Mechanisms of body-temperature regulation in honeybees, *Apis mellifera*. II. Regulation of thoracic temperature at high air temperatures. J Exp Biol 85:73–87

Heinrich B (1981a) The mechanisms and energetics of honeybee swarm temperature regulation. J Exp Biol 91:25–55

Heinrich B (1981b) Insect thermoregulation. Wiley, New York

Heinrich B (1985) The social physiology of temperature regulation in honeybees. In: Hölldobler B, Lindauer M (eds) Experimental behavioral ecology and sociobiology. G Fischer, Stuttgart, pp 393–406

Heinrich B (1987) Thermoregulation of individual honeybees. In: Menzel R, Mercer A (eds) Neurobiology and behavior of honeybees. Springer, Berlin Heidelberg New York, pp 102–111

Heinrich B, Casey TM (1973) Metabolic rate and endothermy in sphinx moths. J Comp Physiol 82:195–206

Heldmaier G (1975) Metabolic and thermoregulatory responses to heat and cold in the Djungarian hamster, *Phodopus sungoras*. J Comp Physiol B102:115–122

Helversen O von, Edrich W (1974) Der Polarisationsempfänger im Bienenauge: ein Ultraviolettrezeptor. J Comp Physiol 94:33–47

Hemmingsen (1960) Energy metabolism as related to body size and respiratory surfaces, and its evolution. Rep Steno Mem Hosp Nord Insulinlab 9:7–110

Hemmling C, Koeniger N, Ruttner F (1979) Quantitative Bestimmung der 9-Oxodecensäure im Lebenszyklus der Kapbiene (*Apis mellifera capensis* Escholtz). Apidologie 10:227–240

Hepburn HR (1986) Honeybees and wax. Springer, Berlin Heidelberg New York

Hepburn HR, Crewe RM (1990) Defining the Cape honeybee: reproductive traits of queenless workers. S Afr J Sci 86:524–527

Hepburn HR, Whiffler LA (1988) Dwarfs and queenballs in swarms of the Cape honeybee. S Afr J Sci 84:851–852

Hepburn HR, Nefdt RJC, Whiffler LA (1988) Queen loss in the Cape honeybee: the interactions of brood, laying workers (false queens?) and queen cells. S Afr J Sci 84:778–780

Heran H (1959) Wahrnehmung und Regelung der Flugeigengeschwindigkeit bei *Apis mellifica* L. Z Vgl Physiol 42:103–163

Hermann HR (1979) Insect sociality – an introduction. In: Hermann HR (ed) Social insects. Academic Press, New York, pp 1–33

Herreid CF, Kessel B (1967) Thermal conductance in birds and mammals. Comp Biochem Physiol 21:405–414

Hertel H, Maronde U (1987) Processing of visual information in the honeybee brain. In: Menzel R, Mercer A (eds) Neurobiology and behavior of honeybees. Springer, Berlin Heidelberg New York, pp 141–157

Heusner A, Roth M (1963) Consommation d'oxygene de l'abeille a differentes temperatures. C R Hebd Sianc Acad Sci Paris 256:284–285

Hill RW, Wyse GA (1989) Animal physiology, 2nd edn. Harper & Row, New York
Hill WG (1976) Order statistics of correlated variables and implications in genetic selection programs. Biometrics 32:889–902
Hill WG (1984) Quantitative genetics I and II. Van Nostrand, New York
Hillesheim E, Koeniger N, Moritz RFA (1989) Colony performance in honeybees (*Apis mellifera capensis* Esch.) depends on the proportion of subordinate and dominant workers. Behav Ecol Sociobiol 24:291–296
Himmer A (1927) Der soziale Wärmehaushalt der Honigbiene, II. Die Wärme der Bienenbrut. Erlanger Jahrb Bienenk 5:1–32
Hoage TR, Kessel RG (1968) An electron microscope study of the process of differentiation during spermatogenesis in the drone honey bee (*Apis mellifera* L.) with special reference to centriole replication and elimination. J Ultrastruct Res 24:6–32
Hoar WS (1983) General and comparative physiology. Prentice-Hall, New York
Hoekstra RF, Bijlsma R, Dolman AJ (1985) Polymorphism from environmental heterogeneity: models are only robust if the heterozygote is close in fitness to the homozygote in each environment. Genet Res 45:299–314
Hölldobler B, Wilson EO (1990) The ants. Springer, Berlin Heidelberg New York
Hogendoorn K, Velthuis HHW (1988) Influence of multiple mating on kin recognition by worker honeybees. Naturwissenschaften 75:412–413
Hoshiba H (1979) Chromosome of diploid and haploid drone honeybee, *Apis mellifera*. Proc 27th Int Congr Apicult, Apimondia, Budcharest, pp 73–74
Hoshiba H, Kusanagi A (1978) Karyological study of honeybee. J Apic Res 17:105–109
Hoshiba H, Okada I (1986) G-banding analysis of male chromosomes in *Apis cerana* and *Apis mellifera ligustuca*. Apidologie 17:101–106
Hoshiba H, Okada I, Kusangi A (1981) The diploid drone of *Apis cerana japonica* and its chromosomes. J Apic Res 20:143–147
Houston A, Schmid-Hempel P, Kacelnik A (1988) Foraging strategy, worker mortality, and the growth of the colony in social insects. Am Nat 131: 107–114
Hubell SP, Johnson LK (1977) Competition and nest spacing in a tropical stingless bee community. Ecology 58:949–963
Huber F (1792) Nouvelles observations sur les abeilles. Paschoud, Geneva
Huber F (1814) Nouvelles observations sur les abeilles, 2. Translation 1926, Dadant & Sons, Hamilton, IL
Ifantidis MD (1978) Wabenorientierung im Nest der Honigbiene (Apis mellifica L.). Apidologie 9:57–73
Imdorf A, Bühlmann G, Gerig L, Kilchenmann V, Wille H (1987) Überprüfung der Schätzmethode zur Ermittlung der Brutfläche und der Anzahl der Arbeiterinnen in freifliegenden Bienenvölkern. Apidologie 18:137–146
Inoue T, Salmah S, Abbas I, Yusuf E (1985) Foraging behavior of individual workers and foraging dynamics of colonies of three Sumatran stingless bees. Res Popul Ecol 27:373–392

Isil S (1983) A study of social behavior in laboratory colonies of the naked mole-rat (*Heterocephalus glaber*): cooperation in colony labor and reproduction. MSc Thesis, Univ California, Santa Cruz, CA
Istomina-Tsvetkova KP (1953) Reciprocal feeding of bees. Apic Abstr 92:54
Istomina-Tsvetkova KP (1960) Contribution to the study of trophic relations in adult worker bees. Proc 18th Int Congr Apicult, Apimondia, Rome, 2:361–368
Jacobs W (1924) Das Duftorgan von *Apis mellifica* und ähnliche Hautdrüsenorgane sozialer und solitärer Apiden. Z Morphol Oekol Tiere 3:1–80
Jacoby R (1949) Das Imker ABC, 2nd edn. Bienenzucht, Bad Segeberg
Jander R (1982) Random and systematic search in foraging insects. In: Breed MD, Michener CD, Evans HE (eds) The biology of social insects. Westview Press, Boulder, CO, pp 28–31
Jarvis, JUM (1978) Energetics and survival in *Heterocephalus glaber* (Rüpell), the naked mole-rat (Rodentia: Bathyergidae). Bull Carnegie Mus Nat Hist 6:81–87
Jarvis JUM (1981) Eusociality in a mammal: cooperative breeding in naked mole-rat colonies. Science 212:571–573
Jarvis JUM, O'Riain MJ, McDaid E (1990) Growth and factors affecting body size in naked mole-rats. In: Sherman PW, Jarvis JUM, Alexander RD (eds) The biology of the naked mole rat. Princeton Univ Press, Princeton, NJ, pp 358–383
Jay SC (1970) The effect of various combinations of immature queen and worker bees on the ovary development of worker honeybees in colonies with and without queens. Can J Zool 48:169–173
Jaycox ER, Parise SG (1980) Homesite selection by Italian honey bee swarms, *Apis mellifera ligustica* (Hymenoptera: Apidae). J Kans Entomol Soc 53:171–178
Jeffree EP (1955) Observations on the decline and growth of honey bee colonies. J Econ Entomol 48:723–726
Jeffree EP (1956) Winter brood and pollen in honeybee colonies. Insect Soc 3:417–422
Jegen G (1920) Zur Geschlechtsbestimmung bei *Apis mellifica*. Festschrift Zschokke 38, Basel
Johansson TSK, Johansson MP (1969) Wintering. Bee World 50:89–100
Jones CE, Buchmann SL (1974) Ultraviolet floral patterns as functional orientation cues in hymenopterous pollination systems. Anim Behav 22: 481–485
Jongbloed J, Wiersma CAG (1934) Der Stoffwechsel der Honigbiene während des Fliegens. Z Vgl Physiol B21:519–533
Jordan RA, Brosemer RW (1974) Characterization of DNA from three bee species. J Insect Physiol 20:2513–2520
Jungmann R, Rothe U, Nachtigall W (1986) Flight of the honey bee I. Thorax surface temperature and thermoregulation during thethered flight. J Comp Physiol B 158:711–718
Jungries AM (1982) Evidence supporting a proposed association between hemolymph gamma-aminobutyric acid and neuorsecretion in lepidopterous insects. Comp Biochem Physiol 71C:169–175

Jungries AM, Omilianowski DR (1980) Gamma-aminobutyric acid and glutamic acid in *Manduca sexta*: proposed roles in insect development. Comp Biochem Physiol 67C:169–175

Juska A (1978) Temporal decline in attractiveness of honeybee queen tracks. Nature (Lond) 276:261

Juska A, Seeley TD, Velthuis HHW (1981) How honeybee queen attendants become ordinary workers. J Insect Physiol 27:515–519

Kaestner A (1969) Lehrbuch der speziellen Zoologie. Band I. Wirbellose 1. Teil. G Fischer, Stuttgart

Kaiser W (1985) Comparative neurobiology of sleep – the honey bee model. In: Koella WP, Rüther E, Schulz H (eds) Sleep '84. G Fischer, Stuttgart, pp 225–227

Kaiser W (1988) Busy bees need rest, too. J Comp Physiol 163:565–584

Kaiser W, Steiner-Kaiser J (1983) Neuronal correlates of sleep, wakefulness and arousal in a diurnal insect. Nature (Lond) 301:707–709

Kaissling KE, Renner M (1968) Antennale Rezeptoren für Queen Substance und Sterzelduft bei der Honigbiene. Z Vgl Physiol 59:357–361

Kalmus H (1939) Der Füllungszustand der Honigblase entscheidet die Flugrichtung der Honigbiene. Z Vgl Physiol 26:76–84

Kalmus H, Ribbands CR (1952) The origin of the odours by which honeybees distinguish their companions. Proc R Soc Lond B 140:50–59

Karlson P, Butenandt A (1959) Pheromones (ectohormones) in insects. Annu Rev Entomol 4:39–58

Katayama E (1971) Observations on the brood development of *Bombus ignitus* (Hymenoptera, Apidae) 1. Egg-laying habits of queens and workers. Kontyu 39:189–203

Kayanja FIB, Jarvis JUM (1971) Histological observations on the ovary, oviduct and uterus of the naked mole-rat. Z Saeugetierk 36:114–121

Keeton WT, Gould JL (1986) Biological science, 4th edn. Norton, New York

Kefuss JA (1978) Influence of photoperiod on the behaviour and brood rearing activities of the honeybee. J Apic Res 17:137–228

Kempthorne O (1969) An introduction to genetic statistics. Iowa State Univ Press, Ames

Kerkut GA, Gilbert LI (1985) Comprehensive insect physiology, biochemistry and pharmacology. Pergamon Press, Oxford

Kerr WE (1950) Genetic determination of castes in the genus Melipona. Genetics 35:143–152

Kerr WE (1951) Bases para o estudo de genetica de populacoes dos Hymenoptera em general e dos Apinae sociais em particular. Anais Esc Super Agric Luiz de Queiroz Univ, Sao Paulo 8:221–354

Kerr WE (1967) The history of the introduction of African bees in Brazil. S Afr Bee J 39:3–5

Kerr WE (1969) Some aspects of the evolution of social bees. Evol Biol 3:119–175

Kerr WE (1972a) Numbers of chromosomes in some species of bees. J Kans Entomol Soc 45:111–122

Kerr WE (1972b) Effect of low temperature on male meiosis in *Melipona marginata*. J Apic Res 11:95–99

Kerr WE (1973) Genetica e biologia de abelhas. Cienc Cult 25:927–934

Kerr WE (1974) Sex determination in bees. III. Caste determination and genetic control in *Melipona*. Insect Soc 21:357–368

Kerr WE (1984) Historia parcial da ciencia apicola no Brasil. In: Goncalves LS, Soares AEE, DeJong D, Steiner J, Martinho MR, Message N (eds) 5th Congr Bras Apicult and 3rd Congr Latino-Ibero-Am Apicult. Imprensa Univ, Vicosa, Brasil, pp 47–60

Kerr WE, Bueno D (1970) Natural crossing between *Apis mellifera adansonii* and *Apis mellifera ligustica*. Evolution 24:145–155

Kerr WE, da Costa Cruz C (1961) Funcoes diferentes tomadas pela glandula mandibular na evolucao das abelhas em general e em "*Trigona* (*Oxytrigona*) *tataira*" em especial. Rev Bras Biol 21:1–16

Kerr WE, Esch H (1965) Comunicao entre as abelhas sociais brasileiras e sua contribucao para o entendinento da sua evolucao. Cienc Cult 17:529–538

Kerr WE, Laidlaw HH (1956) General genetics of bees. Adv Genet 8:109–153

Kerr WE, Nielsen RA (1966) Evidences that genetically determined Melipona queens can become workers. Genetics 54:859–866

Kerr WE, Nielsen RA (1967) Sex determination in bees (Apinae). J Apic Res 6:1–9

Kerr WE, Silveira ZV (1972) Karyotype evolution and corresponding taxonomic implications. Evolution 26:197–202

Kerr WE, Zucchi R, Nakadaira JT, Botolo JE (1962) Reproduction in social bees. J N Y Entomol Soc 70:265–276

Kerr WE, Martinho MR, Goncalves LS (1980) Kinship selection in bees. Rev Bras Genet 3:339–343

King GE (1933) The large glands in the worker honey-bee: a correlation of activity with age and with physiological functioning. PhD Thesis, Univ Illinois, Urbana

Kirchner WH (1990) Learned and spontaneous reactions to airborne sound in honeybees. In: Veeresch GK, Mallik B, Viraktamath CA (eds) Social insects and the environment. Oxford Univ Press & IBM, New Delhi, India, pp 200–201

Kirchner WH, Dreller C, Towne WF (1991) Hearing in honeybees: operant conditioning and spontaneous reactions to airborne sound. J Comp Physiol A 168:85–89

Koenig S (1739) Solution du problème des cellules hexagones des ruches des abeilles. Mem Acad Sci Paris, vol Hist p 30

Koeniger G (1984) Funktionsmorphologische Befunde bei der Kopulation der Honigbiene (*Apis mellifera* L.). Apidologie 15:189–204

Koeniger G (1986) Mating sign and multiple mating in the honeybee. Bee World 67:141–150

Koeniger G (1987) Mating behavior of honeybees. In: Needham GR, Page RE, Delfinado-Baker M, Bowman CE (eds) Africanized honeybees and bee mites. Ellis Horwood, Chichester, pp 167–172

Koeniger G (1988a) Mating flights of honey bee drones (*Apis mellifera* L.). A film documentation. In: Nachtigall W (ed) The flying honeybee. G Fischer, Stuttgart, pp 29–34

Koeniger G (1988b) Discrimination between different coloured queen models by free flying drones (*Apis mellifera* L.). In: Nachtigall W (ed) The flying honeybee. G Fischer, Stuttgart, pp 35–42

Koeniger G (1990) The role of the mating sign in honey bees, *Apis mellifera* L.: does it hinder or promote multiple mating? Anim Behav 39:444–449

Koeniger G, Koeniger N (1990a) Unterschiedliche Genese der Polyandrie bei *Apis*-Arten. Mitt Dtsch Zool Ges 83:618

Koeniger G, Koeniger N (1990b) Evolution of reproductive behavior in honey bees. In: Veeresh GK, Mallik B, Viraktamath CA (eds) Social insects and the environment. Oxford & IHB, New Delhi, pp 101–102

Koeniger G, Ruttner F (1989) Mating behaviour and anatomy of the reproductive organs. in: Moritz RFA (ed) The instrumental insemination of the queen bee. Apimondia, Bucharest, pp 19–34

Koeniger G, Koeniger N, Mardan M, Punichewa RWK, Otis GW (1990) Numbers of spermatozoa in queens and drones indicate multiple mating of queens in *Apis andreniformis* and *Apis dorsata*. Apidologie 21:281–286

Koeniger N (1970) Factors determining the laying of drone and worker eggs by the queen honeybee. Bee World 51:166–169

Koeniger N (1976) Neue Aspekte der Phylogenie innerhalb der Gattung *Apis*. Apidologie 7:257–266

Koeniger N (1982) Interactions among the four species of the genus *Apis*. In: Breed MD, Michener CD, Evans HE (eds) The bioloy of social insects. Westview Press, Boulder, CO, pp 59–64

Koeniger N, Koeniger G (1980) Observations and experiments on migration and dance comminucation of *Apis dorsata* in Sri Lanka. J Apic Res 19:21–39

Koeniger N, Veith HJ (1983) Glyceryl-1,2-dioleate-3-palmitate, a brood pheromone of the honey bee (*Apis mellifera* L.). Experientia 39:1051–1052

Koeniger N, Wijayagunasekera HNP (1976) Time of drone flight in three Asiatic honeybee species (*Apis cerana*, *Apis florea*, *Apis dorsata*). J Apic Res 15:67–71

Koeniger N, Weiss J, Maschwitz U (1979) Alarm pheromones of the sting in the genus *Apis*. J Insect Physiol 25:467–476

Koeniger N, Koeniger G, Tingek S, Mardan M, Rinderer TE (1988) Reproductive isolation by different time of drone flight between *Apis cerana* Fabricius 1793 and *Apis vechti* Maa 1953. Apidologie 19:103–106

Koeniger N, Koeniger G, Wongsiri S (1989) Mating and sperm transfer in *Apis florea*. Apidologie 20:413–418

Kolmes SA (1985) A quantitative study of the division of labour among worker honey bees. Z Tierpsychol 68:287–302

Kolmes SA, Winston ML, Fergusson LA (1989) The division of labour among worker honey bees (Hymenoptera: Apidae): the effects of multiple patrilines. J Kans Entomol Soc 62:80–95

Korst PJAM, Velthuis HHW (1982) The nature of trophallaxis. Insect Soc 29:209–221

Kovac H, Crailsheim K (1988) Lifespan of *Apis mellifera carnica* Pollm. in fested by *Varroa jacobsoni* Oud. in relation to season and extent of infestation. J Apic Res 27:230–238

Kramer E (1976) The orientation of walking honeybees in odour fields with small concentration gradients. Physiol Entomol 1:27–37

Kreil G (1973) Structure of mellitin isolated from two species of honeybees. FEBS Lett 33:241–244

Kreil G (1975) The structure of *Dorsata* mellitin: phylogenetic relationships between honeybees as deducted from sequence data. FEBS Let 54:100–102

Kronenberg F, Heller HC (1982) Colonial thermoregulation in honey bees (*Apis mellifera*). J Comp Physiol 148:65–76

Kryger P, Moritz RFA P (1991) Hygienic behaviour of honeybees and genotypic intragroup variation. Behav Ecol Scoiobiol (in press)

Kukuk PF (1989) Evolutionary genetics of a primitively eusocial halictine bee, *Dialictus zephyrus*. In: Breed MD, Page RE (eds) The genetics of social evolution. Westview Press, Boulder, CO, pp 183–202

Kukuk PF, Decelles PC (1986) Relatedness-dependent spatial distribution of female *Lasioglossum zephyrum* as estimated from aggressive interactions. Behav Ecol Sociobiol 19:233–239

Kukuk PF, Eickwort GC, May BP (1987) Multiple maternity and multiple paternity in first generation brood from single foundress colonies of the sweat bee *Dialictus zephyrus*. Insect Soc 34:131–135

Kumbkarni CG (1964) Cytological studies in Hymenoptera. Part I. Cytology of parthenogenesis in the honeybees-*Apis dorsata*. Indian J Exp Biol 2:65–68

Kuwabara W (1957) Bildung des bedingten Reflexes von Pavlovs Typus bei der Honigbiene, *Apis mellifica*. J Fac Sci Hokkaido Univ Ser 6 13:458

Labhart T (1980) Specialized photoreceptors at the dorsal rim of the honey bee's compound eye: polarization and angular sensitivity. J Comp Physiol 141:19–30

Lacey EA, Sherman PW (1991) Social organization of naked mole-rat colonies: evidence for divisions of labor. In: Sherman PW, Jarvis JUM, Alexander RD (eds) The biology of the naked mole-rat. Princeton Univ Press, Princeton, NJ, pp 275–336

Lachaud JP, Fresneau D (1987) Social regulation in ponerine ants. In: Pasteels JM, Deneubourg JL (eds) From individual to collective behavior in social insects. Birkhäuser, Basel, pp 197–218

Lacher V (1964) Elektrophysiologische Untersuchungen an einzelnen Rezeptoren für Geruch, Luftfeuchtigkeit und Temperatur auf den antennen der Arbeitsbienen und der Drohne. Z Vgl Physiol 48:587–623

Lack D (1954) The natural regulation of animal numbers. Oxford Univ Press, Oxford

Lacy RC (1980) The evolution of eusociality in termites: a haplodipoid analogy? Am Nat 116:449–451

Laidlaw HH (1974) Relationships of bees within a colony. Apiacta 9:49–52

Laidlaw HH, Eckert JE (1950) Queen rearing. Dadant & Sons, Hamilton, IL

Laidlaw HH, Page RE (1984) Polyandry in honey bees (*Apis mellifera* L.): sperm utilization and intracolony genetic relationships. Genetics 108:985–997

Laidlaw HH, Page RE (1986) Mating designs. In: Rinderer TE (ed) Bee genetics and breeding. Academic Press, Orlando, pp 323–344

Laidlaw HH, El-Banby MA, Tucker KW (1964) Five new eye-color mutants in the honey bee. J Hered 54:39–41

Le Conte Y, Arnold G, Trouiller J, Masson C, Chappe B, Ourisson G (1989) Attraction of the parasitic mite *Varroa* to the drone larvae of honey bees by simple aliphatic esters. Science 245:638–639

Lee PC, Winston ML (1985) The effect of swarm size and date of issue on comb construction in newly founded colonies of honeybees (*Apis mellifera* L.). Can J Zool 63:524–527

Legewie H (1925a) Zur Theorie der Staatenbildung. I. Teil: Die Biologie der Furchenbiene *Halictus malachurus* K. Z Morphol Oekol Tiere 3:619–684

Legewie H (1925b) Zur Theorie der Staatenbildung (II. Teil) Z Morphol Oekol Tiere 4:246–300

Leinaas HP (1983) A sex-linked esterase locus and translocation heterozygosity in a termite. Heredity 46:315–320

Lensky Y, Slabezki Y (1981) The inhibiting effect of the queen bee (*Apis mellifera* L.) foot-print pheromone on the construction of swarming queen cups. J Insect Physiol 27:313–323

Lensky Y, Baehr J-C, Pocheron P (1978) Dosages radio-immunologiques des ecdysones et des hormones juveniles au cours du developpement post-embryonnaire chez les ouvrières et les reines d'abeille (*Apis mellifera* L. var. ligustica). C R Acad Sci (Paris) 287D:821–824

Lensky Y, Finkel A, Schlesinger R, Delorme-Joulie C, Levinsohn M (1984) Les glandes tarsales de l'abeille mellifique (*Apis mellifica* L.) reines, ouvrieres et faux-bourdons (Hymenoptera, Apidae) II. Role biologique. Ann Sci Nat Zool 6:165–175

Lensky Y, Cassier P, Notkin M, Delorme-Joulie C, Levinsohn M (1985) Pheromonal activity and fine structure of the mandibular glands of honeybee drones (*Apis mellifera* L.) (Insecta, Hymenoptera, Apidae). J Insect Physiol 31:265–276

Lensky Y, Finkel A, Cassier P, Teeshbee A, Schlesinger R (1987) The tarsal glands of honeybee (*Apis mellifera* L.) queens, workers, and drones – chemical characterization of footprint secretion. Honeybee Sci 8:97–102

Lenz M, Barret RA (1982) Neotenic formation in field colonies of *Coptotermes lacteus* (Froggatt) in Australia, with comments on the role of neotenics in the genus Coptotermes. Sociobiology 7:47–60

Lerer H, Dixon, SE (1975) Polyploidy in fat-body of honeybees during larval growth and development. J Apic Res 14:105–111

Levene H (1953) Genetic equilibrium when more than one ecological niche is available. Am Nat 87:331–333

Levin MD (1959) Distribution patterns of young and experienced honey bees foraging on alfalfa. J Econ Entomol 52:969–971

Lin N, Michener CD (1972) Evolution of sociality in insects. Q Rev Biol 47:131–159

Lindauer M (1951) Bee dances in the clustered swarm. Naturwissenschaften 22:509–513

Lindauer M (1952) Ein Beitrag zur Frage der Arbeitsteilung im Bienenstaat. Z Vgl Physiol 34:299–345

Lindauer M (1954) Temperaturregulierung und Wasserhaushalt im Bienestaat. Z Vgl Physiol 36:391–432

Lindauer M (1955) Schwarmbienen auf Wohnungssuche. Z Vgl Physiol 37:263–324

Lindauer M (1956) Über die verständigung bei indischen Bienen. Z Vgl Physiol 38:521–557

Lindauer M (1961) Communication among social bees. Harvard Univ Press, Cambridge, MA

Lindauer M, Kerr WE (1960) Communication between the workers of stingless bees. Bee World 41:29–41, 65–71
Lindauer M, Martin H (1963) Über die Orientierung der Biene im Duftfeld. Naturwissenschaften 50:509–514
Lindauer M, Martin H (1968) Die Schwereorientierung der Bienen unter dem Einfluß des Erdmagnetfelds. Z Vgl Physiol 60:219–243
Livy T (1919) Ab urbe condita. Liber II:32.9–32.12 (translated by Foster BO) Heinemann, London
Lloyd DG (1988) Benefits and costs of biparental and uniparental repuroduction in plants. In: Michod RE, Levin BR (eds) The evolution of sex. Sinauer, Sunderland, MA, pp 233–252
Lobo JA, Del Lama MA, Mestriner MA (1989) Population differentiation and racial admixture in the Africanized honeybee (*Apis mellifera* L.). Evoltion 43:794–802
Loper GM (1985) Influence of age on the fluctuation of iron in the oenocytes of honey bee (*Apis mellifera*) drones. Apidologie 16:181–184
Loper GM, Wolf WW, Taylor OR (1987) Detection and monitoring of honeybee drone congregation areas by radar. Apidologie 18:163–172
Loper GM, Wolf WW, Taylor OR (1988) The use of radar to document honey-bee (*Apis mellifera*) drone flight behavior. In: Needham GR, Page RE, Delfinado-Baker M, Bowman CD (eds) Africanized honeybees and bee mites. Ellis Horwood, Chichester, pp 193–198
Lovelock JE (1986) Gaia, the world as living organism. New Sci 12:25–28
Lovelock JE (1988) Ages of Gaia. Norton, New York
Lumsden CJ (1982) The social regulation of physical caste: the superorganism revived. J Theor Biol 95:749–781
Lundie AE (1925) The flight activities of the honeybee. U S Dep Agric Bull 1328:1–37
Lüscher M (1951) Beobachtungen über die Koloniegründung bei verschiedenen afrikanischen Termitenarten. Acta Tropica 8:36–43
Luykx P, Syren RM (1979) The cytogenetics of *Incistermes schwarzi* and other Florida termites. Sociobiology 4:191–209
Maa TC (1953) An inquiry into the systematics of the tribus Apidini or honeybees (Hymenoptera; Apidae). Treubia 21:525–640
MacArthur R, Wilson EO (1967) The theory of island biogeography. Princeton Univ Press, Princeton, NJ
Macevicz S (1979) Some consequences of Fisher's sex ratio principle for social Hymenoptera that reproduce by colony fission. Am Nat 113: 363–371
Mackensen O (1951) Viability and sex determination in the honey bee (*Apis mellifera* L.). Genetics 36:500–509
Maeta Y, Sakagami SF, Michener CD (1985) Laboratory studies on the life cycle and nesting biology of *Braunsapis sauteriella*, a social xylocopine bee. J Sociobiol 10:17–41
Maeterlinck M (1901) The life of the bee. Dodd, Mead, New York
Makela ME, Sames IV WJ, Rowell GA (1990) Population dynamics of Africanized, European and hybrid honey bees: simulations based on inherited behavioral characteristics. Am Bee J 130:805
Malecot G (1948) Les mathématiques de l'heredité. Masson, Paris
Malecot G (1969) The mathematics of heredity. Freeman, San Francisco

Maraldi JP (1712) Observation sur les abeilles. Mem Acad Sci Paris 14: 297–336

Mark EL, Copeland M (1906) Some stages in the spermatogenesis of the honey bee. Proc Am Acad Arts Sci 42:103–111

Mark EL, Copeland M (1907) Maturation stages in the spermatogenesis of Vespa. Proc Am Acad Arts Sci 43:69–74

Markl H (1962) Borstenfelder an den Gelenken als Schweresinnesorgane bei Ameisen und anderen Hymenopteren. Z Vgl Physiol 45:475–569

Martin H (1963) Die Steuerung der Volksteilung beim Schwärmen der Bienen: zugleich ein Beitrag zum Problem der Wanderschwärme. Insect Soc 10:13–42

Martin H, Lindauer M (1973) Orientierung im Erdmagnetfeld. Fortschr Zool 21:211–228

Martin H, Lindauer M (1977) Der Einfluß des Erdmagnetfeldes auf die Schwereorientierung der Honigbiene (*Apis mellifica*). J Comp Physiol A 122:145–187

Martins E, Mestriner MA, Contel EPB (1977) Alcohol dehydrogenase polymorphism in *Apis mellifera*. Biochem Genet 15:357–366

Maschwitz UW (1963) Gefahrenalarmstoffe und Gefahrenalarmierung bei sozialen Hymenopteren. Z Vgl Physiol 47:596–655

Maschwitz UW (1964a) Alarm substances and alarm behaviour in social Hymenoptera. Nature (Lond) 204:324–327

Maschwitz UW (1964b) Gefahrenalarmstoffe und Gefahrenalarmierung bei sozialen Hymenopteren. Z Vgl Physiol 47:596–655

Masson C, Arnold G (1987) Organization and plasticity of the olfactory system of the honeybee, *Apis mellifera*. In: Menzel R, Mercer A (eds) Neurobiology and behavior of honeybees. Springer, Berlin Heidelberg New York, pp 280–295

Matessi C, Jayakar SD (1973) A model for the evolution of altruistic behavior. Genetics 74:174

Matessi C, Jayakar SD (1976) Conditions for the evolution of altruism under Darwinian selection. Theor Popul Biol 9:360–387

Matsuura M (1985) Biology of hornets – why do they attack honeybee colonies. Honeybee Sci 6:53–64

Matsuura M, Sakagami SF (1973) A bionomic sketch of the giant hornet, *Vespa mandarina*, a serious pest for Japanese apiculture. J Fac Hokkaido Univ Ser VI Zool 19:125–162

Maurizio A (1950) The influence of pollen feeding and brood rearing on the length of life and physiological condition of the honeybee. Bee World 31:9–12

May ML (1985) Thermoregulation. In: Kerkut GA, Gilbert LI (eds) Comprehensive insect physiology, biochemistry and pharmacology. Pergamon Press, New York, pp 507–552

Maynard Smith J (1964) Group selection and kin selection: a rejoinder. Nature (Lond) 201:1145–1147

Maynard Smith J (1978) The evolution of sex. Cambridge Univ Press, Cambridge

Maynard Smith J (1987a) How to model evolution. In: Dupre J (ed) The latest on the best. Essays on evolution and optimality. MIT Press, Boston, pp 119–131

Maynard Smith J (1987b) Reply to Sober. In: Dupre J (ed) The latest on the best. Essays on evolution and optimality. MIT Press, Boston, pp 147–149

McGinnis W, Levine M, Hafen E, Juroiwa A, Gehring WJ (1984a) A conserved DNA sequence in homeotic genes of the *Drosophila* antennapedia and bithorax complexes. Nature (Lond) 308:428–433

McGinnis W, Garber RL, Wirz J, Kuroiwa A, Gehring WJ (1984b) A homologous protein-coding sequence in *Drosophila* homeotic genes and its conservation in other metazoans. Cell 37:403–408

McGregor SE (1976) Insect pollination of cultivated crop plants. USDA Handbook 496:64–75

McMasters JH (1989) The flight of the bumblebee and related myths of entomological engineering. Am Sci 77:164–169

Mello MLS, Takahashi CS (1971) DNA content and nuclear volume in larval organs of *Apis mellifera*. J Apic Res 10:125–132

Menzel R (1983) Neurobiology of learning and memory: the honey bee as a model system. Naturwissenschaften 70:504–511

Menzel R (1990) Color vision in flower visiting insects. KFA, Jülich

Menzel R, Erber J (1978) Learning and memory in honey bees. Sci Am 239:102–111

Menzel R, Mercer A (1987) Neurobiology and behavior of honeybees. Springer, Berlin Heidelberg New York

Menzel R, Erber J, Masuhr T (1973) Learning and memory in the honey bee. In: Browne LB (ed) Experimental analysis of insect behavior. Springer, Berlin Heidelberg New York, pp 195–217

Menzel R, Hammer M, Braun G, Mauelshagen J, Sugawa M (1991) Neurobiology of learning and memory in honeybees. In: Goodman LJ, Fisher RC (eds) The behaviour and physiology of honeybees. CAB International. Oxon, UK, pp 323–353

Merriam RW, Ris H (1954) Size and DNA content of nuclei in various tissues of male, female and worker honeybees. Chromosoma 6:522–538

Mesteriner MA, Contel EPB (1972) The P-3 and Est loci in the honeybee *Apis mellifera*. Genetics 72:733–738

Meusel MS, Moritz RFA (1992) Transfer of paternal mitochondrial DNA in fertilization of honeybee (*Apis mellifera* L.) eggs. Genetics (submitted)

Meves F (1903) Über "Richtungskörper" im Hoden von Hymenopteren. Anat Anz 24:29–32

Meves F (1907) Die Spermatocytenteilung bei der Honigbiene, Apis mellifica L. nebst Bemerkungen über Chromatinreduction. Arch Mikrosk Anat 24:29–32

Meyer J (1966) Essai d'application de certains modèles cybernetiques à la coordination chez les insectes sociaux. Insect Soc 13:127–138

Meyer W (1956) "Propolis bees" and their activities. Bee World 37:25–36

Michelsen A, Kirchner WH, Lindauer M (1986) Sound and vibrational signals in the dance language of the honeybee, *Apis mellifera*. Behav Ecol Sociobiol 18:207–212

Michelsen A, Towne WF, Kirchner WH, Kryger P (1987) The acoustic near field of the dancing honeybee. J Comp Physiol 161:633–643.

Michelsen A, Andersen BB, Kirchner WH, Lindauer M (1989) Honeybees can be recruited by a mechanical model of a dancing bee. Naturwissenschaften 76:277–280

Michener CD (1944) Comparative external morphology, phylogeny and a clsssification of the bees (Hymenoptera). Bull Am Mus Nat Hist 82: 151–326
Michener CD (1958) The evolution of social behavior in bees. Proc 10th Int Congr Entomol, Montreal 2:441–447
Michener CD (1962) Biological observations on the primitively social bee of the genus "*Allodapula*" in the Australian region. Insect Soc 9:355–373
Michener CD (1971) Biologies of African allodapine bees. Bull Am Mus Nat Hist 145:219–302
Michener CD (1972) Direct food transferring behavior in bees. J Kans Entomol Soc 45:373–376
Michener CD (1974) The social behavior of the bees. Harvard Univ Press, Cambridge, MA
Michener CD (1977) Discordant evolution and the classification of allodapine bees. Syst Zool 26:32–56
Michener CD (1979) Biogreography of the bees. Ann Mo Bot Gard 66: 277–347
Michener CD (1985) From solitary to eusocial: need there be a series of intervening species? In: Hölldobler B, Lindauer M (eds) Experimental behavioral ecology and sociobiology. G Fischer, Stuttgart, pp 293–305
Michener CD (1990) Castes in xylocopine bees. In: Engels W (ed) Social insects. An evolutionary approach to castes and reproduction. Springer, Berlin Heidelberg New York, pp 123–146
Michener CD, Grimaldi DA (1988) The oldest fossil bee: apoid history, evolutionary statsis, and antiquity of social behavior. Proc Natl Acad Sci USA 85:6424–6426
Michod R (1982) The theory of kin selection. Annu Rev Ecol Syst 13:23–55
Milde JJ (1987) The ocellar system of the honeybee. In: Menzel R, Mercer A (eds) Neurobiology and behavior of honeybees. Springer, Berlin Heidelberg New York, pp 191–200
Miller EM (1969) Caste differentiation in lower termites. In: Krishna K, Weesner FM (eds) Biology of termites I. Academic Press, New York, pp 283–310
Milne CP (1982) Laboratory measurement of brood disease resistance in the honeybee. 1. Uncapping and removing of freeze-killed brood by newly emerged workers in laboratory test cages. J Apic Res 21:111–114
Milne CP (1983) Honey bee (Hymenoptera: Apidae) hygienic behavior and resistance to chalkbrood. Ann Entomol Soc Am 76:384–387
Milne CP (1985a) A heritability estimate for honey bee hoarding behavior. Apidologie 16:413–420
Milne CP (1985b) An estimate of the heritability of the corbicular area of the honeybee. J Apic Res 24:137–139
Milne CP (1985c) An estimate of the heritability of worker longevity or length of life in the honeybee. J Apic Res 24:140–143
Milne CP (1985d) Laboratory tests of honey bee hygienic behavior and resistance to European foulbrood. Am Bee J 125:578–580
Milne CP (1985e) Estimates of the heritabilities of and genetic correlation between two components of honey bee (Hymenoptera: Apidae) hygienic behavior: uncapping and removing. Ann Entomol Soc Am 78:841–844

Milne CP, Friars GW (1984) An estimate of the hertiability of honeybee pupal weight. J Hered 75:509–510
Milum VG (1955) Honey bee communication. Am Bee J 95:97–104
Mindt B (1962) Untersuchungen über das Leben der Drohnen, insbesondere Ernährung und Geschlechtsreife. Z Bienenforsch 6:9–33
Minsky M (1986) The society of mind. Simon and Schuster, New York
Moebus R (1980) A new theoretical model of the winter cluster. 27th Int Congr Apicult, Apimondia, Bucharest, p 286
Moody Institute of Science (no date) Mathematics of the honeycomb. Revised version. Video. Moody Inst Sci, Whittier, CA
Moore D, Dankin MA (1985) Circadian locomotor rhythms in individual honeybees. J Physiol Entomol 10:191–197
Mordue W, Goldsworthy GJ, Brady J, Blaney WM (1980) Insect physiology. Wiley, New York
Moreno JL (1934) Who shall survive? Foundations of sociometry, group psychotherapy and sociodrama. Beacon House, New York
Moritz RFA (1983) Homogenous mixing of honeybee semen by centrifugation. J Apic Res 22:249–255
Moritz RFA (1984a) Heritabilitätsschätzungen bei der Honigbiene. 30th Colloq Int Biomet Soc, Dortmund pp 76–77
Moritz RFA (1984b) Selection in small populations of the hoenybee (*Apis mellifera* L.). Z Tierz Zuechtungsbiol 101:394–400
Moritz RFA (1985a) Heritability of the postcapping stage and its relation to varroatosis resistance in the honeybee (*Apis mellifera* L.). J Hered 76:267–270
Moritz RFA (1985b) The effects of multiple mating on the worker-queen conflict in *Apis mellifera* L. Behav Ecol Sociobiol 16:375–377
Moritz RFA (1986a) Intracolonial worker relationship and sperm competition in the honeybee (*Apis mellifera* L.). Experientia 42:445–448
Moritz RFA (1986b) Estimating genetic variance of group characters: social behaviour of honeybees (*Apis mellifera* L.). Theor Appl Genet 72:513–517
Moritz RFA (1986c) Two parthenogenetical strategies of laying workers in populations of the honeybee *Apis mellifera* (Hymenoptera: Apidae). Entomol Gen 11:159–164
Moritz RFA (1988a) Group relatedness and kin discrimination in honey bees (*Apis mellifera* L.). Anim Behav 36:1334–1340
Moritz RFA (1988b) A reevaluation of the two-locus model for hygienic behavior in honeybees (*Apis mellifera*). J Hered 79:257–262
Moritz RFA (1989) colony level and within colony level selection in honeybees. A two allele population model for *Apis mellifera capensis*. Behav Ecol Sociobiol 25:437–444
Moritz RFA (1991) Kin recognition in honeybees: experimental artefact or biological reality? In: Goodman LJ, Fisher RC (eds) The behaviour and physiology of bees. CAB International, Oxon, UK, pp 48–59
Moritz RFA (1992) Nourishment and sociality in honeybees. In: Hunt JH, Nalepa CA (eds) Nourishment and the evolution of sociality. Westview Press, Boulder, CO (in press)

Moritz RFA, Brandes C (1987) Behavior genetics of honeybees (*Apis mellifera* L.). In: Menzel R, Mercer A (eds) Neurobiology and behavior of Honeybees. Springer, Berlin Heidelberg New York, pp 21–35

Moritz RFA, Crewe RM (1988a) Chemical signals of queens in kin recognition of honeybees, *Apis mellifera* L. J Comp Physiol A 164:83–89

Moritz RFA, Crewe RM (1988b) Air ventilation in two African stingless bees *Trigona denoiti* and *Trigona gribidoi*. Experientia 44:1024–1027

Moritz RFA, Heisler T (1992) Trophallaxis and polyandry in kin recognition of honeybee workers (*Apis mellifera* L.). Insect Soc in press

Moritz RFA, Hillesheim E (1985) Inheritance of dominance in honeybees (*Apis mellifera capensis*). Behav Ecol Sociobiol 17:87–89

Moritz RFA, Hillesheim E (1989) Genotypic intragroup variance and hoarding behaviour in honeybees (*Apis mellifera* L.). Apidologie 20: 383–390

Moritz RFA, Hillesheim E (1990) Trophallaxis and genetic variance for kin recognition in honeybees (*Apis mellifera* L.). Anim Behav 40:641–647

Moritz RFA, Jordan M (1991) Selection of varroatosis resistance across caste and sex in honeybees (*Apis mellifera* L.). Exp Appl Acarol 9 (in press)

Moritz RFA, Kauhausen D (1984) Hybridization between Apis mellifera capensis and adjacent races of *Apis mellifera*. Apidologie 15:211–222

Moritz RFA, Kühnert M (1984) Seasonal effects on artificial insemination of honeybee queens (*Apis mellifera* L.). Apidologie 15:223–231

Moritz RFA, Meusel MS (1992) Mitochondrial DNA polymorphism in Africanized honeybees *Apis mellifera* L.: theoretical model and empirical evidence. J Evol Biol 5:71–81

Moritz RFA, Sakofski F (1991) The role of the queen in circadian rhythms of honeybees (*Apis mellifera* L.). Behav Ecol Sociobiol 29:361–365

Moritz RFA, Southwick EE (1987) Phenotype interactions in group behaviour of honey bee workers (*Apis mellifera* L.). Behav Ecol Sociobiol 21:53–57

Moritz RFA, Southwick EE, Breh M (1985) A metabolic test for the quantitative analysis of alarm behavior of honeybees (*Apis mellifera* L.) J Exp Zool 235:1–5

Moritz RFA, Hawkins CF, Crozier RH, Mackinlay AG (1986) A mitochondrial DNA polymorphism in honeybees (*Apis mellifera* L.). Experientia 42:322–324

Moritz RFA, Southwick EE, Harbo JB (1987a) Genetic analysis of defensive behaviour of honeybee colonies (*Apis mellifera* L.) in a field test. Apidologie 18:27–42

Moritz RFA, Southwick EE, Harbo JB (1987b) Maternal and preeclosional factors affecting alarm behaviour in adult honey bees. Insect Soc 34: 298–307

Moritz RFA, Crozier RH, Pamilo P (1987c) Sociogenetics in insects: introduction. In: Eder J, Rembold H (eds) Chemistry and biology of social insects. Peperny, Munich, pp 331–332

Moritz RFA, Kirchner M, Crewe RM (1991a) Chemical camouflage of the death head hawkmoth, *Acherontia atropus*. Naturwissenschaften 78: 179–182

Moritz RFA, Sakofski F, Kainz D (1991b) Circadian rhythms in social groups of the honeybee. In: Binnenbruck H, Richter J (eds) Forschung unter Weltraumbedinungen. DARA, Aachen, pp 403–408

Moritz RFA, Meusel MS, Haberl M (1991c) Oligonucleotide DNA fingerprinting discriminates super- and half-sisters in honeybee colonies (*Apis mellifera* L.). Naturwissenschaften 78:422–424
Morris W (1976) The American heritage dictionary of the English language. Houghton Mifflin, Boston
Morse DH (1982) Behavior and ecology of bumblebees. In: Hermann HR (ed) Social insects, vol III. Academic Press, New York, pp 245–322
Morse RA (1956) Florida beekeeping. State Plant Bd FL 11:Bull 10
Morse RA (1963) Swarm orientation in honeybees. Science 141:357–358
Morse RA (1972) Environmental control in the beehive. Sci Am 226:93–98
Morse RA (1978) Honey bee pests, predators, and diseases. Comstock, Cornell Univ Press, Ithaca, NY
Morse RA (1988) The complete guide to beekeeping. R Hale, London
Morse RA, Boch R (1971) Pheromone concert in swarming honey bees. Ann Entomol Soc Am 64:1414–1417
Morse RA, Shearer DA, Boch R, Benton AW (1967) Observations on alarm-substances in the genus *Apis*. J Apic Res 6:113–118
Mousseau TA, Roff DA (1986) Natural selection and the heritability of fitness components. Heredity 59:181–197
Myles TG (1988) Resource inheritance in social evolution from termites to man. In: Slobodchikoff CN (ed) The ecology of social behavior. Academic Press, New York, pp 379–423
Myser WC (1952) Ingestion of eggs by honey bee workers. Am Bee J 92:67
Nachtigall W, Rothe U, Feller P, Jungmann R (1989) Flight of the honey bee III. Flight metabolic power calculated from gas analysis, thermoregulation and fuel consumption. J Comp Physiol 158B:729–737
Nachtsheim H (1914) Cytologische Studien über die Geschlechtsbestimmung bei der Honigbiene (*Apis mellifica* L.). Arch Zellforsch 2:169–241
National Geographic Society (US) (1990) National geographic atlas of the world, 6th edn. Nat Geogr Soc, Washington, DC
Naumann K, Winston ML, Slessor KN, Prestwich GD, Webster FX (1991) Production and transmission of honey bee queen (*Apis mellifera* L.) mandibular gland pheromone. Behav Ecol Sociobiol 29:321–332
Neese V (1965) Zur Funktion der Augenborsten bei der Honigbiene. Z Vgl Physiol 49:543–585
Nelson JA (1915) The embryology of the honey bee. Princeton Univ Press, Princeton, NJ
Neukirch A (1982) Dependence of life span of the honeybee (*Apis mellifera*) upon flight performance and energy consumption. J Comp Physiol A 146:35–40
Newell W (1915) Inheritance in the honeybee. Science 41:218–219
Newsholme EA, Crabtree B, Higgins SJ, Thorton SD, Start C (1972) The activities of fructose diphosphatase in flight muscles for the bumblebee and the role of this enzyme in heat generation. Biochem J 128:897–898
Nijhout HF (1975) A threshold size for metamorphosis in the tobacco worm *Manduca sexta* (L.). Biol Bull 149:214–225
Nixon HL, Ribbands CR (1952) Food transmission within the honeybee community. proc R Soc B140:43–50
Noirot C (1985) Pathways of caste development in lower termites. In: Watson JAL, Okot-Kober BM, Noirot C (eds) Caste differentiation in social insects. Pergamon Press, Oxford, pp 177–186

Noirot C (1989) Social structure in termite societies. Ethol Ecol Evol 1: 1–17
Noirot C (1990) Sexual castes and reproductive strategies in termites. In: Engels W (ed) Social insects. An evolutionary approach to castes and reproduction. Springer, Berlin Heidelberg New York, pp 5–35
Noirot C, Pasteels JM (1987) Ontogenetic development and evolution of the worker caste in termites. Experientia 43:851–860
Nolan WJ (1925) The brood-rearing cycle of the honeybee. Bull US Dep Agric 1349:1–56
Nonacs P (1986) Sex-ratio determination within colonies of ants. Evolution 40:199–204
Noonan KC (1986) Recognition of queen larvae by worker honey bees. Ethology 73:295–306
Nunamaker RA (1980) Subspecies determination in the honeybee (*Apis mellifera*) based on isoelectric focusing of malate dehydrogenase. PhD Thesis, Univ Wyoming, Laramie
Nunamaker RA, Wilson WT (1981) Comparison of MDH allozyme patterns in the African honey bee (*Apis mellifera adansonii* L.). J Kans Entomol Soc 54:794–710
Nunez JA (1966) Quantitative Beziehungen zwischen den Eigenschaften von Futterquellen und den Verhalten von Sammelbienen. Z Vgl Physiol 53:142–164
Nunez JA (1967) Sammelbienen markieren versiegte Futterquellen durch Duft. Naturwissenschaften 54:322–323
Nunez JA (1970) The relationship between sugar flow and foraging and recruiting behaviour of honey bees (*Apis mellifera* L.), Anim Behav 18:527–538
Nunez JA (1982) Honeybee foraging strategies at a food source in relation to its distance from the hive an the rate of sugar flow. J Apic Res 21:139–150
Nüsslein-Volhard C, Wieschaus E (1980) Mutations affecting segment number and polarity in *Drosophila*. Nature (Lond) 287:801
Nutting WL (1969) Flight and colony foundation. In: Krishna K, Weesner FM (eds) Biololgy of termites, vol 1. Academic Press, New York, pp 233–282
O'Brien (1956) Desoxyribose nucleic acid in a haplo-diploid species (*Steatococcus tuberculatus* Morrison). Chromosoma 39:381–401
Odum EP (1971) Fundamentals of ecology. Saunders, Philadelphia
Oertel E (1944) Variation in the sugar concentration of some southern nectars. J Econ Entomol 37:525–527
Oettingen-Spielberg T (1949) Über das Wesen der Suchbienen. Z Vgl Physiol 31:454–489
O'Keefe KJ, Schwarz PM (1990) Pheromones are implicated in reproductive differentiation in a primitively social bee. Naturwissenschaften 77:83–86
Oldroyd BP (1988) Are queens or workers more important to honey production? In: Needham G, Page RE, Delfinado-Baker M, Bowman E (eds) Africanized honey bees and bee mites. Ellis Horwood, Chichester, pp 91–96
Oldroyd BP, Goodman RD (1988) Heterosis and inbreeding in queen bees in relation to brood area and honey production. Aust J Agric Res 39: 959–964

Oldroyd BP, Goodman RD (1990) On the relative importance of queens and workers to honey production. Apidologie 21:153–159
Oldroyd BP, Moran C (1983) Heritability of worker characters in the honeybee (*Apis mellifera*). Aust J Biol Sci 36:323–332
Oldroyd BP, Rinderer TE, Buco SM (1990) Nepotism in honey bees, computer programmes and scientific hypotheses. Nature (Lond) 346:707
Oldroyd BP, Rinderer TE, Buco SM (1991) Does patriline recognition really occur in honey bees? (unpubl manuscript)
Olmstead AL, Wooten DB (1987) Bee pollination and productivity growth: the case of alfalfa. Am J Agric Econ 87/2:56–63
Omholt SW (1987) Thermoregulation in the winter cluster of the honeybee *Apis mellifera*. J Theor Biol 128:219–231
Onions GW (1912) South African "fertile-worker bees". Agric J Union S Afr 3:720–728
Ono M, Okada I, Sasaki M (1987) Heat production by balling in the Japenese honey bees, *Apis cerana japonica* as a defensive behavior against the hornet, *Vespa simillima xanthoptera* (Hymenoptera: Vespidae). Experientia 43:1031–1032
Orians GH, Pearson NE (1979) On the theory of central place foraging. In: Horn DJ, Mitchell RD, Stairs GR (eds) Analysis of ecological systems. Ohio State Univ Press, Columbus, pp 155–177
Oster GF, Wilson EO (1978) Caste and ecology in the social insects. Princeton Univ Press, Princeton
Otis GW (1982) Population biology of the Africanized honey bee. In: Jaisson P (ed) Social insects in the tropics, vol 1. Univ Paris-Nord, Paris, pp 209–219
Owen RE (1986) Colony-level selection in the social insects: single locus additive and nonadditive models. Theor Popul Biol 29:198–234
Owen RE (1989a) Differential size variation of male and female bumblebees. J Hered 80:39–43
Owen RE (1989b) The genetics of colony-level selection. In: Breed MD, Page RE (eds) The genetics of social evolution. Westview Press, Boulder, CO, pp 31–59
Owen RE, Plowright RC (1982) Worker-queen conflict and male parentage in bumble bees. Behav Ecol Sociobiol 11:91–99
Owen RE, Rodd FH, Plowright RC (1980) Sex ratios in bumble bee colonies: complications due to orphaning? Behav Ecol Sociobiol 7: 287–291
Owens CD (1971) The thermology of wintering honey bee colonies. US Dep Agric Tech Bull 1429:1–32
Packer L (1986a) The biology of a subtropical population of *Halictus ligatus* IV. A cuckoo-like caste. J NY Entomol Soc 94:458–466
Packer L (1986b) The social organization of *Halictus ligatus* (Hymenoptera: Halictidae) in southern Ontario. Can J Zool 64:2317–2324
Page RE (1980) The evolution of multiple mating behavior by honey bee queens (*Apis mellifera* L.). Genetics 96:263–273
Page RE (1981) Protandrous reproduction in honey bees. Environ Entomol 10:359–361
Page RE (1986) Sperm utilization in social insects. Annu Rev Entomol 31:297–320

Page RE, Erickson EE (1986a) Kin recognition and virgin queen acceptance by worker honey bees. Anim Behav 34:1061–1064
Page RE, Erickson EE (1986b) Kin recognition during emergency queen rearing by worker bees (Hymenoptera: Apidae). Ann Entomol Soc Am 79:460–467
Page RE, Erickson EE (1988) Reproduction by worker honey bees (*Apis mellifera*). Behav Ecol Sociobiol 23:117–126
Page RE, Laidlaw HH (1988) Full sisters and super sisters: a terminological paradigm. Anim Behav 36:944–945
Page RE, Metcalf RA (1982) Multiple mating, sperm utilization, and social evolution. Am Nat 119:263–218
Page RE, Metcalf RA (1984) A population investment sex ratio for the honey bee (*Apis mellifera* L.) Am Nat 124:680–702
Page RE, Robinson GE (1991) The genetics of division of labour in honey bee colonies. Adv Insect Physiol 23:117–171
Page RE, Kimsey RB, Laidlaw HH (1984) Migration and dispersal of spermatozoa in spermathecae of queen honeybees (*Apis mellifera* L.). Experientia 40:182–184
Page RE, Post DC, Blum MS (1987) A honeybee queen pheromone that repels workers. In: Eder J, Rembold H (eds) Chemistry and biology of social insects. Peperny, Munich, pp 435–436
Page RE, Robinson GE, Fondrk MK (1989a) Genetic specialists, kin recognition and nepotism in honey-bee colonies. Nature (Lond) 338: 576–579
Page RE, Robinson GE, Calderone NW, Rothenbuhler WC (1989b) Genetic structure, division of labor, and the evolution of insect societies. In: Breed MD, Page RE (eds) The genetics of social evolution. Westview Press, Boulder, CO, pp 15–29
Page RE, Robinson GE, Fondrk MK (1990) Nepotism in the honeybee. Nature (Lond) 346:707–708
Pain J (1961) Sur la pheromone des reines d'abeilles et ses effets physiologiques. Ann Abeille 4:73–153
Pain J (1973) Pheromones and Hymenoptera. Bee World 54:11–24
Pain J, Barbier M (1981) The pheromone of the queen honeybee. Naturwissenschaften 68:429–430
Painter TS (1945) Nuclear phenomena associated with secretion in certain gland cells with especial reference to the origin of cytoplasmic acid. J Exp Zool 100:523–541
Palmer EL, Fowler HS (1975) Fieldbook of natural history. McGraw-Hill, New York
Palmquist-Momot J, Rothenbuhler WC (1971) Behavior genetics of nest cleaning in honeybees VI. Interactions of age and genotypes of bees, and nectar flow. J Apic Res 10:11–21
Pamilo P (1984) Genotypic correlation and regression in social groups: multiple alleles, multiple loci and subdivided populations. Genetics 107:307–320
Pamilo P (1988) Genetic variation in heterogeneous environments. Ann Zool Fenn 25:99–106
Pamilo P (1990) Comparison of relatedness estimators. Evolution 44: 1378–1382

Pamilo P, Crozier RH (1982) Measuring genetic relatedness in natural populations: methodology. Theor Popul Biol 21:171–193

Pamilo P, Rosengren R (1983) Sex ratio strategies in Formica ants. Oikos 40:199–204

Pamilo P, Varvio-Aho SL, Pekkarinen A (1978) Low enzyme gene variability in Hymenoptera as a consequence of haplodiploidy. Hereditas 88:93–99

Parhon M (1909) Les échanges nutritifs chez les abeilles pendant les quatre saisons. Ann Sc Nat Zool 9:1–59

Park OW (1937) Testing for resistance to American foulbrood in honeybees. J Econ Entomol 30:504–512

Parker FD, Torchio PF (1980) Management of wild bees. In: Beekeeping in the United States. Handb 335. US Dep Agric, Washington, DC, pp 144–160

Parker RL (1926) The collection and utilization of pollen by the honeybee. Mem Cornell Agric Exp Stn 98:1–55

Payne SF (1982) A study of social behavior in laboratory colonies of the naked mole-rat (*Heterocepahlus glaber*): cooperation in colony labor and reproduction. MSc Thesis, Univ California, Santa Cruz, CA

Peer DE (1957) Further studies on the mating range of the honeybee, *Apis mellifera*. Can Entomol 89:108–110

Peng Y-S (1988) The resistance mechanism of the Asian honey bee (*Apis cerana*) to the mite *Varroa jacobsoni*. In: Needham GR, Page RE, Delfinado-Baker M, Bowman CE (eds) Africanized honey bees and bee mites. Ellis Horwood, Chichester, pp 426–429

Peng Y-S, Fang Y, Xu Sh, Ge L (1987) Resistance mechanism of the Asiatic honeybee *Apis cerana* Fabr. to an ectoparasitic mite *Varroa jacobsoni* Oud. J Invertebr Pathol 49:54–60

Percival MS (1961) Types of nectar in angiosperms. New Phytol 60:235–281

Perepelova L (1928) Materials concerning the biology of the bee: the work of the bees in the hive. Opuit Pas 11:492–502

Petrunkewitsch (1901) Die Richtungskörper und ihr Schicksal im befruchteten und unbefruchteten Bienenei. Zool Jahrb 14:573–608

Pham MH, Roger B, Pain J (1982) Variation en fonction de l'âge des ouvrières d'abeilles (*Apis mellifica ligusta* S.) du pouvoir d'attraction d'un extrait de pheromones royales. Apidologie 13:143–155

Phillips EF, Demuth GS (1914) The temperature of the honeybee cluster in winter. Bull US Dep Agric 93:1–16

Phillips WS (1991) Desert. In: Lorimer LT (ed) Encyclopedia Americana, vol 9. Grolier, Danbury, CT, pp 1–11

Pickard RS (1988) From bee brain to biosensors. Sci Tech Rev 4:59–64

Pickett JA, Williams IH, Martin AP, Smith MC (1980) Nasonov pheromone of the honeybee *Apis mellifera* L. (Hymenoptera, Apidae), I. Chemical characterization. J Chem Ecol 6:425–434

Pickett JA, Williams IH, Smith MC, Martin AP (1981) Nasonov pheromone of the honeybee *Apis mellifera* L. (Hymenoptera, Apidae). III. Regulation of pheromone comosition and production. J Chem Ecol 7:543–554

Pimentel D, Pimentel M (1979) Food, energy and society. Arnold, London

Pirchner F (1979) Populationsgenetik in der Tierzucht. Parey, Hamburg

Plass F (1953) Inzuchtwirkung und Heterosiseffekt bei der Honigbiene. Allg Inf Dienst, Bonn 66:49–68
Pleasants JM (1980) Competition for bumblebee pollinators in Rocky Mountain plant communities. Ecology 61:1446–1459
Pleasants JM (1981) Bumblebee response to variation in nectar availability. Ecology 62:1648–1661
Plowright RC, Jay SC (1968) Caste differentiation in bumblebees (Bombus Latr.: Hym.). I-The determination of female size. Insect Soc 15:171–192
Plowright RC, Jay SC (1977) On the size determination of bumble bee castes (Hymenoptera: Apidae). Can J Zool 55:1133–1138
Plowright RC, Laverty TM (1984) The ecology and sociobiology of bumble bees. Annu Rev Entomol 29:175–199
Plowright RC, Plowright CMS (1990) The laying of male eggs by bumble bee queens: an experimental reappraisal and a new hypothesis. Can J Zool 68:493–497
Polhemus MS, Lush JL, Rothenbuhler WC (1950) Mating systems in honey bees. J Hered 41:151–155
Pomeroy N (1981) Reproductive dominance interactions and colony development in bumblebees. Thesis, Univ Toronto, Ontario
Pomeroy N, Plowright RC (1982) The relation between worker numbers and the production of males and queens in the bumble bee *Bombus perplexus*. Can J Zool 60:954–957
Pompolo SG, Takahashi CS, Kerr WE (1989) Cytogenetic analysis of 21 stingless bee species (Meliponinae, Apidae, Hymenoptera). Proc 32nd Int Congr Apicult, Rio de Janeiro, pp 70–71
Post DC, Page RE, Erickson EH (1986) Honey bee (*Apis mellifera* L.) queen feces: source of a pheromone that repels worker bees. J Chem Ecol 13:583–592
Prestwich GD, Blomquist GJ (1987) Pheromone biochemistry. Academic Press, New York
Prosser CL (1986) Adaptational biology: molecules to organisms. Wiley, New York
Pulliam HR, Caraco T (1984) Living in groups: is there an optimal group size? In: Krebs JR, Davies NB (eds) Behavioural ecology: an evolutionary approach. Blackwell, Oxford, pp 122–147
Pulliam HR, Millikan (1982) Social organization in the nonreproductive season. In: Farner DS, King JR (eds) Avian biology, vol 6. Academic Press, New York, pp 169–197
Pyke G (1978) Optimal foraging: movement patterns of bumble bees between inflorescences. Theor Popul Biol 13:72–98
Pyke G, Blazer L (1982) The effects of the introduced honey bee *Apis mellifera* on Australian native bees. Rep NSW Nat Parks Wildlife Serv, Sydney
Pyke GH, Pulliam HR, Charnov EL (1977) Optimal foraging: a selective review of theory and tests. Q Rev Biol 52:137–154
Queller DC (1989) The evolution of eusociality: reproductive head starts of workers. Proc Natl Acad Sci USA 86:3224–3226
Queller DC, Goodnight KF (1989) Estimating relatedness using genetic markers. Evolution 43:258–275

Ragland SS, Sohal RS (1975) Ambient temperature, physical activity and ageing in the house fly, *Musca domestica*. Exp Gerontol 10:279–290

Ratnieks FLW (1988) Reproductive harmony via mutual policing by workers in eusocial Hymenoptera. Am Nat 132:217–236

Ratnieks FLW (1990a) The evolution of polyandry by queens in social Hymenoptera: the significance of the timing of removal of diploid males. Behav Ecol Sociobiol 26:343–348

Ratnieks FLW (1990b) Assessment of queen mating frequency by workers in social Hymenoptera. J Theor Biol 142:87–93

Ratnieks FLW (1991) Africanized bees: natural selection for colonizing ability. In: Spivak M, Fletcher DJC, Breed MD (eds) The "African" honey bee. Westview Press, Boulder, CO, pp 119–135

Ratnieks FLW, Reeve HK (1991) The evolution of queen-rearing nepotism in social Hymenoptera: effects of discrimination costs in swarming species. J Evol Biol 4:93–115

Ratnieks FLW, Visscher PK (1989) Worker policing in honeybees. Nature (Lond) 342:796–797

Real LA, Ott J, Silverfine E (1982) On the trade-off between the mean and the variance in foraging: effect of spatial distribution and color preference. Ecology 63:1617–1623

Reaumur RAF (1759) Oeconomische Abhandlungen von den Bienen. Martin Göbhards Erben, Frankfurt

Reeve HK, Westneat DF, Noon WA, Sherman PW, Aquadro CF (1990) DNA "fingerprinting" reveals high levels of inbreeding in colonies of the eusocial naked mole-rat. Proc Natl Acad Sci USA 87:2496–2500

Rembold H (1976) The role of the determinator in caste formation in the honey bee. In: Lüscher M (ed) Phase and caste determination in insects. Pergamon Press, Oxford pp 21–34

Rembold H (1986) Control of imaginal development by juvenile hormone. Adv Invertebr Reprod 4:59–68

Renner M (1957) Neue Versuche über den Zeitsinn der Honigbiene. Z Vgl Physiol 40:85–118

Renner M, Baumann M (1964) Über Komplexe von subepidermalen Drüsenzellen (Duftdrüsen?) der Bienenkönigin. Naturwissenschaften 51:68–69

Renner M, Vierling G (1977) Die Rolle des Taschendrüsenpheromons beim Hochzeitsflug der Bienenkönigin. Behav Ecol Sociobiol 2:329–338

Rhein W von (1933) Über die Entstehung des weiblichen Dimorphismus im Bienenstaate. Wilhelm Roux's Arch Entwicklungsmech Org 129D: 601–665

Ribbands CR (1949) The foraging method of individual honeybees. J Anim Ecol 18:47–66

Ribbands CR (1952) Division of labour in the honeybee community. Proc R Soc Lond B140:32–42

Ribbands CR (1953) The behaviour and social life of honeybees. Bee Res Assoc, London

Ribbands CR, Spiers N (1953) The adaptability of the homecoming honeybee. Br J Anim Behav 1:59–66

Richards KW (1977) Ovarian development of queen and worker bumble bees (Hymenoptera: Apidae) in southern Alberta. Can Entomol 109:109–116

Rinderer TE (1977) Measuring the heritability of characters of honeybees. J Apic Res 16:95–98
Rinderer TE (1986) Bee genetics and breeding. Academic Press, Orlando
Rinderer TE, Baxter JR (1978) The effect of empty comb on hoarding behavior and honey production of the honey bees. J Econ Entomol 71:757–759
Rinderer TE, Collins AM (1986) Behavioral genetics. In: Rinderer TE (ed) Bee genetics and breeding. Academic Press, Orlando, pp 155–176
Rinderer TE, Hellmich RL (1991) The processes of Africanization. In: Spivak M, Fletcher DJC, Breed MD (eds) The "African" honey bee. Westview Press, Boulder, CO, pp 95–117
Rinderer TE, Collins AM, Brown MA (1983) Heritabilities and correlations of the honeybee: response to *Nosema apis*, longevity and alarm response to isopentyl acetate. Apidologie 14:79–85
Rinderer TE, Hellmich RL, Danka RG, Collins AM (1985) Male reproductive parasitism. A factor in the Africanization of honeybee populations. Science 228:1119–1121
Rinderer TE, Collins AM, Hellmich RL, Danka RG (1987a) Differential drone production by Africanized and European honey bee colonies. Apidologie 18:61–68
Rinderer TE, Wright JE, Shimanuki H, Parker F, Erickson EH, Wilson WT (1987b) The proposed honey-bee regulated zone in Mexico. Am Bee J 127:160–164
Rinderer TE, Koeniger N, Tingek S, Mardan M, Koeniger G (1989) A morphological comparison of the cavity dwelling honeybees of Borneo *Apis koschevnikovi* (Buttel-Reepen, 1906) and *Apis cerana* (Fabricius, 1973). Apidologie 20:405–411
Risler H (1954) Die somatische Polyploidie in der Entwicklung der Honigbiene (*Apis mellifera*) und die Wiederherstellung der Diploidie bei den Drohnen. Z Zellforsch Mikrosk Anat 41:1–78
Risler H, Romer F (1968) Chromosomenzahlen und DNS-Gehalt der Epidermiszellen während der Entwicklung der Honigbiene. Dtsch Zool Ges Verh Suppl 32:153–160
Ritter W (1978) Influence of the brood on the change of heat production in honeybee-colonies (*Apis mellifera carnica*). Verh Dtsch Zool Ges 1978:220
Ritter W (1982) Experimenteller Beitrag zur Thermoregulation des Bienenvolkes (*Apis mellifera* L.). Apidologie 13:169–185
Roberts WC (1944) Multiple mating of queen bees proved by progeny and flight tests. Glean Bee Cult 72:281–283
Robertson A (1962) Selection for heterozygotes in small populations. Genetics 47:1297–1300
Robinson GE (1987a) Hormonal regulation of age polyethism in the honey bee, *Apis mellifera*. In: Menzel R, Mercer A (eds) Neurobiology and behaviour in honeybees. Springer, Berlin Heidelberg New York, pp 266–280
Robinson GE (1987b) Regulation of honey bee age polyethism by juvenile hormone. Behav Ecol Sociobiol 120:329–338
Robinson GE (1991) Hormonal and genetic control of division of labor in the honey bee colony. In: Goodman LJ, Fisher RC (eds) The behaviour and physiology of bees. CAB International, Oxon, UK, pp 14–27

Robinson GE, Page RE (1988) Genetic determination of guarding and undertaking in honeybee colonies. Nature (Lond) 333:356–358

Robinson GE, Page RE (1989a) Genetic basis for division of labour in an insect society. In: Breed MD, Page RE (eds) The genetics of social evolution. Westview Press, Boulder, CO, pp 61–80

Robinson GE, Page RE (1989b) Genetic determination of nectar foraging, pollen foraging, and nest site scouting in honeybee colonies. Behav Ecol Sociobiol 24:317–323

Robinson GE, Page RE, Strambi C, Strambi A (1989a) Hormonal and genetic control of behavioral integration in honey bee colonies. Science 246:109–112

Robinson GE, Page RE, Strambi C, Strambi A (1989b) Hormonal and genetic control of behavioral integration in honey bee colonies. Science 246:109–112

Robinson GE, Page RE, Fondrk MK (1990) Intracolonial behavioral variation in oviposition, oophagy, and larval care in queenless honeybee colonies. Behav Ecol Sociobiol 26:315–323

Rockstein M (1972) The role of molecular genetic mechanisms in the ageing process. In: Rockstein M, Baker GT (eds) Molecular genetic mechanisms in development and ageing. Academic Press, New York, pp 1–10

Rockstein M (1978) Biochemistry of insects. Academic Press, New York

Rockstein M (1987) The physiology of insects. Academic Press, New York

Roegl F, Steininger F (1984) Neogene Paratethys, Mediterranean and Indopacific seaways. In: Brenchley P (ed) Fossils and climate. Wiley, Chichester, pp 171–200

Rogers EM, Agarwala-Rogers R (1976) Communication in organizations. Free Press, New York

Rogers EM, Kincaid DL (1981) Communication networks. Toward a new paradigm for research. Free Press, New York

Rösch GA (1927) Untersuchungen über die Arbeitsteilung im Bienenstaat. I. Die Tätigkeiten im normalen Bienenstaate und ihre Beziehungen zum Alter der Arbeitsbiene. Z Vgl Physiol 6:264–298

Rösch GA (1930) Untersuchungen über die Arbeitsteilung im Bienenstaat. II. Die Tätigkeiten der Arbeitsbienen unter experimentell veränderten Bedingungen. Z Vgl Physiol 12:1–17

Röseler P-F (1970) Unterschiede in der Kastendetermination zwischen den Hummelarten *Bombus hypnorum* und *Bombus terrestris*. Z Naturforsch 25b:543–548

Röseler P-F (1974) Größenpolymorphismus, Geschlechtsregulation und Stabilisierung der Kasten im Hummelvolk. In: Schmidt GH (ed) Sozialpolymorphismus bei Insekten. Wiss Verlagsgesellschaft, Stuttgart, pp 298–335

Röseler P-F, Röseler I (1977) Dominance in bumblebees. Proc 8th Int Congr IUSSI, Cent Agric Publ Doc, Wageningen, pp 22–23

Röseler P-F, Röseler I, van Honk CGJ (1981) Evidence for inhibition of corpora allata activity in workers of *Bombus terrestris* by a pheromone from the queens mandibular glands. Experientia 37:348–351

Rosengren RW, Fortelius W, Lindström K, Luther A (1986) Phenology and causation of nest heating and thermoregulation in red wood ants of the *Formica rufa* group studied in coniferous forest habitats in southern Finland. Ann Zool Fenn 24:147–155

Rossel S, Wehner R (1984) Celestial orientation in bees: the use of spectral cues. J Comp Physiol A 155:605–613
Rothe U, Nachtigall W (1989) Flight of the honey bee IV. Respiratory quotients and metabolic rates during sitting, walking and flying. J Comp Physiol B 158:739–749
Rothenbuhler WC (1957) Diploid male tissue as new evidence on sex determination in honey bees. J Hered 48:160–168
Rothenbuhler WC (1960) A technique for studying genetics of colony behaviour in honey bees. Am Bee J 100:176, 198
Rothenbuhler WC (1964a) Behavior genetics of nest cleaning in honeybees I. Responses of four inbred lines to disease-killed brood. Anim Behav 12:578–583
Rothenbuhler WC (1964b) Behavior genetics of nest cleaning in honeybees IV. Responses of F_1 and backcross generations to disease-killed brood. Am Zool 4:111–123
Rothenbuhler WC, Page RE (1989) Genetic variability for temporal polyethism in colonies consisting of similarly-aged worker honey bees. Apidologie 20:433–437
Rothenbuhler WC, Kulincevic JM, Thompson VC (1979) Successful selection of honeybees for fast and slow hoarding of sugar syrup in the laboratory. J Apic Res 18:272–278
Roubaud E (1916) Recherches biologiques sur les guêpes solitaires et sociaux d'Afrique. La genèse de la vie sociale et l'évolution de l'instinct maternel chez les vespides. Ann Sci Nat Zool 1:1–160
Roubik DW (1978) Competitive interactions between neotropical pollinators and Africanized honeybees. Science 201:1030–1032
Roubik DW (1983) Experimental community studies: time-series tests of competition between African and neotropical bees. Ecology 64: 971–978
Roubik DW (1987) Long-term consequences of the Africanized honey bee invasion: implications for the United States. In: Sorensen AA (ed) Proc Africanized Honey Bee Symp, Am Farm Bureau, Park Ridge, IL, pp 46–54
Roubik DW (1989) Ecology and natural history of tropical bees. Cambridge Univ Press, New York
Roubik DW, Boreham MM (1990) Learning to live with Africanized honeybees. Interciencia 15:146–153
Roubik DW, Buchmann SL (1982) Africanized bees confirmed in Panama. Am Bee J 122:322
Roubik DW, Buchmann SL (1984) Nectar selection by *Melipona* and *Apis mellifera* (Hymenoptera: Apidae) and the ecology of nectar intake by bee colonies in a tropical forest. Oecologia 61:1–10
Rowell GA, Taylor OR (1988) Some computer simulations using the neutral mating model for the honey bee, *Apis mellifera*. In: Needham GR, Page RE, Delfinado-Baker M, Bowman CE (eds) Africanized honey bees and bee mites. Ellis Horwood, Chichester, pp 184–192
Rüschkamp PF (1921) Wheelers Trophallaxis und der Ursprung der Insektenstaaten. Biol Zentralbl 41(11)
Ruttner F (1968) Les races des abeilles. In: Chauvin R (ed) Traité de biologie de l'abeille, vol 1. Masson, Paris, pp 27–44

Ruttner F (1973) Die Bienenrassen des mediterranen Beckens. Apidologie 4:171–172

Ruttner F (1977) The problem of the Cape bee (*Apis mellifera capensis* Escholtz): parthenogenesis, size of population, evolution. Apidologie 8:281–294

Ruttner F (1986) Geographical variation and classification. In: Rinderer TE (ed) Bee genetics and breeding. Academic Press, Orlando, pp 23–56

Ruttner F (1988) Biogeography and taxonomy of honeybees. Springer, Berlin Heidelberg New York

Ruttner F, Hesse B (1981) Rassenspezifische Unterschiede in Ovarentwicklung und Eiablage von weisellosen Arbeiterinnen der Honigbiene Apis mellifera L. Apidologie 12:159–183

Ruttner F, Maul V (1983) Experimental analysis of the reproductive interspecific isolation of *Apis mellifera* L. and *Apis cerana* Fabr. Apidologie 14:309–327

Ruttner F, Ruttner H (1963) Untersuchungen über die Flugaktivität und das Paarungsverhalten der Drohnen. 1. Proc 19th Int Congr Apicult, Apimondia, Prague, pp 604–605

Ruttner F, Ruttner H (1965) Untersuchungen über die Flugaktivität und das Paarungsverhalten der Drohnen. 2. Beobachtungen an Drohnensammelplätzen. Z Bienenforsch 8:1–8

Ruttner F, Ruttner H (1966) Untersuchungen über die Flugaktivität und das Paarungsverhalten der Drohnen. 3. Flugweite und Flugrichtung der Drohnen. Z Bienenforsch 8:332–354

Ruttner F, Koeniger N, Veith HJ (1967) Queen substance bei eierlegenden Arbeiterinnen der Honigbiene (*Apis mellifera* L.). Naturwissenschaften 63:434

Ruttner F, Woyke J, Koeniger N (1973) Reproduction in *Apis cerana*. 1. Mating behaviour. J Apic Res 11:141–146

Ruttner F, Wilson R, Snelling G, Vorwohl G, Kauhausen D (1986) The evolution of the honeybee wing venation. Apidologie 17:349

Ruttner F, Kauhausen D, Koeniger N (1989) Position of the red honey bee, *Apis koshevnikovi* (Buttel-Reepen 1906), within the genus Apis. Apidologie 20:395–404

Ruttner H, Ruttner F (1972) Untersuchungen über die Flugaktivität und das Paarungsverhalten der Drohnen. 5. Drohnensammelplätze und Paarungsdistanz. Apidologie 3:203–232

Rutz WL, Gerig L, Wille H, Lüscher M (1976) The function of juvenile hormone in adult worker honeybees, *Apis mellifera*. J Insect Physiol 22:1485–1491

Safranek L, Williams CM (1984) Critical weights for metamorphosis in the tobacco hornworm, *Manduca sexta*. Biol Bull 167:555–567

Sagan D (1990) Biospheres. Metamorphosis of planet earth. McGraw-Hill, New York

Sakagami SF (1953) Untersuchungen über die Arbeitsteilung in einem Zwergvolk der Honigbiene: Beiträge zur Biologie des Bienenvolkes, *Apis mellifera* L. I. Jpn J Zool 11:117–185

Sakagami SF (1976) Specific differences in the bionomic characters of bumblebees. A comparative review. J Fac Sci Hokkaido Univ Zool 20:390–447

Sakagami SF (1982) Stingless bees. In: Hermann, HR (ed) Social insects, vol 3. Academic Press, New York, pp 361–423
Sakagami SF, Laroca S (1963) Additional observations on the habits of the cleptophobic stingless bee, the genus *Lestrimelitta* Friese. J Fac Sci Hokkaido Univ Ser VI Zool 15:319–339
Sakagami SF, Maeta Y (1985) Multifemale nests and rudimentary castes in the normally solitary bee *Ceratina japonica*. J Kans Entomol Soc 57: 639–656
Sakagami SF, Montenegro MJ, Kerr WE (1965) Behavior studies of the stingless bee, with special reference to the ovipostion process. V. *Melipona quadrifasciata anthidioides* Lepeletier. J Fac Sci Hokkaido Univ Ser VI Zool 15:578–607
Sakagami SF, Matsumura T, Ito K (1980) *Apis laboriosa* in Himalaya, the little known world largest honeybee (Hymenoptera, Apidae). Insecta Matsumurana 19:47–77
Sanderson AR (1932) The cytology of parthenogenesis in Tenthredinidae (Hymeoptera: Symphyta). Genetica 14:29–36
Sanderson AR, Hall DW (1948) The cytology of the honey bee *Apis mellifica* L. Nature (Lond) 162:34–35
Sanderson AR, Hall DW (1951) Sex in the honey-bee. Endeavour 10:33–39
Santschi F (1911) Observations et remarques critiques sur le mécanisme de l'orientation chez les fourmis. Rev Suisse Zool 19:303–338
Sasagawa H, Sasaki M, Okada I (1986) Experimental induction of the division of Labor in worker *Apis mellifera* L. by juvenile hormone (JH) and its analog. 30th Int Congr Apicult, Apimondia, Nagoya, pp 130–132
Schaffer WM, Zeh DW, Buchmann SL, Kleinhaus S, Schaffer MV, Antrim J (1983) Competition for nectar between introduced honey bees and native North American bees and ants. Ecology 64:564–577
Schmid-Hempel P (1984) The importance of handling time for flight directionality in bees. Behav Ecol Sociobiol 15:303–309
Schmid-Hempel P (1987) Efficient nectar-collection by honeybees. I. Economic models. J Anim Ecol 56:209–218
Schmid-Hempel P, Wolf T (1988) Foraging effort and life span of workers in a social insect. J Anim Ecol 57:509–522
Schmid-Hempel P, Kacelink A, Houston A (1985) Honeybees maximize efficiency by not filling their crop. Behav Ecol Sociobiol 17:61–66
Schmidt GH (1974) Sozialpolymorphismus bei Insekten. Wiss Verlagsgesellschaft, Stuttgart
Schmidt JO, Buchmann SL (1985) Pollen digestion and nitrogen utilization by *Apis mellifera* L. (Hymenoptera: Apidae). Comp Biochem Physiol 82A:499–503
Schmidt RS (1964) *Apticotermes* nests. Am Zool 4:221–225
Schmidt-Nielsen K (1985) Animal physiology. Cambridge Univ Press, Cambridge
Schneider P, Kloft W (1971) Beobachtungen zum Gruppenverteidigungsverhalten der östlichen Honigbiene *Apis cerana* Fabr. Z Tierpsychol 29: 337–342
Schneider SS, Stamps JA, Gary NE (1986a) The vibration dance of the honey bee. I. Communication regulating foraging on two time scales. Anim Behav 34:366–385

Schneider SS, Stamps JA, Gary NE (1986b) The vibration dance of the honey bee. II. The effect of foraging success on daily patterns of vibration activity. Anim Behav 34:386–391

Schreffler AK (1983) Energy costs of foraging by honey bees on artificial flower patches of variable and constant nectar distributions. Diss, Dep Biol Sci, State Univ New York, Brockport, NY

Schua L (1952) Untersuchungen über den Einfluß meteorologischer Elemente auf das Verhalten der Honigbiene. A Vgl Physiol 34:258–277

Schulze BR (1974) Climate of South Africa. Part 8. General Survey, Weather Bur Dep Transp, Pretoria

Schürmann F-W, Elekes K (1987) Synaptic connectivity in the mushroom bodies of the honeybee brain: electron microscopy and immunocytochemistry of neuroactive compounds. In: Menzel R, Mercer A (eds) Neurobiology and behavior of honeybees. Springer, Berlin Heidelberg New York, pp 225–234

Schwarz HF (1948) Stingless bees (Meliponinae) of the western hemisphere. Bull Am Mus Nat Hist 90:1–546

Schwarz MP (1986) Persistent multifemale nests in an Australian allodapine bee *Exoneura bicolor* (Hymenopytera; Anthophoridae). Insect Soc 33: 258–277

Schwarz MP (1987) Intra-colony relatedness and sociality in the allodapine bee *Exoneura bicolor*. Behav Ecol Sociobiol 21:387–392

Schwarz MP (1988a) Local resource enhancement and sex ratios in a primitively social bee. Nature (Lond) 331:346–348

Schwarz MP (1988b) Intra-specific mutualism and kin-association of cofoundresses in allodapine bees (Hymenoptera Antophoride). Monit Zool Ital 22:245–254

Schwarz MP (1988c) Some notes on cofounded nests of three species of social bees in the genus *Exoneura*. Victorian Nat 105:212–215

Schwarz MP, O'Keefe KJ (1991) Order of eclosion and reproductive differentiation in a social allodapine bee. Ethol Ecol Evol 3:233–246

Schwarz MP, Scholz O, Jensen G (1987) Ovarian inhibition among nestmates of *Exoneura bicolor* Smith. J Aust Entomol Soc 26:355–359

Scott CD (1986) Biology and management of wild bee and domesticated honey bee pollinators for tree pollination. PhD Thesis, Simon Fraser Univ, Burnaby, BC

Seeley TD (1974) Atmosheric carbon dioxide regulation in honey-bee (*Apis mellifera*) colonies. J Insect Physiol 20:2301–2305

Seeley TD (1977) Measurement of nest cavity volume by the honey bee (*Apis mellifera*). Behav Ecol Sociobiol 2:201–227

Seeley TD (1978) Life history strategy of the honey bee, *Apis mellifera*. Oecologia 32:109–118

Seeley TD (1979) Queen substance dispersal by messenger workers in honeybee colonies. Behav Ecol Sociobiol 5:391–415

Seeley TD (1982a) Adaptive significance of the age polyethism schedule in honeybee colonies. Behav Ecol Sociobiol 11:287–293

Seeley, TD (1982b) How honey bees find a home. Sci Am 247:158–168

Seeley TD (1983) Division of labour between scouts and recruits in honeybee foraging. Behav Ecol Sociobiol 11:287–293

Seeley TD (1985) Honeybee ecology – a study of adaptation in social life. Princeton Univ Press, Princeton, NJ

Seeley TE (1986) Social foraging by honey bees: how colonies allocate foragers among patches of flowers. Behav Ecolo Sociobiol 19:343–354
Seeley TD (1989) The honey bee colony as a superorganism. Am Sci 77: 546–553
Seeley TD, Heinrich B (1981) Regulation of temperature in the nests of social insects. In: Heinrich B (ed) Insect thermoregulation. Wiley, New York, pp 159–234
Seeley TD, Morse RA (1976) The nest of the honeybee (*Apis mellifera*). Insect Soc 23:495–512
Seeley TD, Visscher PK (1985) Survival of honeybees in cold climates: the critical timing of colony growth and reproduction. Ecol Entomol 10:81–88
Seeley TD, Morse RA, Visscher PK (1979) The natural history of the flight of honey bee swarms. Psyche 86:103–113
Seeley TD, Seeley RH, Akratanakul P (1982) Colony defense strategies of the honeybees in Thailand. Ecol Monogr 52:43–63
Seeley TD, Camazine S, Sneyd J (1991) Collective decision-making in honey bees: how colonies choose among nectar sources. Behav Ecol Sociol 28:277–290
Seger J (1989) Who are the drone police? Nature (Lond) 342:741–742
Severson DW, Aiken JM, Marsh RF (1988) Molecular analyses of North American and Africanized honey bees. In: Needham GR, Page RE, Delfinado-Baker M, Bowman CE (eds) Africanized honey bees and bee mites. Ellis Horwood, Chichester, pp 294–302
Sharma GP, Gupta CG, Kumbkarni CG (1961) Cytology of spermatogenesis in the honey-bee, *Apis indica* (F.). J R Micros Soc 79:337–351
Shehata SM, Townsend GF, Shuel RW (1981) Seasonal physiological changes in queen and worker honeybees. J Apic Res 20:69–78
Sheppard WS, Berlocher SH (1984) Enzyme polymorphism in *Apis mellifera* from Norway. J Apic Res 23:64–69
Sheppard WS, Berlocher SH (1985) New allozyme variability in Italian honey bees. J Hered 76:45–48
Sheppard WS, Berlocher SH (1989) Allozyme variation and differentiation among four *Apis* species. Apidologie 20:419–432
Sheppard WS, Rinderer TE, Mazzoli JA, Stelzer JA, Shimanuki H (1991) Gene flow between African- and European-derived honey bee populations in Argentina. Nature (Lond) 349:782–784
Sherman PW, Seeley TD, Reeve HK (1988) Parasites, pathogens, and polyandry in social hymenoptera. Am Nat 131:602–610
Sherman PW, Jarvis JUM, Alexander RD (1990) The biology of the naked mole rat. Princeton Univ Press, Princeton, NJ
Shuel RW (1975) The production of nectar. In: Dadant CP (ed) The hive and the honey bee. Dadant & Sons, Hamilton, IL, pp 283–302
Shykoff JA, Schmid-Hempel P (1991a) Parasites and the advantage of genetic variability within social insect colonies. Proc R Soc Lond B 243:55–58
Shykoff JA, Schmid-Hempel P (1991b) Genetic relatedness and eusociality: parasite-mediated selection on the genetic composition of groups. Behav Ecol Sociobiol 28:371–376
Silberrad REM (1976) Beekeeping in Zambia. Apimondia, Bucharest

Silva DLN da (1972) Consideracoes em torno de um caso de substituicao de rainha em *Plebeia* (*Plebeia*) *droryana* (Friese, 1900). In: Homenagem a Warwick Kerr, Rio Claro SP, Brazil, pp 267–273
Silva DLN da, Zucchi R, Kerr WE (1972) Biological and behavioural aspects of the reproduction in some species of *Melipona*. Anim Behav 20:123–132
Silveira ZV (1971) Numero de cromosomos em Meliponideos barsileiros. Cienc Cult (Sao Paulo) Suppl 23:105–106
Silveira ZV (1972) Numero de cromosomos em Meliponideos barsileiros II. Cienc Cult (Sao Paulo) Suppl 24:160–161
Silverstein RM, Rodin JO, Wood DL (1966) Sex attractants in frass produced by male *Ips confussis* in ponderosa pine. Science 154:509–510
Simmel G (1964) The web of group affiliations. Free Press, New York
Simpson J (1950) Humidity in the winter cluster of honey-bees. Bee World 31:41–44
Simpson J (1961) Nest climate regulation in honeybee colonies. Science 133:1327–1333
Simpson J (1964) The mechanism of honeybee queen piping. Z Vgl Physiol 48:277–282
Simpson J (1979) The existence and physical properties of pheromones by which worker honeybees recognize queens. J Apic Res 18:233–249
Singh S (1962) Beekeeping in India. Indian Counc Agric Res, New Delhi
Skaife SH (1954) African insect life. Longmans, London
Sladen FWL (1901) A scent organ in the bee. Br Bee J 29:142, 143, 151–153
Sladen FWL (1905) Queen rearing in England. Houlson, London
Sladen FWL (1912) The bumble-bee, its life history and how to domesticate it. Macmillan, London
Slessor KN, Kaminski LA, King GGS, Borden JH, Winston ML (1988) Semiochemical basis of the retinue response to queen honey bees. Nature (Lond) 332:354–356
Slifer EH, Sekhon SS (1961) Fine structure of the sense organs on the antennal flagellum of the honey bee *Apis mellifera* L. J Morphol 109: 351–362
Smeathman H (1781) Some account of the termites which are found in Africa and other hot climates. Philos Trans 71:139–192
Smith BH (1987) Effects of genealogical relationship and colony age on the dominance hierarchy in the primitively eusocial bee *Lasioglossum zephyrum*. Anim Behav 35:211–217
Smith DR (1988) Mitochondrial DNA polymorphisms in five Old World subspecies of honey bees and in New World hybrids. In: Needham GR, Page RE, Delfinado-Baker M, Bowman CE (eds) Africanized honey bees and bee mites. Ellis Horwood, Chichester, pp 303–312
Smith DR (1990) Mitochondrial DNA and biogeography of *Apis* species. In: Veeresh GK, Mallik B, Viraktamath CA (eds) Social insects and the environment. Oxford & IBH, New Delhi, pp 113–114
Smith DR (1991) Diversity in the genus *Apis*. Westview Press, Boulder, CO
Smith DR, Brown WM (1988) Polymorphisms in mitochondrial DNA of European and Africanized honeybees (*Apis mellifera*). Experientia 44:257–260

Smith DR, Taylor OR, Brown WM (1989) Neotropical Africanized honey bees have African mitochondrial DNA. Nature (Lond) 339:213–215

Smith IC, Peacock AD (1957) The cytology of Pharaoh's ant, *Monomorium pharaonis* (L.). Proc R Soc Edinb B 66:235–261

Smith M (1969) Optimum selection procedures in animal breeding. Anim Prod 11:433–442

Smugor D (1991) Specific activity of phosphofructokinase in flight muscle of honeybees (*Apis mellifera* L.) in winter, summer and in a flight room. MSc Thesis, State Univ New York, Brockport, NY

Sneyd J, Camazine A (1991) Selection of high-quality nectar sources by honey bees: self-organized decision-making. In: Goodman LJ, Fisher RC (eds) The behaviour and physiology of bees. CAB International, Oxon, UK pp 102–115

Snodgrass RE (1956) Anatomy of the honeybee. Cornell Univ Press, Ithaca, NY

Sober E (1984) The nature of selection. MIT Press, Cambridge, MA

Sohal RS (1975) Mitochondrial changes in flight muscle of normal and flightless *Drosophila melanogaster* with age. J Morphol 145:337–354

Sohal RS (1976) Metabolic rate and life span. In: Cutler RG (ed) Cellular ageing, part I, vol. 9. Interdisciplinary topics in gerontology. Karger, Basel, pp 25–40

Sohal RS (1985) Aging in insects. In: Kerkut GA, Gilbert LI (eds) Comprehensive insect physiology and pharmacology, vol 10. Pergamon Press, Oxford, pp 595–531

Sohal RS (1986) The rate of living theory: a contemporary interpretation. In: Collatz KG, Sohal RS (eds) Insect aging: strategies and mechanisms. Springer, Berlin Heidelberg New York, pp 23–44

Southwick AK, Southwick EE (1983) Aging effect on nectar production in *Asclepias syriaca*. Oecologia 56:121–125

Southwick EE (1982) Metabolic energy of intact honey bee colonies. Comp Biochem Physiol 71A:277–281

Southwick EE (1983a) The honey bee cluster as a homeothermic superorganism. Comp Biochem Physiol 75A:641–645

Southwick EE (1983b) Nectar biology and nectar feeders of common milkweed, *Asclepias syriaca* L. Bull Torrey Bot Club 110:324–345

Southwick EE (1984) Photosynthate allocation to floral nectar: a neglected energy investment. Ecology 65:1775–1779

Southwick EE (1985a) Allometric relations, metabolism and heat conductance in clusters of honey bees at cool temperatures. J Comp Physiol B 156:143–149

Southwick EE (1985b) Bee hair structure and the effect of hair on metabolism at cool temperature. J Apic Res 24:144–149

Southwick EE (1985c) Thermal conductivity of wax comb and its effect on heat balance in colonial honey. bees (*Apis mellifera* L.). Experientia 41:1486–1487

Southwick EE (1986a) Colonial synergism of rhythms in honey bees. Am Bee J 126:834

Southwick EE (1986b) Effect of amino acids in flower nectar on honey bee feeding. Am Bee J 126:34

Southwick EE (1987a) Cooperative metabolism in honey bees: an alternative to antifreeze and hibernation. J Therm Biol 12:155–158
Southwick EE (1987b) Regulation of respiration in honey bee colonies. In: Rembold H, Eder J (eds) Chemistry and biology of social insects. Peperny, Munich, p 240
Southwick EE (1987c) Selection in social insects: the superorganism concept. In: Rembold H, Eder J (eds) Chemistry and biology of social insects. Peperny, Munich, p358
Southwick EE (1988) Thermoregulation in honey-bee colonies. In: Needham G, Page RE, Delfinado-Baker M, Bowman CE (eds) Africanized honey bees and bee mites. Wiley, New York, pp 223–236
Southwick EE (1989) Social temperature control and heat production in honey bees. In: Mercer JB (ed) Thermal physiology. Elsevier, Amsterdam, pp 755–759
Southwick EE (1990) How will Africanized honey bees behave? Am Bee J 130:200–201
Southwick EE (1991a) Overwintering in honey bees: implications for apiculture. In: Lee RE Jr, Denlinger DL (eds) Insects at low temperature. Chapman Hall, New York, pp 446–460
Southwick EE (1991b) The colony as a thermoregultaing superorganism. In: Goodman LJ, Fisher RC (eds) The behaviour and physiology of bees. CAB International, Oxon, UK, pp 28–47
Southwick EE, Heldmaier G (1987) Temperature control in honey bee colonies. Bio Sci 37:395–399
Southwick EE, Moritz RFA (1985) Metabolic response to alarm pheromone in honeybees. J Insect Physiol 31:389–392
Southwick EE, Moritz RFA (1986) Effects of meteorological factors on defensive behaviour of honeybees. Int J Biometeorol 31:259–265
Southwick EE, Moritz RFA (1987a) Social control of air ventilation in colonies of honey bees, *Apis mellifera*. J Insect Physiol 33:623–626
Southwick EE, Moritz RFA (1987b) Effects of meteorological factors on defensive behavior of honey bees. Int J Biometeorol 31:259–265
Southwick EE, Moritz RFA (1987c) Social synchronization of circadian rhythms in honeybees. Physiol Entomol 12:209–212
Southwick EE, Mugaas JN (1971) A hypothetical homeotherm: the honey bee hive. Comp Bioch Physiol 40A:935–944
Southwick EE, Pimentel D (1981) Energy efficiency of honey production by bees. Bioscience 31:730–732
Southwick EE, Schreffler AK (1984) Nectar supply, foraging and pollination by bees. Colloq INRA 21:85–88
Southwick EE, Southwick AK (1986) Nectar characteristics and phenology of spring bee plants in northwestern New York. Agric Ecosyst Environ 16:55–62
Southwick EE, Southwick L Jr (1989a) Pollination puzzle. Glean Bee Cult 117:498
Southwick EE, Southwick L Jr (1992) Economic value of honey bees (Hymenoptera: Apidae) in the United States. J Econ Entomol (in press)
Southwick EE, Loper GM, Sadwick SE (1981) Nectar production, composition, energetics and pollinator attractiveness in spring flowers of western New York. Am J Bot 68:994–1002

Southwick EE, Roubik DW, Williams J (1990) Cooparative energy balance in groups of Africanized and European honey bees: ecological implications. Comp Biochem Physiol 97A:1–7
Southwick L Jr, Southwick EE (1989b) A comment on "Value of honey bees as pollinators of U.S. crops." Am Bee J 129:805–807
Southwood TRE (1966) Ecological methods: with particular reference to the study of insect populations. Methuen, London
Spangler HG (1986) High-frequency sound production by honeybees. J Apic Res 25:213–219
Spangler HG (1991) Do honey bees encode distance information into wing vibrations of the waggle dance? J Insect Behav 4:15–20
Spencer H (1862) First principles. Williams Norgate, London
Sperlin A, Campbell R, Brosemer RW (1975) The hybridization of DNA from two species of honeybees. J Insect Physiol 21:373–376
Spiess EB (1977) Genes in populations. Wiley, New York
Spinola M (1840) Observations sur les apiaries meliponides. Ann Sci Nat Zool 13:116–140
Spivak M (1989) Identification and relative success of Africanized and European honey bees in Costa Rica. PhD Thesis, Univ Kans, Lawrence
Spivak M (1991) The Africanization process in Costa Rica. In: Spivak M, Fletcher DJC, Breed MD (eds) The "African" honey bee. Westview Press, Boulder, CO, pp 137–155
Spivak M, Fletcher DJC, Breed MD (1991a) The "African" honey bee. Westview Press, Boulder, CO
Spivak M, Zeltzer A, DeGrandi-Hoffman G, Martin JH (1991b) The influence of temperature on the rate of development and color patterns of queen honey bees (*Apis mellifera* L.). Environ Entomol (in press)
Stanley RG, Linskens HF (1974) Pollen: biology, biochemistry and management. Springer, Berlin Heidelberg New York
Starr CK (1979) Origin and evolution of insect sociality: a review of modern theory. In: Hermann HR (ed) Social insects, vol 1. Academic Press, New York, pp 38–61
Starr CK (1984) Sperm competition, kinship and sociality in the aculeate Hymenoptera. In: Smith RL (ed) Sperm competition and the evolution of animal mating systems. Academic Press, New York, pp 427–463
Stearns SC (1976) Life-history tactics: a review of the ideas. Q Rev Biol 51:3–47
Stearns (1988) Why sex evolved and the difference it makes. In: Stearns (ed) The evolution of sex and its consequences. Birkhäuser, Basel, pp 15–31
Stekolshchikov MG (1971) Soderzhanie DNK v yadrakh kletok pautinnykh zhelez lichinok pchel. Tsitol Genet 5:536–540
Stephens DW, Krebs JR (1986) Foraging theory. Princeton Univ Press, Prinecton, NJ
Stöckhert E (1923) Über Entwicklung und Lebensweise der Bienengattung *Halictus* Latr. und ihrer Schmarotzer (Hym.). Zugleich ein Beitrag zur Stammesgeschichte des Bienenstaates. Konowia 2:48–247
Storey KB (1985) Metabolic biochemistry of insect flight. In: Gilles R (ed) Circulation, respiration and metabolism. Springer, Berlin Heidelberg New York, pp 193–207

Stort AC (1974) Genetic study of the aggressiveness of two subspecies of *Apis mellifera* in Brazil. 1. Some tests to measure aggressiveness. J Apic Res 13:32–38

Stort AC (1975a) Genetic study of the aggressiveness of two subspecies of *Apis mellifera* in Brazil. 2. Time at which the first sting reached the leather ball. J Apic Res 14:171–175

Stort AC (1975b) Genetic study of the aggressiveness of two subspecies of *Apis mellifera* in Brazil. 4. Number of stings in the gloves of the observer. Behav Genet 5:269–274

Stort AC (1975c) Genetic study of the aggressiveness of two subspecies of *Apis mellifera* in Brazil. 5. Number of stings in the leather ball. J Kans Entomol Soc 48:381–387

Stort AC (1976) Genetic study of the aggressiveness of two subspecies of *Apis mellifera* in Brazil. 3. Time taken for the colony to become aggressive. Cienc Cult 28:1128–1185

Stort AC (1978) Genetic study of the aggressiveness of two subspecies of *Apis mellifera* in Brazil. 7. Correlation of the various aggressiveness characters among each other and with the gene for abdominal color. Cienc Cult 30:491–496

Stussi T (1972) L'heterotherme de l'abeille. Arch Sci Physiol 26:131–159

Sylvester HA (1976) Allozyme variation in honeybees (*Apis mellifera* L.). PhD Thesis, Univ Calif, Davis

Sylvester HA (1986) Biochemical genetics. In: Rinderer TE (ed) Bee genetics and breeding. Academic Press, Orlando, pp 177–203

Symth T Jr (1985) Muscle systems. In: Blum MS (ed) Fundamentals of insect physiology. Wiley, New York, pp 227–252

Syren RM, Luykx P (1977) Permanent segmental interchange complex in the termite *Incisistermes schwarzi*. Nature (Lond) 266:167–168

Syren RM, Luykx P (1981) Geographic variation of sex-linked genes in males of the termite *Kalotermes approximatus* Snyder (Insecta: Isoptera). Chromosoma 82:65–88

Szabo TI (1987) Outdoor wintering of honeybees in multiple-nucleus and 4-colony packs of honeybees. J Apic Res 26:238–239

Taber S (1955) Sperm distribution in the spermathecae of multiple-mated queen honey bees. J Econ Entomol 48:522–525

Taber S (1980) A population of feral honeybee colonies. Am Bee J 119: 842–847

Taber S, Owens CD (1970) Colony founding and initial nest design of honey bees, *Apis mellifera* L. Anim Behav 18:625–632

Takenaka T, Kaatz HH (1987) Protein synthesis by hypopharyngeal glands of worker honey bees. In: Eder J, Rembold H (eds) Chemistry and biology of social Insects. Peperny, Munich, pp 166–167

Tauber MJ, Tauber CA, Masaki S (1986) Seasonal adaptations of insects. Oxford Univ Press, New York

Taylor OR (1977) The past and the possible future spread of Africanized honeybees in the Americas. Bee World 58:19–30

Taylor OR (1985) African bees: potential impact in the United States. Bull Entomol Soc Am 34:14–24

Taylor OR (1988) Ecology and economic impact of African and Africanized honey bees. In: Needham GR, Page RE, Delfinado-Baker M, Bowman CE (eds) Africanized honey bees and bee mites. Ellis Horwood, Chichester, pp 29–41

Taylor OR, Rowell GA (1988) Drone abundance, queen flight distance, and the neutral mating model for the honey bee, *Apis mellifera*. In: Needham GR, Page RE, Delfinado-Baker M, Bowman CE (eds) Africanized honey bees and bee mites. Ellis Horwood, Chichester, pp 173–183

Taylor OR, Spivak M (1984) Climatic limits of tropical African honeybees in the Americas. Bee World 65:38–47

Thakar CV, Deodikar WB (1966) Chromosome number in *Apis florea* Fabr. Curr Sci Bangalore 7:186

Thompson VC (1964) Behavior genetics of nest cleaning in honeybees III. Effect of age of bees of a resistant line on their responses to disease-killed brood. J Apic Res 3:25–30

Thornhill R, Alcock J (1983) Evolution of insect mating systems. Harvard Univ Press, Cambridge, MA

Tiesler FK (1972) Die Inselbelegstellen aus dem Norden der BRD. In: Ruttner F, Ruttner H, Harnaj V (eds) Paarungskontrolle und Selektion bei der Honigbiene. Apimondia, Bucharest, pp 52–56

Tingek S, Mardan M, Rinderer TE, Koeniger N, Koeniger G (1988) Rediscovery of *Apis vechti* (Maa 1953): the Saban honeybee. Apidologie 19:97–102

Towne WF, Kirchner WH (1989) Hearing in honey bees: detection of air-particle oscillations. Science 244:686–688

Traniello JFA (1987) Social and individual responses to environmental factors in ants. In: Pasteels JM, Deneubourg JL (eds) From individual to collective behavior in social insects. Birkhäuser, Basel, pp 63–80

Tribe GD (1983) What is the Cape bee? S Afr Bee J 55:77–87

Trivers RL, Hare H (1976) Haplodiploidy and the evolution of social insects. Science 191:249–263

Trout WE, Kaplan WD (1970) A relationship between longevity, metabolic rate, and activity in shaker mutants of *Drosophila melanogaster*. Exp Gerontol 5:89–92

Trump RF, Thompson VC, Rothenbuhler WC (1967) Behavior genetics of nest cleaning in honeybees V. Effect of previous experience and composition of mixed colonies on response to disease-killed brood. J Apic Res 6:127–131

Tucker KW (1958) Automictic parthenogenesis in the honeybee. Genetics 43:299–316

Underwood BJ (1990) Bee cool. Nat Hist Dec 90:50–57

Valli E, Summers D (1988) Honey hunters of Nepal. Nat Geogr 174:660–671

Van der Blom J (1986) Reproductive dominance within colonies of *Bombus terrestris*. Behaviour 97:37–49

Van der Blom J, Velthuis HHW (1988) Social behaviour of the carpenter bee (*Xylocopa pubescens* Spinola). Ethology 79:281–294

Van der Have TM, Boosma JJ, Menken SBJ (1988) Sex investment ratios and relatedness in the monogynous ant *Lasius niger* L. Evolution 42: 160–172

Van Doorn A (1988) Reproductive dominance in bumblebees: an ethophysiological study. PhD Thesis, Rijksuniv Utrecht, Utrecht
Van Doorn A, Heringa J (1986) The ontogeny of a dominance hierarchy in colonies of the bumblebee, *Bombus terrestris* (Hymenoptera, Apidae). Insect Soc 33:3–25
Van Doorn A, Hogeweg P (1985) Die Entwicklung des agonistischen Verhaltens innerhalb der Arbeiterinnenkaste und zwischen Arbeiterinnen und der Königin während der Volksentwicklung bei der Erdhummel, *Bombus terrestris*. Mitt Dtsch Ges Allg Angew Entomol 4:328–331
Van Honk CGJ, Hogeweg P (1981) The ontogeny of the social structure in a captive *Bombus terrestris* colony. Behav Ecol Sociobiol 9:111–119
Van Honk CGJ, Velthuis HHW, Röseler PF, Malotaux ME (1980) The mandibular glands of *Bombus terrestris* queens as a source of queen pheromones. Entomol Exp Appl 28:191–198
Van Honk CGJ, Röseler PF, Velthuis HHW, Hoogeveen JC (1981) Factors influencing the egg laying behaviour in a captive *Bombus terrestris* colony. Behav Ecol Sociobiol 9:9–14
Van Praagh JP, Arendse MC, Ruttner F (1976) Measurements and characterization of light or radiance fields at drone congregation areas (*Apis mellifera* L.). Verh Dtsch Zool Ges 69:273
Van Praagh JP, Ribi W, Wehrmann C, Wittmann D (1980) Drone bees fixate the queen with the dorsal-frontal part of their compound eyes. J Comp Physiol A 136:263–266
Velthuis HHW (1970) Queen substances from the abdomen of the honeybee queen. Z Vgl Physiol 70:210–222
Velthuis HHW (1972) Observations on the transmission of queen substance n the honeybee colony by the attendants of the queen. Behaviour 41: 105–129
Velthuis HHW (1976) Environmental, genetic and endocrine influences in stingless bee caste determination. In: Lüscher (ed) Phase and caste determination in insects. Pergamon Press, Oxford, pp 63–70
Velthuis HHW (1985) The honeybee queen and the social organization of her colony. In: Hölldobler B, Lindauer M (eds) Experimental behavioral ecology. G Fischer, Stuttgart, pp 343–357
Velthuis HHW (1987) The evolution of sociality: ultimate and proximate factors leading to primitive social behavior in carpenter bees. In: Pasteels JM, Deneubourg J-L (eds) From individual to collective behavior in social insects. Birkhäuser, Basel, pp 405–430
Velthuis HHW (1990) Regulation of honey bee reproduction at colony level. In: Hoshi M, Yamashita O (eds) Advances in invertebrate reproduction, vol 5. Elsevier, Amsterdam, pp 489–494
Velthuis HHW, Es J van (1964) Some functional aspects of the mandibular glands of the queen honeybee. J Apic Res 3:11–16
Velthuis HHW, Hogendoorn K (1989) Das fakultative Sozialverhalten der Holzbiene *Xylocopa pubescens*: Brutmortalität in sozialen und solitären Nestern. Proc 12th Cong IUSSI Germ Sect, Bern, p 55
Velthuis HHW, Velthuis-Kluppel FM (1975) Caste differences in a stingless bee, *Melipona quadrifasciata* Lep., influenced by juvenile hormone application. Proc K Ned Akad Wet 78(C):81–84

Velthuis HHW, Ruttner F, Crewe RM (1990) Differentiation in reproductive physiology and behaviour during the development of laying worker honey bees. In: Engels W (ed) Social insects: an evolutionary approach to castes and reproduction. Springer, Berlin Heidelberg New York, pp 231–243

Verheijen-Voogd C (1959) How worker bees perceive the presence of their queen. Z Vgl Physiol 41:527–582

Verma S, Ruttner F (1983) Cytological analysis of the thelytokous parthenogenesis in the Cape honeybee (*Apis mellifera capensis* Escholtz). Apidologie 14:41–58

Verma S, Ruttner F, Verma LR (1982) Cytological mechanisms of syngamy in the automictic laying workers of *Apis mellifera capensis*. In: Breed MD, Michener CD, Evans HE (eds) The biology of social insects. Westview Press, Boulder, CO, pp 412–413

Vierling G, Renner M (1977) Die Bedeutung des Sekretes der Tergittaschendrüsen für die Attraktivität der Bienenkönigin gegenüber jungen Arbeiterinnen. Behav Ecol Sociobiol 2:185–200

Villa JD (1987) Africanized and European colony conditions at different elevations in Colombia. Am Bee J 127:53–57

Villa JD, Koeniger N, Rinderer TE (1991) Overwintering of Africanized, European, and hybrid honey bees in Germany. Environ Entomol 20: 39–43

Vincke PP, Tilquin JP (1978) A sex-linked ring quadrivalent in Termitidae (Isoptera). Chromosoma 67:151–156

Visscher PK (1982) Foraging strategy of honey bee colonies in a temperate deciduous forest. MSc Thesis, Cornell Univ, Ithaca, NY

Visscher PK (1983) The honey bee way of death: necrophoric behaviour in *Apis mellifera* colonies. Anim Behav 31:1790–1801

Visscher PK (1986) Kinship discrimination in queen rearing by honey bees (*Apis mellifera*). Behav Ecol Sociobiol 18:453–460

Visscher PK (1989) A quantitative study of worker reproduction in honey bee colonies. Behav Ecol Sociobiol 25:247–254

Visscher PK, Seeley TD (1982) Foraging strategy of honeybee colonies in a temperate deciduous forest. Ecology 63:1790–1801

Vlasak I, Burgschwaiger S, Kreil G (1987) Nucleotide sequence of the large ribosomal RNA of honeybee mitochondria. Nucleic Acids Res 15:2388

Von Frisch K (1949) Die Polarisation des Himmelslichtes als orientierender Faktor bei den Tänzen der Biene. Experientia 5:142–148

Von Frisch K (1950) Bees: their vision, chemical senses, and language. Cornell Univ Press, Ithaca, NY

Von Frisch K (1965) Tanzsprache und Orientierung der Bienen. Springer, Berlin Heidelberg New York

Von Frisch K (1967) The dance language and orientation of bees. Harvard Univ Press, Cambridge, MA

Von Frisch K (1974) Animal architecture. Harcourt Brace Jovanovich, New York

Von Frisch, Rösch GA (1926) Neue Versuche über die Bedeutung von Duftorgan und Pollenduft für die Verständigung im Bienenvolk. Z Vgl Physiol 4:1–21

Voogd S (1955) Inhibition of ovary development in worker bees by extraction fluid of the queen. Experientia 11:181–182

Waddington KD (1980) Flight patterns of foraging bees relative to density of artificial flowers and distribution of nectar. Oecologia 44:199–204
Waddington KD (1982) Honey bee foraging profitability and round dance correlates. J Comp Physiol 148:292–301
Waddington KD (1985) Cost-intake information used in foraging. J Insect Physiol 31:891–897
Waddington KD (1987) Perception of foraging costs and intakes, and foraging decisions. In: Menzel R, Mercer A (eds) Neurobiology and behavior of honeybees. Springer, Berlin Heidelberg New York, pp 66–75
Waddington KD, Holden LR (1979) Optimal foraging: on flower selection by bees. Am Nat 114:179–186
Waller GD (1972) Evaluating responses of sugar solutions using an artificial-flower feeder. Ann Entomol Soc Am 65:857–862
Watson JAL, Abbey HM (1985) Development of neotenice in *Mastotermes darwiniensis* Froggatt: an alternative strategy. In: Watson JAL, Okot-Kotber BM, Noirot C (eds) Caste differentiation in social insects. Pergamon Press, Oxford, pp 107–124
Weber E (1978) Mathematische Grundlagen der Genetik. VEB G Fischer, Jena
Wedmore EB (1947) The ventilation of bee-hives. Bee Craft, Petts Wood, UK
Wehner R, Rossel S (1985) The bee's celestial compass – a case study in behavioural neurobiology. In: Hölldobler B, Lindauer M (eds) Experimental behavioural ecology. G Fischer, Stuttgart, pp 11–53
Weiss K (1962) Untersuchungen über die Drohnenerzeugung im Bienenvolk. Arch Bienenkd 39:1–7
Weiss K (1984) Regulierung des Proteinhausthaltes im Bienenvolk (*Apis mellifera* L.) durch Brutkannibalismus. Apidologie 15:339–354
Wells PH, Wenner AM (1971) The influence of food scent on behavior of fraging honey bees. Physiol Zool 44:191–209
Wells PH, Wenner AM (1973) Do honey bees have a language? Nature (Lond) 241:171–175
Wenner AM (1964) Sound communication in honey bees. Sci Am 210: 116–124
Wenner AM, Wells PH (1990) Anatomy of a controversy: the question of a "language" among bees. Columbia Univ Press, New York
Wenner AM, Wells PH, Rohlf FJ (1967) An analysis of the waggle dance and recruitment in honey bees. Physiol Zool 40:317–344
Wenner AM, Alcock, JE, Meade D (1990) Removal of feral honey bee colonies from Santa Cruz Island, California. Am Bee J 130:818–819
Werner A, Menzel R, Wehrhalm C (1988) Color constancy in the honey bee. J Neuroscience 8:156–159
West-Eberhard MJ (1975) The evolution of social behavior by kin selection. Q Rev Biol 50:1–33
West-Eberhard MJ (1978) Polygyny and the evolution of social behavior in wasps. J Kans Entomol Soc 51:832–856
Westrich P (1988a) Die Wildbienen Baden-Württembergs. Allgemeiner Teil. Ulmer, Stuttgart
Westrich P (1988b) Die Wildbienen Baden-Württembergs. Spezieller Teil. Ulmer, Stuttgart

Wheeler WM (1911) The ant-colony as an organism. J Morphol 22:307–325
Wheeler WM (1918) A study of some ant larvae, with a consideration of the origin and meaning of the social habit among social insects. Proc Natl Acad Sci USA 57:293–343
Wheeler WM (1928) The social insects: their origin and evolution. Kegan Paul, Trench, Trubner & Co, London
White JW Jr, Riethof ML, Subers MH, Kushnir I (1962) Composition of American honeys. US Dep Agric Tech Bull 1261
Whiting PW (1939) Sex determination and reproductive economy in *Habrobracon*. Genetics 24:110–111
Whiting PW (1940) Multiple alleles in sex determination of *Habrobracon*. J Morphol 66:323–355
Whiting PW (1943) Multiple alleles in complementary sex determination of *Habrobracon*. Genetics 28:365–382
Wigglesworth VB (1984) Insect physiology. Chapman Hall, London
Wilde J de (1976) Juvenile hormone and caste differentiation in the honey bee (*Apis mellifera*). In: Lüscher M (ed) Phase and caste determination in insects. Pergamon Press, New York, pp 5–20
Wille H, Imdorf A, Bühlmann G, Kilchenmann V, Wille M (1985) Beziehungen zwischen Polleneintrag, Brutaufzucht und mittlere Lebenserwartung der Arbeiterinnen in Bienenvölkern (*Apis mellifica*). Mitt Schweiz Entomol Ges 58:205–213
Williams GC (1966) Adaptation and Natural Selection. Princeton Univ Press, Princeton, NJ
Williams CM (1976) Juvenile hormone... In retrospect and prospect. In: Gilbert LI (ed) The juvenile hormones. Plenum Press, New York, pp 1–14
Williams IH, Pickett JA, Martin AP (1981) The Nasanov pheromone of the honeybee, *Apis mellifera* L. (Hymenoptera: Apidae), II. Bioassay of the components using foragers. J Chem Ecol 7:225–237
Williams IH, Pickett JA, Martin AP (1982) The Nasanov pheromone of the honeybee, *Apis mellifera* L. (Hymenoptera: Apidae), IV. Comparative electroantennogram responses. J Chem Ecol 8:567–574
Wilson DS (1975) A theory of group selection. Proc Natl Acad Sci USA 72:143–146
Wilson DS (1977) Structured demes and the evolution of group-advantageous traits. Am Nat 111:157–185
Wilson DS, Sober E (1989) Reviving the superorganism. J Theor Biol 136:337–356
Wilson EO (1968) The superorganism concept and beyond. In: Chauvin M (ed) L'effet de group chez les animaux. CNRS, Paris, pp 27–39
Wilson EO (1971) The insect societies. Harvard Univ Press, Cambridge, MA
Wilson EO (1975) Sociobiology: the new synthesis. Harvard Univ Press, Cambridge, MA
Wilson EO (1980) Caste and division of labor in leaf-cutter ants (Hymnoptera: Formicidae: *Atta*), I. The overall pattern in *A. sexdens*. Behav Ecol Sociobiol 7:143–156
Wilson EO (1984) Biophilia. Harvard Univ Press, Cambridge, MA
Wilson EO (1985) The sociogenesis of insect colonies. Science 228: 1489–1495

Wilson EO (1990) Habitable earth. In: Graves W (ed) National Geographic atlas of the world, 6th edn. Natl Geogr Soc, Washington, DC, p 5
Winston ML (1980) Swarming, afterswarming, and reproductive rate of unmanaged honeybee colonies (*Apis mellifera*). Insect Soc 27:391–398
Winston ML (1987) The biology of the honey bee. Harvard Univ Press, Cambridge, MA
Winston ML, Katz SJ (1981) Longevity of cross-fostered honey bees workers (*Apis mellifera*) of European and Africanized races. Can J Zool 59: 1571–1575
Winston ML, Katz SJ (1982) Foraging differences between cross-fostered honey bees workers (*Apis mellifera*) of European and Africanized races. Behav Ecol Sociobiol 10:125–129
Winston ML, Michener CD (1977) Dual origin of highly social behavior among bees. Proc Natl Acad Sci USA 74:1135–1137
Winston ML, Slessor KN, Smirle MJ, Kandil AA (1982) The influence of a queen-produced substance, 9-HDA, on swarm clustering behavior in the honeybee *Apis mellifera* L. J Chem Ecol 8:1283–1288
Winston ML, Taylor OR, Otis GW (1983) Some differences between temperate European and tropical African and South American honeybees. Bee World 64:12–21
Wirtz P (1973) Differentiation in the honey bee larvae. Med Landbouwhoogesch Wageningen 73-5:1–155
Wirtz P, Beetsma J (1972) Induction of caste differentiation in the honey bee (Apis mellifera L.) by juvenile hormone. Entomol Exp Appl 15:517–520
Wittmann D, Engels W (1987) Welche Diät ergibt Arbeiterinnen bei in vitro Aufzucht von Honigbienen? Apidologie 18:279–288
Wolf BE (1960) Eine Nachuntersuchung zur Cytologie der Honigbiene (*Apis mellifica* L.). Zool Beitr 5:373–391
Wolf E (1926) Über das Heimkehrvermögen der Bienen. Z Vgl Physiol 3:615–691
Wolf E (1927) Über das Heimkehrvermögen der Bienen II. Z Vgl Physiol 6:221–254
Wolf TJ, Schmidt-Hempel P (1989) Extra loads and foraging life span in honeybee workers. J Anim Ecol 58:943–954
Wongsiri S, Limbipichai K, Tangkanasing P, Mardan M, Rinderer T, Sylvester HA, Koeniger G, Otis G (1990) Evidence of reproductive isolation confirms that *Apis andreniformis* (Smith, 1858) is a separate species from sympatric *Apis florea* (Fabricius, 1787). Apidologie 21:47–52
Woyke J (1955) Multiple mating of the honeybee queen (*Apis mellifera* L.) in one nuptial flight. Bull Acad Pol Sci II 3:175–180
Woyke J (1956) Anatomo-physiological changes in queenbees returning from mating flights and the process of multiple mating. Bull Acad Pol Sci Cl II 4:81–87
Woyke J (1962) The hatchability of "lethal" eggs in a two sex allele fraternity of honeybee. J Apic Res 1:6–13
Woyke J (1963a) Drone larvae from fertilized eggs of the honeybee. J Apic Res 2:19–24
Woyke J (1963b) Contribution of successive drones to the insemination of a queen. 19th Int Congr Apicult, Apimondia, Prague, pp 715–718

Woyke J (1965) Genetic proof of the origin of drones from fertilized eggs of the honeybee. J Apic Res 4:7–11

Woyke J (1966) Wovon hängt die Zahl der Spermien in der Samenblase der auf natürlichem Wege begatteten Königin ab? Z Bienenforsch 8:236–247

Woyke J (1967) Diploid drone substance – cannibalism substance. Proc 21st Int Congr Apicult, Apimondia, Maryland, pp 232–234

Woyke J (1969) A method for rearing diploid drones in a honeybee colony. J Apic Res 8:65–71

Woyke J (1973) Experiences with *Apis mellifera adansonii* in Brazil and Poland. Apiacta 8:115–116

Woyke J (1975) Natural and artificial insemination of *Apis cerana* in India. J Apic Res 14:153–159

Woyke J (1983) Dynamics of entry of spermatozoa into the spermatheca of instrumentally inseminated queen honeybees. J Apic Res 22:150–154

Woyke J, Hillesheim E (1989) Genetic aspects of instrumental insemination. In: Moritz RFA (ed) The instrumental insemination of the queen bee. Apimondia, Bucharest, pp 105–123

Woyke J, Skowronek W (1967) Spermatogenesis in diploid drones. Proc 21st Int Congr Apicult, Apimondia, Maryland, pp 470–471

Wright S (1922) Coefficients of inbreeding and relationship. Am Nat 56: 330–338

Wright S (1933) Homozygosis and inbreeding. Proc Natl Acad Sci Wash 19:411–420

Wright S (1969) Evolution and the genetics of populations, vol 2. The theory of gene frequencies. Univ Chicago Press, Chicago

Wykes GR (1952) An investigation of the sugars present in the nectar of flowers of various species. New Phytol 51:210–215

Wynne-Edwards VC (1962) Animal dispersion in relation to social behaviour. Oliver & Boyd, Edinburgh

Wynne-Edwards VC (1963) Intergroup selection in the evolution of social systems. Nature (Lond) 200:623–626

Yokoyama S, Felsenstein JA (1978) A model of kin selection for an altruistic trait considered as a quantitative character. Proc Natl Acad Sci USA 75:420–422

Yokoyama S, Nei M (1979) Population dynamics of sex determining alleles in honey bees and self-incompatibility alleles in plants. Genetics 91: 609–626

Zaltman G, Duncan R, Holbeck J (1973) Innovations and organizations. Wiley, New York

Zeuner FE, Manning FJ, Morris SF (1976) A monograph on fossil bees (Hymenoptera: Apoidea). Bull Br Mus Nat Hist 27:149–268

Zucchi R, Sakagami SF (1972) Capacidade termo-reguladora em *Trigona spinipes* e em algumas outras especies de abelhas sem ferrao. In: Homenagem a Warwick E Kerr. Rio Claro, SP, Brazil pp 301–309

Subject Index

The manufacturer's authorised representative in the EU is Springer
Nature Customer Service Centre.GmbH, Europaplatz 3, 69115 Heidelberg,
Germany. If you have any concerns regarding our products, please
contact ProductSafety@springernature.com

Printed and bound by CPI Group (UK) Ltd, Croydon, CR0 4YY
22/09/2026
02220988-0004